Upstream Dynamics

The upstream petroleum sector is a cornerstone of the global energy landscape. As the world transitions toward more sustainable and diversified energy sources, the role of upstream petroleum economics becomes increasingly complex and dynamic. Blending analytical rigor with real-world case studies from various basins, *Upstream Dynamics* serves as a professional guide to the economics that govern oil and gas exploration and production. It covers the full asset life cycle—exploration, appraisal, development, production, and abandonment—linking engineering realities with financial decision-making. It provides insights into exploration, drilling, production, reservoir management, risk mitigation, regulatory compliance, and technological innovation.

- Details project screening, cash flow modeling, fiscal regimes, reserve classification, risk and uncertainty, field development planning, cost estimation, production optimization, and portfolio management
- Goes beyond traditional discounted cash flow by integrating Monte Carlo simulation, real-options thinking, and scenario planning, with clear step-by-step illustrations and spreadsheet-ready examples
- Addresses the commercial impact of regulation, ESG/sustainability, digitalization/AI, and price volatility, translating these into practical decisions on timing, sequencing, and capital allocation
- Features real-world case studies from across the Middle East, the North Sea, the Gulf of Mexico, West Africa, and Southeast Asia
- Emphasizes what practitioners can do on Monday morning: build better economic cases, manage risk credibly, and communicate value to executives, partners, and regulators

Written by an experienced and practicing senior petroleum economist, this text is aimed at industry professionals and stakeholders, including petroleum economists and engineers, as well as asset managers and analysts, to help them successfully navigate the complexities of the upstream oil and gas sector.

Sulaiman F. Al-Sajari is a Senior Petroleum Economist at the Kuwait Foreign Petroleum Exploration Company (KUFPEC), an international oil and gas company with a global portfolio. He has extensive experience in upstream planning and petroleum economics, covering prospect screening, fiscal analysis, reserves and resources classification, field development planning, and portfolio evaluation across projects in the Middle East, Africa, Europe, and Asia. He earned a BSc in mechanical engineering at Florida Atlantic University and has led economic assessments and decision-support analyses for major exploration and development assets under diverse fiscal regimes. His expertise spans traditional economic evaluation methods—such as discounted cash flow—alongside advanced techniques including Monte Carlo simulation, real options valuation, and risk-adjusted modeling.

Upstream Dynamics

Navigating the Economics of Petroleum Exploration and Production

Sulaiman F. Al-Sajari

CRC Press
Taylor & Francis Group
Boca Raton London New York

CRC Press is an imprint of the
Taylor & Francis Group, an **informa** business

Designed cover image: © 2026 Shutterstock

First edition published 2027
by CRC Press
2385 NW Executive Center Drive, Suite 320, Boca Raton FL 33431

and by CRC Press
4 Park Square, Milton Park, Abingdon, Oxon, OX14 4RN

CRC Press is an imprint of Taylor & Francis Group, LLC

ISBN: 978-1-041-26443-9 (hbk)
ISBN: 978-1-041-26444-6 (pbk)
ISBN: 978-1-003-75034-5 (ebk)

DOI: 10.1201/9781003750345

Typeset in Times
by KnowledgeWorks Global Ltd.

Contents

Preface

The upstream petroleum sector, encompassing oil and gas exploration, extraction, and production, is a cornerstone of the global energy landscape. As the world transitions toward more sustainable and diversified energy sources, the role of upstream petroleum economics becomes increasingly complex and dynamic. This book, *Upstream Dynamics: Navigating the Economics of Petroleum Exploration and Production*, aims to provide a comprehensive exploration of the economic principles, strategies, and challenges that define this critical sector.

Over the past decades, the upstream petroleum industry has faced numerous challenges, including fluctuating commodity prices, technological advancements, geopolitical shifts, and environmental concerns. Despite these challenges, the sector has continuously adapted and evolved, driven by innovation, strategic investments, and a commitment to meeting global energy demands.

This book is structured to provide a comprehensive overview of the various aspects of upstream petroleum economics. Each chapter is meticulously crafted to provide insights into key areas, including exploration, drilling, production, reservoir management, risk mitigation, regulatory compliance, and technological innovation. Through detailed explanations, real-world case studies, and economic analyses, we aim to equip industry professionals, students, and stakeholders with the knowledge and tools necessary to navigate the complexities of the upstream oil and gas sector.

As we embark on this exploration, it is essential to recognize the pivotal role of the upstream petroleum industry in the global energy ecosystem. While the industry faces significant challenges, it also presents immense opportunities for those willing to innovate, adapt, and lead. By understanding the economic forces at play and embracing a forward-thinking approach, we can ensure the sustainable and profitable development of our hydrocarbon resources.

We hope this book serves as a valuable resource, fostering a deeper understanding of upstream petroleum economics and inspiring innovative ideas and strategies for the future. Your journey through these pages will offer a blend of theoretical knowledge and practical insights, reflecting the dynamic nature of the upstream oil and gas industry.

Finally, a note on process: as a non-native English speaker, I relied extensively on editorial tools such as Grammarly and AI-based language refinement software to strengthen clarity, identify gaps, and improve readability. These tools served only as aids for expression—much like a digital proofreader—while all research, models, analysis, and figures were fully developed and authored independently. Every intellectual framework and dataset originates from my professional experience in petroleum economics, ensuring originality and integrity throughout.

Preface

Introduction

In the dynamic world of the upstream petroleum industry, understanding the intricacies of exploration and production economics is crucial for making informed decisions and ensuring the sector's sustainability and profitability. This book, *Upstream Dynamics: Navigating the Economics of Petroleum Exploration and Production*, delves into the fundamental aspects of upstream petroleum economics, offering insights into the various stages of the upstream process, from exploration to production, and the economic principles that guide these activities.

The upstream sector, commonly referred to as Exploration and Production (E&P), plays a crucial role in meeting global energy demands. Despite the growing prominence of renewable energy sources, hydrocarbons remain a cornerstone of the world's energy mix, powering transportation, industrial processes, and electricity generation. The sector's significance extends beyond energy supply, as it contributes substantially to economic growth, job creation, and government revenues.

This book is structured into twenty comprehensive chapters, each focusing on several aspects of upstream petroleum economics. From the initial stages of exploration, reservoir management, and drilling operations to production optimization and risk management, each chapter provides detailed insights and case studies that illustrate the challenges and opportunities in the upstream sector. The book also explores the evolving landscape of upstream economics, driven by technological advancements, regulatory changes, and the global energy transition.

This book examines real-world examples and employs economic analysis techniques to equip industry professionals, students, and stakeholders with the knowledge and tools needed to navigate the complexities of upstream petroleum economics. Whether you are an oil and gas professional seeking to optimize operations, an investor evaluating potential projects, or a student exploring the field of petroleum economics, this book offers valuable perspectives and practical guidance.

As we explore upstream dynamics, it is essential to recognize the sector's challenges and embrace the opportunities presented by innovation, sustainability, and strategic foresight. The future of upstream petroleum economics lies in striking a balance between economic viability, environmental stewardship, and social responsibility, ensuring the industry thrives in an ever-changing energy landscape.

Part I

Core Operations

1 Introduction to Upstream Petroleum Economics

1.1 WHAT IS THE UPSTREAM PETROLEUM SECTOR?

The upstream petroleum sector, a high-stakes, capital-intensive industry, forms the bedrock of the oil and gas industry. It encompasses all activities related to discovering, evaluating, and extracting crude oil and natural gas from subsurface reservoirs. Often referred to as Exploration and Production (E&P), it represents the crucial first step in the journey of hydrocarbons from their geological origins to their eventual use as energy. This sector, distinct from the other petroleum industry segments, is a testament to the oil and gas industry's high-risk, high-reward nature.

> The oil industry is not for the faint of heart. It's a business where risks are taken in the billions, often miles below the surface.
>
> **Daniel Yergin**
> *Energy Historian and Author of the Prize*

Upstream Activities:

- **Exploration:** This is the process of identifying potential hydrocarbon accumulations. It involves geological and geophysical studies, as well as seismic surveys and advanced technologies, to locate oil and gas reserves beneath the Earth's surface.
- **Drilling:** Once potential reserves are identified, exploratory wells are drilled to confirm their presence. Drilling operations are complex and require sophisticated equipment and expertise.
- **Production:** If commercially viable reserves are found, production facilities are developed and operated to extract the oil and gas. This involves constructing infrastructure such as wells, pipelines, and processing facilities to bring the hydrocarbons to the surface.

Midstream Activities:

- **Transportation:** Midstream activities transport crude oil and natural gas from production sites to refineries and distribution networks. This includes pipelines, tankers, and rail transport.
- **Storage:** Crude oil and natural gas are stored in various facilities, including tanks and underground storage sites, to ensure a steady supply to refineries and consumers.
- **Processing:** The initial processing of natural gas involves removing impurities and separating valuable components, such as natural gas liquids (NGLs), from the gas stream.

DOI: 10.1201/9781003750345-2

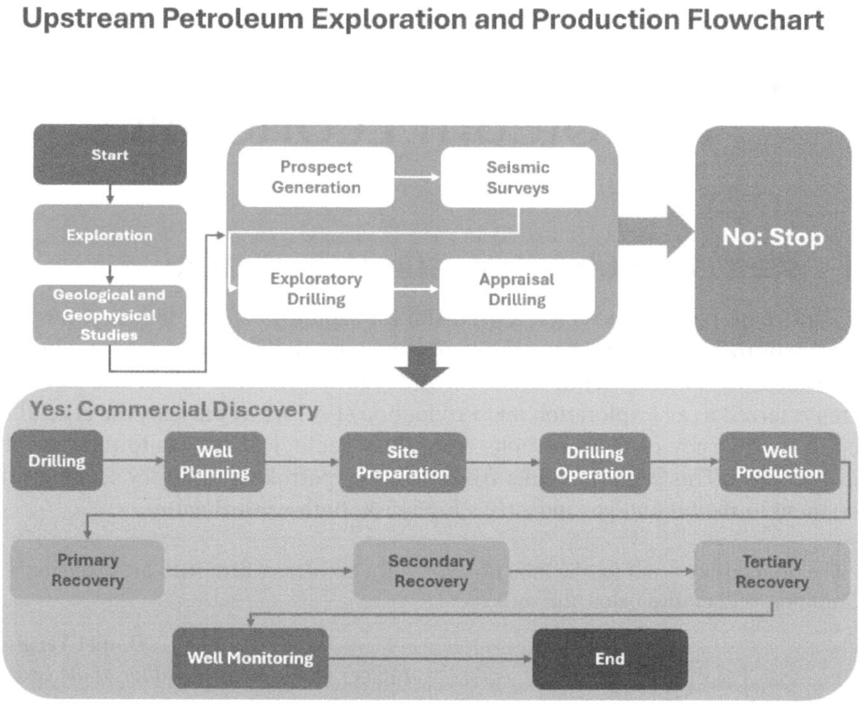

FIGURE 1.1 Upstream petroleum exploration and production flowchart sequential stages from seismic survey to tertiary recovery, illustrating the high-risk lifecycle of upstream operations.

This flowchart outlines the typical sequence of activities in upstream oil and gas operations (Figure 1.1). It begins with geological studies and seismic surveys, followed by exploratory and appraisal drilling. If a commercial discovery is made, the project proceeds to development phases, including well planning, drilling, and production. Over time, recovery efforts evolve from primary to secondary and eventually tertiary methods. If no viable discovery is made, the process is halted. This visual helps illustrate the high-risk, multi-stage nature of upstream investment decisions.

WHO COINED "UPSTREAM"?

The terms "upstream," "midstream," and "downstream" originated in the early days of vertical integration in the petroleum industry. "Upstream" refers metaphorically to the origin of the oil's journey—deep underground (Figure 1.2).

Downstream Activities:

- **Refining:** Downstream activities involve refining crude oil into finished products such as gasoline, diesel, jet fuel, and petrochemicals. Refineries use complex processes to convert crude oil into these valuable products.
- **Marketing and distribution:** Finished petroleum products are marketed and distributed through retail stations, pipelines, and transportation systems to deliver them to consumers.

upstream vs midstream vs downstream

Sector	Main Activities	Key Players	Examples
Upstream	Exploration, drilling, production	E&P companies (e.g., ExxonMobil, Shell)	North Sea oil, Permian Basin
Midstream	Transportation, storage, pipelines	Pipeline operators, terminal operators	Trans-Alaska Pipeline
Downstream	Refining, marketing, distribution	Refiners, marketers, retail networks	BP retail stations, PetroChina refineries

FIGURE 1.2 Comparison of upstream, midstream, and downstream sectors: Overview of key activities, players, and examples across the oil and gas value chain, highlighting the distinct roles of each segment.

Key Players in the Upstream Sector:

- **Oil and gas companies:** These companies, ranging from multinational integrated oil companies (IOCs) like ExxonMobil, Shell, and BP to smaller independent operators, are the driving force behind upstream activities. They invest in exploration, acquire licenses and concessions, manage drilling and production operations, and bear the financial risks associated with these ventures.
- **Service providers:** A diverse array of specialized service companies supports the operations of oil and gas companies. These include seismic survey companies (e.g., Schlumberger, Halliburton), drilling contractors (e.g., Baker Hughes, Nabors Industries), well completion and intervention specialists (e.g., Weatherford International), and reservoir engineering consultants (e.g., RPS Group). These companies provide the technical expertise and equipment necessary for upstream operations.
- **Governments**: Governments play a pivotal role in the upstream sector, acting as resource owners, regulators, and often as partners. They grant E&P licenses, establish environmental regulations, collect royalties and taxes, and, in many cases, participate in upstream projects through National Oil Companies (NOCs) like Saudi Aramco, Petrobras, and Gazprom. Their involvement underscores the sector's strategic and economic importance.

1.2 IMPORTANCE OF EXPLORATION AND PRODUCTION

E&P of oil and gas are fundamental to meeting global energy demands. Despite the growing prominence of renewable energy sources, hydrocarbons are crucial in powering the world's economies. Oil and natural gas fuel transportation, drive industrial processes, generate electricity, and provide feedstock for numerous petrochemical products.

> ExxonMobil and Chevron are plowing ahead with billions of dollars of spending on fossil-fuel projects, betting that the world will still need abundant oil and gas for decades despite calls to limit consumption and shift to cleaner energy.
>
> *The Wall Street Journal, March 2024*

Economic contributions: The upstream sector's significance extends beyond energy supply. It makes substantial local, national, and global economic contributions, significantly impacting the financial landscape. Upstream activities generate numerous high-paying jobs both directly within the industry and indirectly in supporting sectors such as manufacturing, construction, and transportation. Furthermore, the industry generates significant government revenue through royalties, taxes, and licensing fees, contributing to national budgets and economic development. This impact highlights the sector's importance and influence, positioning it as a key player in the global economy.

Geopolitical significance: Oil and gas resources hold significant geopolitical weight, shaping international relations and influencing political power dynamics. Access to and control over these resources can even trigger conflicts. The geographic distribution of oil and gas reserves, concentrated in some areas of the world, creates dependencies and strategic considerations for consuming nations, highlighting the sector's influence on global politics.

Environmental impact: While the upstream sector is crucial for energy supply, it also has significant ecological implications. E&P activities can result in habitat disruption, oil spills, and greenhouse gas emissions. As a result, the industry is under increasing pressure to adopt sustainable practices and reduce its environmental footprint. This includes measures to minimize habitat disruption, prevent and respond to oil spills, and reduce greenhouse gas emissions through the use of improved technology and operational practices. The industry's response to these challenges can significantly impact its economic viability and social acceptance.

1.3 BASIC PRINCIPLES OF UPSTREAM PETROLEUM ECONOMICS

The upstream petroleum sector operates within a complex economic environment characterized by substantial capital investments, significant technical and geological risks, and volatile commodity prices. Financial decision-making in this sector requires a sophisticated understanding of these factors.

Key Economic Drivers:

- **Oil and gas prices:** Fluctuations in global oil and gas prices are perhaps the most influential factors impacting the profitability and viability of upstream projects. A complex interplay of supply and demand dynamics, geopolitical events, and market sentiment determines these prices.
- **Costs:** Upstream projects entail substantial capital expenditure (CAPEX) and operating expenses (OPEX). CAPEX includes exploration and development costs (such as building production facilities) and infrastructure costs. OPEX encompasses ongoing expenses, including maintenance, labor, and transportation.
- **Risks:** The upstream sector is inherently risky. Exploration ventures carry the risk of not discovering commercial quantities of hydrocarbons. Drilling operations face technical challenges and potential hazards. Reservoir characteristics are often uncertain, impacting production forecasts. These risks must be carefully assessed and managed.

Economic metrics: Several key economic metrics are employed to evaluate the economic viability of upstream projects:

- **Net present value (NPV):** NPV[1] calculates the present value of all future cash flows associated with a project, discounted at an appropriate rate that reflects the time value of money and the project's risk. A positive NPV indicates that the project is expected to generate a positive return on investment.
- **Internal rate of return (IRR):** The IRR is the discount rate that makes a project's NPV equal to zero. It represents the project's expected rate of return. Projects with an IRR higher than the company's hurdle rate (minimum acceptable rate of return) are typically considered financially attractive.
- **Payback period:** The payback period measures the time it takes for a project to recoup its initial investment. It provides a simple measure of project profitability but does not consider the time value of money.
- **Profitability index (PI):** The PI is the ratio of the present value of future cash flows to the initial investment. A PI greater than one indicates a profitable project.

Decision-making frameworks: Upstream companies employ sophisticated decision-making frameworks, often incorporating discounted cash flow analysis, risk assessment, and scenario planning, to assess the economic potential of upstream projects. These frameworks help companies navigate the industry's uncertainties and complexities.

1.4 THE EVOLVING LANDSCAPE OF UPSTREAM ECONOMICS

The future of oil isn't just in the ground—it's in the data.

Patrick Pouyanné
CEO, TotalEnergies

The economics of the upstream petroleum sector are in constant flux, influenced by a complex interplay of numerous factors, including geopolitical tensions, global oil demand, advancements in extraction technologies, and regulatory changes. Fluctuations in crude oil prices, driven by global market trends and OPEC's production decisions, further complicate the landscape. Environmental considerations and the shift toward renewable energy sources are also reshaping the industry's investment strategies and operational practices. As a result, stakeholders must continuously adapt to these evolving dynamics to optimize their E&P strategies.

Technological advancements: Technological innovation is transforming upstream operations in the oil and gas industry. Advances in seismic imaging, horizontal drilling, hydraulic fracturing, enhanced oil recovery (EOR) techniques, and digital technologies are improving exploration success rates, enhancing production efficiency, and extending the lifespan of mature fields. For instance, 3D and 4D seismic imaging technologies provide more accurate mapping of subsurface reservoirs, thereby reducing the risk of drilling dry wells. Additionally, technologies such as artificial intelligence (AI), digital twins, and predictive analytics have significantly reduced OPEX in specific fields by over 20%. According to research by McKinsey & Company, the adoption of digital technologies, such as predictive maintenance and AI-driven analytics, is significantly improving efficiency. For example, some companies have seen a 15% reduction in maintenance costs and a 20% improvement in production uptime by using these technologies to predict equipment failures before they occur.

Sustainability considerations: The growing awareness of climate change and environmental concerns is driving the industry toward greater sustainability. This includes reducing greenhouse gas emissions, minimizing environmental impact, investing in carbon capture and storage (CCS) technologies, and exploring opportunities in renewable energy. Companies adopt best practices and innovations to reduce their carbon footprint and enhance environmental stewardship.

Geopolitical factors and market volatility: Geopolitical events, including political instability, resource nationalism, international sanctions, and conflicts, can significantly disrupt oil and gas markets, leading to price volatility and influencing investment decisions. For example, political upheaval in major oil-producing regions can lead to supply disruptions, affecting global prices and market stability. The Russian invasion of Ukraine in 2022 highlighted how upstream constraints can destabilize global energy markets almost overnight. With sanctions and supply chain disruptions becoming increasingly common, companies must now consider risks that extend beyond geopolitical factors.

Energy transition: The global shift toward cleaner energy sources presents challenges and opportunities for the upstream sector. Companies are adapting their strategies to navigate this energy transition by investing in renewable energy and developing technologies to reduce the carbon footprint of oil and gas production. Integrating renewable energy sources, such as wind and solar, into oil and gas operations is becoming increasingly common. Additionally, natural gas—particularly in liquefied natural gas (LNG) form—is viewed by many international oil companies (IOCs) as the "transition fuel." In 2024, more than 30% of new upstream investments are expected to focus on gas.

1.4.1 Energy Transition and the Role of LNG

As the world transitions toward cleaner energy, natural gas is increasingly positioned as a bridge fuel between traditional fossil fuels and renewable energy sources. LNG plays a crucial role in reducing carbon emissions by replacing coal in power generation. Many oil and gas companies are expanding their LNG portfolios to align with global energy transition goals while maintaining a strong revenue stream from hydrocarbon-based resources.

Technology's Role in Upstream Operations:

- 3D and 4D seismic imaging for accurate reservoir mapping
- AI-driven predictive maintenance for drilling equipment
- EOR techniques to optimize output.

1.4.2 Sustainability and the Energy Transition

The oil and gas industry is facing pressure to reduce its carbon footprint. Companies are:

- Investing in CCS
- Integrating renewable energy sources into their operations
- Reducing methane emissions through advanced monitoring systems

Regulatory changes: Governments worldwide are implementing new regulations to address environmental concerns and promote sustainable practices. These regulations can impact the economics of upstream projects by imposing additional costs or requirements. Companies must stay informed about regulatory developments and adjust their operations accordingly.

1.4.3 Regulatory Factors and ESG Considerations

Environmental, social, and governance (ESG) policies are increasingly shaping investment decisions in the upstream sector. Investors and financial institutions now demand more substantial commitments to:

- Reducing greenhouse gas emissions
- Implementing sustainable drilling practices
- Enhancing transparency in corporate governance and ethical labor practices

Companies that fail to meet ESG standards may face difficulties in securing funding, obtaining regulatory approvals, and maintaining stakeholder trust, making compliance a critical factor in achieving long-term success.

Market dynamics: Supply and demand, technological advancements, and geopolitical events significantly influence the dynamics of the oil and gas market. Companies must continuously monitor market trends and adjust their strategies to remain competitive. For example, the emergence of shale oil and gas has transformed the energy landscape, affecting global supply and pricing (Figure 1.3).

FAST FACT: SHALE VS. CONVENTIONAL DECLINE

Shale oil wells can decline by **70–90% within the first three years**, requiring constant reinvestment and new drilling. In contrast, **conventional reservoirs** often maintain production for decades with modest intervention, making them more capital efficient over time.

Market Dynamics and Global Policy Shifts:

- Political instability can cause supply disruptions.
- Environmental regulations influence project costs and viability.
- The rise of electric vehicles and renewables impacts long-term demand.

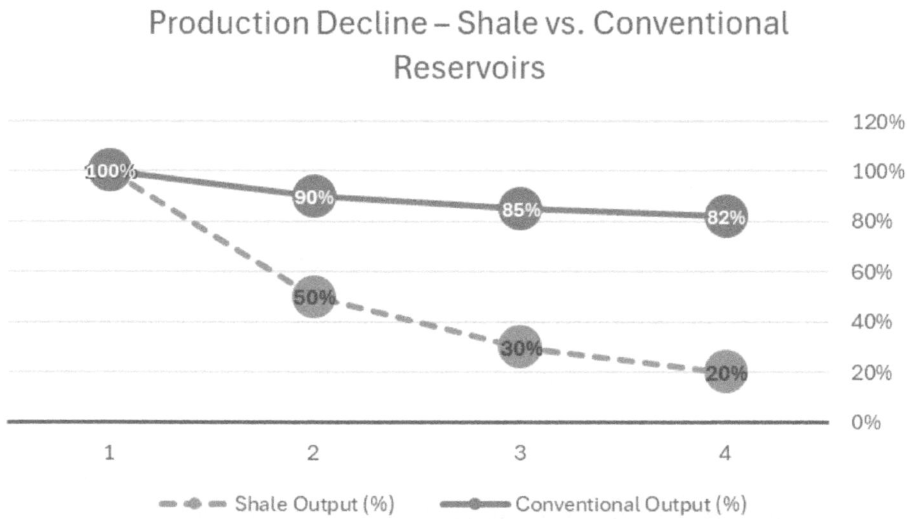

FIGURE 1.3 Production decline: Shale vs. Conventional Reservoirs Comparison of output decline rates over time, highlighting the rapid depletion of shale wells versus the sustained performance of conventional reservoirs.

1.5 CHAPTER SUMMARY

This chapter has provided a foundational understanding of the upstream petroleum sector, its critical role in meeting global energy needs, and the key economic principles that guide its operations. We explored the distinct activities of the upstream segment, differentiating it from midstream and downstream. We also examined the diverse players involved, from multinational oil companies to specialized service providers and government entities.

The chapter highlighted the economic importance, contribution to global economies, and geopolitical significance of E&P. Furthermore, we explored the fundamental financial principles, including key drivers, metrics, and decision-making frameworks used in upstream project evaluations. Finally, we discussed the sector's evolving landscape, emphasizing the impact of technological advancements, sustainability considerations, geopolitical factors, and the energy transition.

The next chapter will examine the specific economic considerations associated with exploration activities.

1.6 CASE STUDY: THE DEVELOPMENT OF A MAJOR OIL FIELD

To gain a deeper understanding of the upstream petroleum sector, it is essential to examine a detailed case study of the development of a major oil field. The discovery and subsequent development of the North Sea oil fields, particularly during the late 20th century, exemplify the significant challenges and notable triumphs associated with upstream oil and gas activities.

The North Sea oil fields, primarily located between the United Kingdom and Norway, were first discovered in the 1960s, with the first significant find at the Ekofisk field in 1969. This marked a pivotal moment for the oil industry, signaling the beginning of large-scale offshore drilling in deeper waters. Companies faced numerous challenges, including harsh weather conditions, complex geology, and the extensive logistics required for offshore operations.

Nonetheless, the successful extraction of oil from these fields led to transformative changes in the energy industry and the surrounding countries. By the late 1970s and 1980s, North Sea oil production had surged, significantly contributing to the UK's economy through job creation, investment, and enhanced energy security. Furthermore, the technological advancements spurred by the need to extract oil in challenging environments efficiently paved the way for future E&P strategies in other offshore locations worldwide.

This case study highlights the technical and economic aspects of upstream petroleum activities, underscoring the importance of regulatory frameworks and environmental considerations that have evolved in response to the industry's impact on ecosystems and communities. The North Sea oil fields are a profound example of how exploration and development in the upstream sector can significantly impact both local and global energy landscapes.

Discovery phase: The North Sea, located between the United Kingdom and continental Europe, was long suspected to contain significant oil and gas reserves. In the 1960s, technological advancements in seismic imaging and drilling made it possible to explore these deepwater areas. The discovery of the Ekofisk field by Phillips Petroleum in 1969 marked a turning point, as it confirmed the presence of commercially viable reserves.

Exploration and drilling: Following the discovery of Ekofisk, a wave of exploration activities ensued. International oil companies, including BP, Shell, and ExxonMobil, invested heavily in drilling exploratory wells across the North Sea. The harsh offshore environment posed significant technical challenges, requiring the development of specialized drilling rigs and techniques to withstand rough seas and deepwater conditions.

Development and production: As more fields were discovered, the focus shifted to developing production facilities. The construction of platforms, subsea infrastructure, and pipelines enabled the extraction and transportation of hydrocarbons to shore. The Forties field, discovered by BP in 1970, became one of the largest producing fields in the North Sea, delivering substantial volumes of oil and gas to the global market.

Economic impact: The North Sea oil boom had a profound impact on the United Kingdom and Norway. It created thousands of jobs, stimulated economic growth, and generated significant government revenues through royalties and taxes. The development of the North Sea oil fields also reduced Europe's dependence on Middle Eastern oil, enhancing energy security.

Environmental considerations: The North Sea oil industry faced ecological challenges, including the risk of oil spills and the need to protect marine ecosystems. To mitigate these risks, the industry implemented stringent safety measures and environmental regulations. Technological advancements, such as improved blowout preventers and oil spill response techniques, have enhanced the industry's ability to respond to potential environmental incidents.

1.7 FUTURE TRENDS IN UPSTREAM PETROLEUM ECONOMICS

As the upstream petroleum sector continues to evolve, several key trends are shaping its future landscape. The advancement of digital technology, including AI and machine learning, is revolutionizing E&P processes, enabling companies to optimize resource extraction and enhance operational efficiency. Additionally, the increasing focus on sustainability and environmental responsibility is pushing the industry to adopt greener practices, such as reducing flaring and investing in carbon capture technologies. The rise of renewable energy sources and shifting consumer preferences are also driving oil and gas companies to diversify their portfolios, resulting in increased investment in alternative energy projects. Furthermore, geopolitical factors and regulatory changes are influencing supply chains and market dynamics, highlighting the need for flexibility and agility in operations. These trends collectively underscore a transitioning industry that is adapting to new challenges and opportunities.

Digital transformation: Integrating digital technologies, such as AI, big data analytics, and the Internet of Things (IoT), revolutionizes upstream operations. These technologies enable real-time monitoring of drilling and production activities, predictive maintenance of equipment, and optimized decision-making. For example, AI-powered algorithms can analyze seismic data to identify potential drilling targets with greater accuracy.

Sustainable practices: The industry is increasingly adopting sustainable practices to reduce environmental impact. This includes the development of green technologies, such as CCS, as well as the use of renewable energy sources to power upstream operations. Companies are also exploring ways to minimize flaring and methane emissions, contributing to global efforts to combat climate change.

Collaboration and partnerships: Effective collaboration among oil and gas companies, service providers, and governments is crucial for addressing the industry's challenges. Joint ventures and partnerships enable the sharing of expertise, resources, and risks. For example, international collaboration in developing the Arctic oil fields has led to significant advancements in cold-weather drilling technologies.

Investment in renewable energy: Many oil and gas companies are diversifying their portfolios by investing in renewable energy projects, including offshore wind farms, solar energy, and hydrogen production. Integrating renewable energy sources into traditional oil and gas operations can enhance sustainability and reduce carbon emissions.

> Saudi Aramco Power, along with other firms, signed $8.3 billion worth of clean energy deals including photovoltaic and wind projects with a total capacity of 15 GW, marking a major step in Saudi Arabia's renewable energy transition.
>
> *Reuters, July 2025*

Regulatory and policy changes: Governments are implementing new policies and regulations to promote sustainability and address environmental concerns. This includes stricter emissions standards, incentives for renewable energy investments, and requirements for carbon reporting. Companies must navigate these regulatory changes while maintaining profitability.

1.8 CONCLUSION

The upstream petroleum sector remains the beating heart of global energy systems, supporting economic prosperity and industrial advancement. This vital segment drives economic growth by generating significant revenue for nations, creating employment opportunities, and fueling industrial development across multiple sectors, including manufacturing, transportation, and technology. Moreover, it plays a crucial role in strategic decision-making at a global scale, influencing geopolitical relations and energy security scenarios worldwide.

However, the challenges facing this sector are intensifying. These challenges manifest below the Earth's surface through geological complexities and resource depletion, and above it, where they become evident in the high-stakes environments of boardrooms, trading floors, and international climate negotiations. Factors such as fluctuating oil prices, regulatory changes, and societal shifts toward sustainable practices are pressing the industry to adapt rapidly.

This chapter laid the groundwork for a deeper understanding of the sector's intricate structure, economic dynamics, and strategic significance. Yet, the true complexity—and growth potential—lies in mastering the multifaceted forces reshaping it: geopolitical disruptions caused by regional conflicts and trade tensions, stringent environmental demands driven by climate change initiatives, and the ongoing digital transformation revolutionizing operational efficiencies and market strategies.

As we progress through this book, we'll explore how upstream players are not merely striving to survive but actively reinvent themselves to achieve leadership in an increasingly volatile landscape. This journey involves leveraging advanced technologies, embracing innovative practices to achieve sustainability goals, and navigating the energy transition toward renewable sources, all while responding to an uncertain future characterized by rapid change and heightened competition.

NOTE

1. John Burr Williams, *The Theory of Investment Value*, 1938.

2 Exploration Economics

2.1 INTRODUCTION TO EXPLORATION

Exploration is the lifeblood of upstream petroleum economics. Without discoveries, reserves decline, production falls, and economic value deteriorates. Exploration strategies define where, when, and how companies deploy capital to discover new hydrocarbon resources. These strategies influence ecological potential, fiscal regimes, technological advances, and market dynamics.

A successful exploration strategy balances risk and reward at its core—navigating between high-risk, high-reward frontier areas and lower-risk, mature basins. With growing investor scrutiny, increasing energy transition pressures, and tightening capital availability, modern exploration strategies must align with profitability goals and consider carbon and ESG factors.

> Headline – "Oil Majors Slash Exploration Budgets Amid Energy Transition"
>
> *Financial Times, June 2022*

As of 2023, exploration spending by international oil companies (IOCs) had dropped nearly 60% from its 2014 peak, reflecting a pivot toward lower-cost, faster-cycle assets and capital discipline.[1]

> We can no longer afford to explore blindly. Data must drive every decision.
>
> **Vicki Hollub**
> *CEO, Occidental Petroleum*

The exploration process: The exploration process is a complex, multi-staged endeavor often involving significant financial investment and technical expertise. It typically begins with regional geological studies and progresses through various phases, culminating in drilling exploratory wells. Let's break down these stages:

1. **Geological and geophysical studies:** Explorationists start by gathering and analyzing existing geological data, including maps, well logs, and seismic surveys. They also conduct new geophysical surveys, such as seismic reflection and gravity surveys, to image the subsurface and identify potential hydrocarbon traps. This stage is crucial for understanding the area's geological history and structural features.
2. **Prospect generation:** Based on the geological and geophysical data, explorationists identify potential areas where hydrocarbons might have accumulated. These areas are called prospects. This stage involves evaluating the likelihood of finding commercially viable hydrocarbon reserves.

DOI: 10.1201/9781003750345-3

3. **Exploratory drilling:** The most critical and expensive exploration stage is drilling exploratory wells, also known as wildcat wells. These wells are drilled to evaluate the prospects and determine whether hydrocarbons are present. Drilling operations require advanced technologies and highly skilled personnel.
4. **Appraisal drilling:** If an exploratory well encounters hydrocarbons, appraisal wells are drilled to delineate the size and extent of the discovery and gather more information about the reservoir characteristics. This stage helps estimate the volume of recoverable hydrocarbons.
5. **Feasibility study:** A feasibility study assesses the economic viability of developing the discovered reserves. This involves evaluating the potential revenue, costs, and risks associated with the project.

STRATEGY SHIFT—FROM VOLUME TO VALUE

In the past, exploration strategies focused on maximizing discovered volumes. Today, the focus has shifted to value-based metrics:

- **Shorter time to first oil**
- **Low breakeven prices (< $40/bbl)**
- **Low carbon intensity barrels**
- **High netback margin per barrel**

Companies increasingly evaluate prospects on geological promise, carbon-adjusted returns, and infrastructure synergies.

Types of exploration activities: Exploration activities can be broadly categorized into:

- **Geophysical exploration:** This involves using seismic, gravity, and magnetic surveys to image the subsurface and identify geological structures that may contain hydrocarbons. Seismic surveys are the most commonly used method, providing detailed information about subsurface formations.
- **Geological exploration:** This focuses on studying an area's geological history, rock types, and structural features to understand the potential for hydrocarbon accumulation. This can include fieldwork, analysis of rock samples, and geological modeling.
- **Geochemical exploration:** This involves analyzing soil, water, and rock samples for the presence of hydrocarbon indicators. Geochemical methods can help identify areas where hydrocarbons may be seeping to the surface (Figure 2.1).

Overview of geophysical, geological, and geochemical exploration methods used to identify hydrocarbon accumulations. Visual representation of subsurface layers and survey tools highlights how seismic, sampling, and modeling techniques guide prospect generation.

FIGURE 2.1 Types of exploration activities and subsurface imaging techniques.

2.2 COST ANALYSIS OF EXPLORATION ACTIVITIES

Exploration activities are costly, and understanding the cost structure is crucial for making informed investment decisions. Key cost components include:

- **Seismic surveys:** Seismic surveys are a significant expense in exploration. Costs vary depending on the survey area's size, the subsurface geology's complexity, and whether the survey is onshore or offshore. Costs include data acquisition, processing, and interpretation. For example, a large-scale 3D seismic survey in a frontier offshore area can cost millions. Seismic imaging advancements, such as full waveform inversion (FWI), allow geoscientists to create more accurate subsurface maps even in complex geological settings.[2]
    - Seismic surveys, particularly in offshore exploration, represent a significant expense, and costs can vary dramatically based on location and survey complexity. While onshore seismic surveys typically cost **between $10,000 and $30,000 per square kilometer**, offshore surveys, especially in **deepwater and ultra-deepwater regions, can exceed $200,000 per square kilometer**. Modern **4D seismic surveys,**

which involve time-lapse monitoring of reservoir changes, **can push the total cost of an offshore survey above $100 million**,[3] particularly in challenging deepwater environments. These costs include data acquisition, processing, and interpretation. Given the high expense, companies often focus on improving **seismic imaging technology and data integration** to minimize the number of dry wells drilled, thereby reducing exploration risks and costs.

- **Geological and geochemical studies:** These studies involve fieldwork, laboratory analysis of samples, and geological modeling. Costs can vary depending on the study's scope and the complexity of the geology. For instance, detailed geological mapping and sampling in a remote area can be expensive due to logistical challenges.
- **Exploratory drilling:** Drilling exploratory wells is the most expensive part of exploration. Well, costs depend on factors such as the depth of the well, the drilling time, the type of drilling rig used, and the complexity of the geological formations. Offshore drilling is significantly more expensive than onshore drilling. The cost of drilling a deepwater well can exceed $100 million.[4]
- Drilling exploratory wells is the most **capital-intensive phase** of upstream petroleum operations, and costs can vary significantly depending on the drilling environment. **Depending on depth, reservoir pressure, and logistical challenges, onshore exploratory wells typically range from $5 million to $15 million**. In contrast, **deepwater exploratory wells can cost over $100 million**, especially in complex geological settings requiring advanced well control and casing designs. **High-pressure, High-Temperature (HPHT) wells** present additional technical challenges, increasing costs even further. Another critical financial consideration is **dry hole costs**, where an exploratory well fails to encounter commercial hydrocarbons. Companies must account for these failures in their economic models, as unsuccessful wells can significantly impact project viability.

UNLOCKING THE UNSEEN—WHAT IS FWI?

Full Waveform Inversion (FWI) is a cutting-edge seismic technique that uses the entire seismic wavefield to build detailed subsurface models. Unlike conventional seismic imaging, FWI captures both travel time and waveform data, enhancing resolution and reducing drilling risk, especially in geologically complex regions like salt domes and deepwater basins.

"DRY HOLES: THE HIGH COST OF NOTHING"

Even with cutting-edge technology, more than 30% of exploratory wells drilled globally result in dry holes, which cost millions and yield no return. The dry hole rate in frontier basins can exceed 60%, underscoring the importance of robust economic screening before drilling.

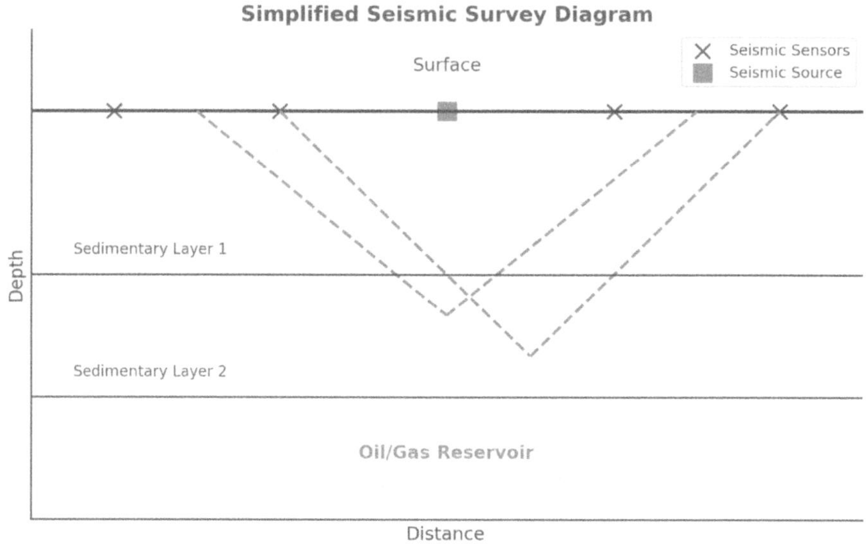

FIGURE 2.2 Seismic survey method for subsurface imaging.

- **Other exploration expenses:** Other expenses include land acquisition costs, permitting fees, environmental impact assessments, and administrative overhead. These costs can add up, especially in regions with stringent regulatory requirements (Figure 2.2).

Illustration of seismic wave propagation and reflection used to detect subsurface geological layers and hydrocarbon reservoirs. Highlights how seismic sensors and source arrays generate data for structural mapping and exploration targeting (Figure 2.3).

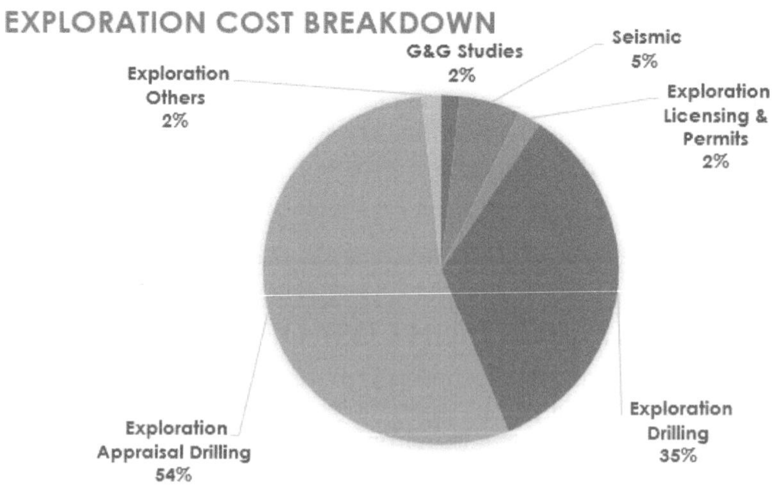

FIGURE 2.3 Exploration cost breakdown by activity type.

Distribution of exploration expenditures across key categories, including drilling, seismic, licensing, and geological studies. Highlights that appraisal and exploratory drilling account for nearly 90% of total costs, underscoring the capital intensity and risk of subsurface evaluation.

2.3 RISK ASSESSMENT AND PROSPECT EVALUATION

Exploration is inherently risky. There is always the possibility that an exploratory well will not find commercial quantities of hydrocarbons. Risk assessment is, therefore, a critical part of the exploration process. Key risks include:

- **Geological risk:** The risk that hydrocarbons are not present in a prospect. This is assessed by analyzing geological and geophysical data to estimate the likelihood of finding hydrocarbons.
  - Assessing geological risk is essential in petroleum exploration, as the likelihood of discovering commercial hydrocarbons varies widely between regions. Companies use the **Probability of Geological Success (PgS)** metric to quantify the chance of making a hydrocarbon discovery. In **frontier basins with limited data, PgS values typically range between 5% and 15%**, indicating a high-risk, high-reward scenario. Conversely, **near-field exploration, which targets areas close to existing discoveries, often has a PgS between 30% and 60%**, due to better geological understanding and data availability. Integrating **seismic data, well logs, and geological modeling** helps refine PgS estimates, allowing companies to make informed exploration decisions and allocate capital more effectively.
- **Technical risk:** Encountering technical challenges during drilling and well completion. This includes risks related to drilling equipment, wellbore stability, and reservoir characteristics.
- **Economic risk:** The risk that the discovered reserves are not economically viable to develop due to factors such as low oil and gas prices or high development costs. This involves evaluating market conditions and cost estimates (Figure 2.4).

Illustrates probabilistic outcomes of a drilling decision, including success, partial success, and dry hole scenarios. Highlights how expected monetary values and branching probabilities guide exploration risk management and capital allocation.

Risk assessment methods: Several methods are used to assess these risks, including:

- **Geological modeling:** Creating 3D subsurface models to better understand the geological structure and potential hydrocarbon traps. Advanced software tools integrate seismic data, well logs, and geological information.
- **Probabilistic analysis:** Using statistical methods to estimate the probability of success in finding hydrocarbons. This involves defining different scenarios and calculating the likelihood of each scenario.
- **Decision Tree Analysis is a decision-making tool that helps evaluate different exploration scenarios and their potential outcomes. It helps identify** the best course of action based on the probability of success and associated costs.

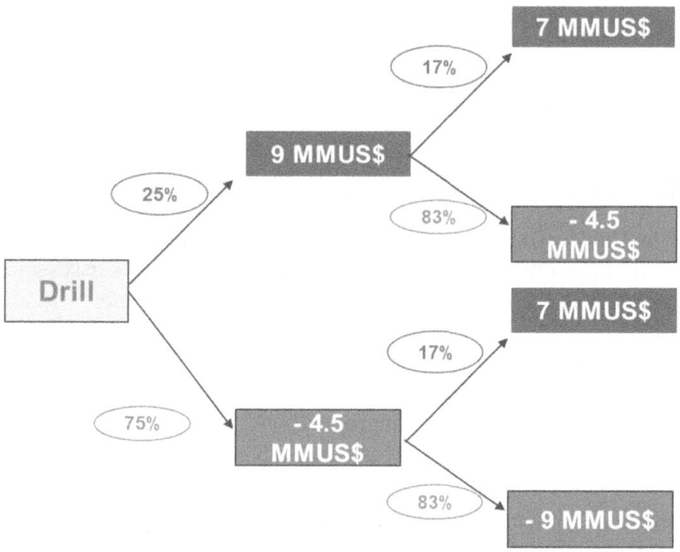

FIGURE 2.4 Decision tree analysis for drilling outcomes.

Exploration is not about striking it rich; it's about managing uncertainty.

Peter Voser
Former CEO, Shell

Economic metrics for prospect evaluation: Exploration prospects are evaluated using economic metrics such as Net Present Value (NPV) and Internal Rate of Return (IRR), considering the associated project risks. These metrics help determine the prospect's potential profitability and attractiveness (Figure 2.5).

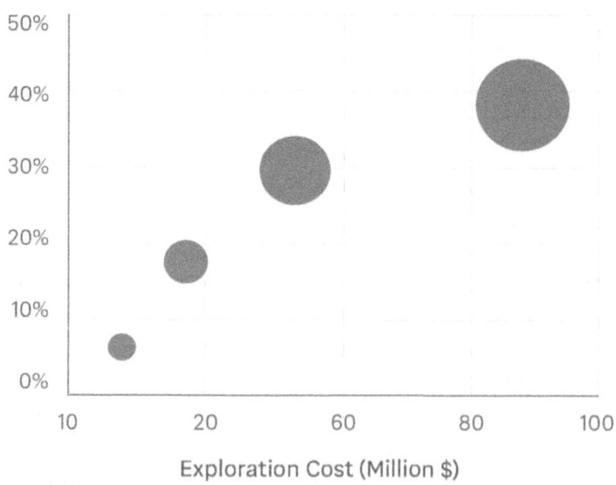

FIGURE 2.5 Probability of geological success vs. exploration cost.

Bubble chart illustrating how increased exploration investment correlates with higher geological success probability. Highlights the capital-risk trade-off in prospect evaluation, with bubble size suggesting project scale or impact.

2.4 EXPLORATION STRATEGIES AND PORTFOLIO MANAGEMENT

Exploration companies employ various strategies to manage their exploration activities. These include:

- **Frontier exploration involves exploring new, underexplored areas. This activity** carries substantial risk but also has the potential for significant discoveries. Companies often target frontier basins with promising geological features.
- **Near-field exploration:** Exploring areas adjacent to existing oil and gas fields. This carries a lower risk as the geology is better understood. Near-field exploration aims to extend the life of existing fields and infrastructure.
- **Focusing on specific geological plays:** Concentrating exploration efforts on specific geological formations that are known to contain hydrocarbons. This involves identifying plays with proven potential and applying advanced techniques to unlock new reserves.
- **Play-based exploration**: In recent years, many companies have shifted toward a **play-based exploration strategy**, focusing on well-defined geological plays rather than isolated prospects. A **play** is a group of petroleum accumulations with similar geological characteristics, such as source rock, migration pathways, reservoir rock, and trapping mechanisms. This approach reduces exploration uncertainty by targeting regions with **proven hydrocarbon potential**, rather than testing individual prospects with limited supporting data. A notable example is the **U.S. shale boom**, where companies systematically targeted **shale plays like the Permian Basin and Eagle Ford**, leading to rapid production growth. Play-based exploration enhances **efficiency, reduces risk, and allows for economies of scale** in development and production.

Portfolio management: Exploration companies also manage their exploration portfolios to diversify risk and maximize returns. This involves investing in a mix of exploration projects with varying levels of risk and potential reward. Companies allocate their exploration budgets across different regions, plays, and project types to balance risk exposure.

Farm-in/farm-out agreements: Farm-in/farm-out agreements are often used to share exploration risks and costs with other companies. A farm-out agreement allows one company (the farmor) to transfer a portion of its interest in an exploration project to another company (the farmee) in exchange for financial or operational commitments. This enables companies to leverage each other's strengths and resources.

Farm-in and farm-out agreements are crucial tools for risk-sharing in exploration, allowing companies to participate in projects while minimizing financial exposure. In a typical **farm-out agreement**, the company holding exploration rights (farmor)

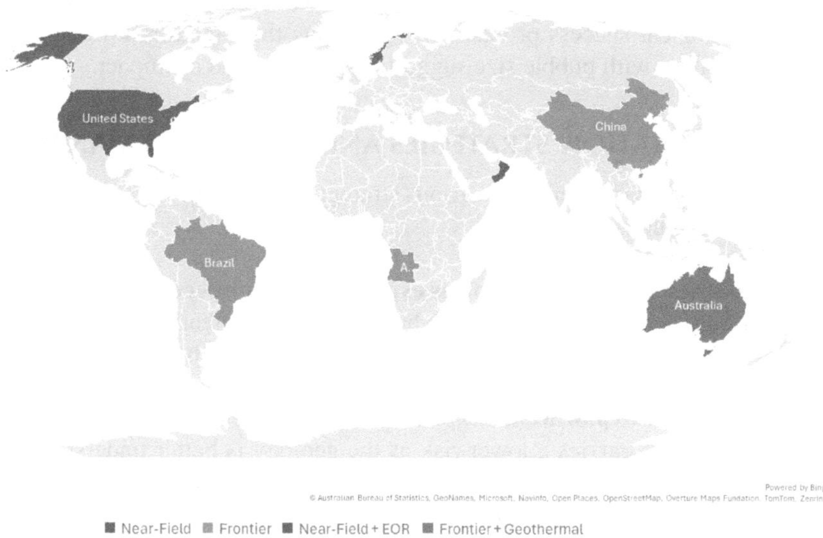

FIGURE 2.6 Frontier vs. near-field discoveries (2010–2024).

transfers a portion of its interest to another company (farmee) in exchange for financial or operational commitments. A key component of such agreements is **carried interest**, where the farmee agrees to **cover the farmor's share of exploration costs**, often up to the completion of a well or discovery phase. This arrangement allows smaller companies to **access capital and technical expertise** while enabling larger firms to **diversify risk across multiple prospects**. Farm-in/farm-out deals are especially common in **frontier exploration**, where the financial risks of drilling wildcat wells are particularly high (Figure 2.6).

This map illustrates significant upstream discoveries over the past decade, differentiating between frontier plays and near-field redevelopment. While mature basins like the North Sea and Permian yield incremental gains, frontier regions like Brazil's pre-salt and China's shale gas zones have attracted high-risk, high-reward investment strategies.

2.5 CASE STUDIES OF EXPLORATION PROJECTS

Case Study 1: The Johan Sverdrup Field (Norway): The financial realities of exploration projects vary significantly depending on location, geological complexity, and development strategies. Some projects achieve high economic returns, while others struggle due to technical challenges or poor reservoir performance. For instance, the Johan Sverdrup field in Norway, discovered by Lundin Petroleum in 2010, is one of the most significant oil discoveries in the North Sea in recent decades. With estimated reserves of 2.2–3.2 billion barrels of oil equivalent, its initial development required an investment of approximately $10 billion. However, its break-even price of $20–30 per barrel made it one of the most economically robust offshore developments. The field's success is attributed to advanced seismic imaging, strategic partnerships, and a stable regulatory framework. Over its projected 50-year production lifespan, Johan Sverdrup is expected to generate substantial revenue for Norway.

Case Study 2: The Chinguetti Field (Mauritania), discovered in 2001, illustrates the risks of frontier exploration. Initially estimated to hold 120 million barrels, the field suffered from poor reservoir performance and high production costs. As a result, its output declined rapidly, leading to its eventual shutdown in 2017. Meanwhile, the Lula field in Brazil's pre-salt formation is a prime example of a high-risk, high-reward deepwater project. Despite holding over 8 billion barrels of recoverable reserves, its development required an investment of more than $50 billion, driven by technical challenges in ultra-deepwater drilling and complex reservoir characteristics. These case studies underscore the importance of thorough feasibility studies, risk management, and technological advancements before committing to large-scale oil and gas exploration investments.

Case Study 3: The Gulf of Mexico—Marine CSEM Exploration: The integration of **3D inversion techniques with seismic imaging** demonstrated the **potential of marine CSEM technology in offshore exploration**, particularly for identifying small hydrocarbon targets in geologically complex regions. CSEM surveys typically cost between **$5–$15 million per survey**, a fraction of the cost of drilling an unsuccessful deepwater well, which can exceed **$100 million**. This case study underscores how **advanced electromagnetic surveying methods can complement traditional seismic techniques**, reducing exploration risk and improving drilling success rates in deepwater environments.

Case Study 4: North America—Surface Geochemical Exploration: By incorporating soil-iodine geochemistry into hydrocarbon migration analysis, explorations could **significantly lower exploration risk and reduce finding costs by 20–40%**. In some areas, geochemical methods have led to **a 30% increase in drilling success rates**. This case highlights the **importance of integrating geochemical data with seismic imaging**, enabling more precise drilling decisions and increasing oil and gas exploration efficiency.

Case Study 5: Australia—Geothermal Exploration (continued): The **Cooper Basin Geothermal Project** models how **traditional petroleum exploration expertise can be leveraged for renewable energy development**. With estimated geothermal potential exceeding **5,000 megawatts (MW)**, the project has demonstrated how enhanced drilling technologies can improve efficiency. Geothermal wells in the Cooper Basin typically cost **$10–$20 million per well**, significantly lower than deepwater oil wells, making this an attractive alternative energy source.

Case Study 6: Africa—Frontier Basin Exploration: The success of the **Kwanza Basin exploration** demonstrates how **advanced seismic imaging and geological modeling** can unlock significant hydrocarbon reserves in previously underexplored regions. Recent discoveries in the Kwanza Basin have added **over 2 billion barrels of recoverable oil** to Angola's reserves, attracting **investment exceeding $3 billion**. The subsequent commercial discoveries have reaffirmed the **economic potential of frontier basin exploration despite its inherent risks**.

Case Study 7: Europe—Near-Field Exploration: The **Buzzard Field discovery** showcases how **near-field exploration in mature basins can maximize production and extend the economic life of existing fields**. Discovered in **2001**, the field has an estimated **recoverable reserve of over 1.5 billion barrels of oil equivalent**, generating more than **$50 billion in revenue since production began in 2007**. Using **advanced seismic techniques and reservoir modeling**, companies have **reduced drilling uncertainty by up to 40%**, proving that **targeted exploration around existing infrastructure remains a cost-effective strategy**.

Case Study 8: South America—Deepwater Exploration: The **Lula Field in Brazil's Santos Basin** highlights the **transformative impact of deepwater exploration**, positioning the country as a **major global oil producer**. The field holds over **8 billion barrels of recoverable reserves** and requires an investment exceeding **$50 billion** for development. However, its break-even price of **$35–45 per barrel** ensures profitability even in volatile market conditions. The success of Lula and other pre-salt discoveries has **attracted more than $100 billion in international investment** into Brazil's deepwater sector.

Case Study 9: Asia—Shale Gas Exploration: China's **Sichuan Basin shale gas development** demonstrates the **scalability and adaptability of hydraulic fracturing and horizontal drilling technologies**. The basin holds estimated **recoverable reserves of 400–500 trillion cubic feet (TCF)** of natural gas, making it one of the largest shale plays outside North America. Since 2010, China has invested **over $50 billion** in shale gas exploration, reducing **LNG import dependency by 15%**. This initiative's success has reinforced the unconventional resources' role **in global energy diversification**.

Case Study 10: Middle East—Enhanced Oil Recovery (EOR): The **Mukhaizna Field in Oman** is a **benchmark for successful EOR applications in heavy oil reservoirs**. By deploying **steam flooding techniques**, Occidental Petroleum has significantly **increased recovery rates from 10% to 40%**, adding **more than 500 million barrels of recoverable reserves**. The total investment in the project has exceeded **$2 billion**, but the improved recovery has generated **over $10 billion in revenue** since implementation. This case study highlights how **EOR can extend the productive lifespan of mature oil fields while maintaining economic viability**.

2.5.1 Summary of Case Studies

These case studies provide diverse exploration scenarios, highlighting various techniques, strategies, and technologies for discovering and developing hydrocarbon reserves. They demonstrate the importance of integrating multiple data sources, advanced technologies, and innovative approaches to reduce risk and improve exploration success rates. These examples underscore the potential for exploration activities to contribute to energy security, economic development, and sustainability.

2.6 CHAPTER SUMMARY

This chapter covers the key aspects of exploration economics, from the exploration process and cost analysis to risk assessment, prospect evaluation, exploration strategies, and portfolio management. We emphasize the exploration's high-risk, high-reward nature and the importance of sound economic decision-making.

By understanding these factors, companies can make informed decisions that balance risk and reward, ultimately ensuring the sustainability and profitability of their exploration activities. The diverse case studies illustrate the successful application of various exploration techniques across regions and geological settings.

The next chapter will discuss reservoir economics, focusing on the evaluation, development, and management of hydrocarbon reservoirs. We will explore the principles and practices governing reservoir engineering and the economic considerations in optimizing hydrocarbon recovery.

NOTES

1. Rystad Energy, *Global Exploration Trends Report*, 2023.
2. Schlumberger, "Full Waveform Inversion: Unlocking the Subsurface," Technical Brief, 2022.
3. Source: Wood Mackenzie, Global Seismic Cost Benchmarks, 2023.
4. Source: Rystad Energy, 2023 Offshore Drilling Report.

3 Reservoir Economics

3.1 INTRODUCTION TO RESERVOIR ENGINEERING AND ECONOMICS

Reservoir engineering is a crucial discipline within the upstream petroleum sector that focuses on efficiently and effectively managing hydrocarbon reservoirs. Its primary goal is to maximize the ultimate recovery of oil and gas while optimizing production rates and minimizing costs. Reservoir economics plays a vital role in this process, providing the framework for evaluating the economic viability of different reservoir management strategies.

A thorough understanding of reservoir properties is essential for accurate economic evaluation. Reservoir engineers collaborate closely with geologists and geophysicists to characterize the reservoir and develop models that predict its performance under various production scenarios. These models are then used to forecast production rates, estimate ultimate recovery, and assess the economic impact of different development and management options. Integrating reservoir engineering and economic analysis is therefore critical for making sound investment decisions and maximizing the value of hydrocarbon assets.

> Reservoirs don't just hold oil; they hold value — if managed right.
>
> **Abdelrahman Al-Omari**
> *Aramco Senior Reservoir Engineer*

Reservoir economics is an interdisciplinary field that combines technical knowledge with financial analysis. It involves assessing the profitability of various reservoir management strategies, taking into account production rates, recovery factors, costs, risks, and market conditions. Reservoir engineers and economists use various tools and techniques to evaluate the economic viability of projects, including discounted cash flow analysis, sensitivity analysis, and probabilistic modeling.

3D surface plot showing hydrocarbon saturation variability across reservoir width and length. Highlights spatial heterogeneity and its implications for well placement, recovery optimization, and economic forecasting.

"DIGITAL RESERVOIRS"

With modern AI and real-time sensors, engineers now simulate entire fields before a single well is drilled—cutting planning time by 40% (Figure 3.1).

3.2 EVALUATION OF RESERVOIR PROPERTIES

Accurately characterizing reservoir properties, including reservoir rock types, fluid saturation, porosity, permeability, and pressure conditions, is fundamental

DOI: 10.1201/9781003750345-4

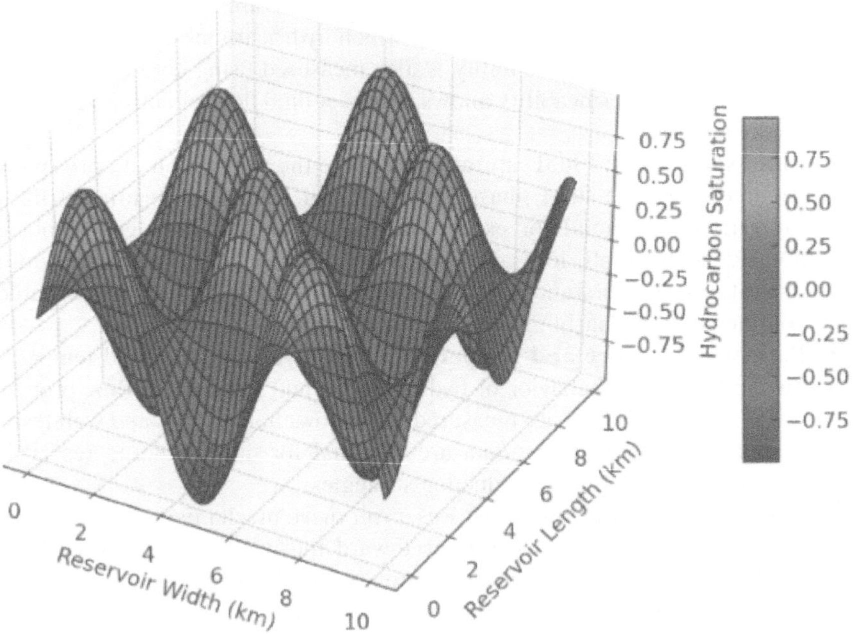

FIGURE 3.1 Reservoir simulation model: Performance and economic impact.

to effective reservoir management. A detailed understanding of these properties enables better prediction of reservoir behavior, optimization of resource recovery, and implementation of tailored production strategies. Additionally, comprehensive characterization informs decision-making regarding well placement, stimulation techniques, and enhanced oil recovery (EOR) methods, maximizing reservoir exploitation's economic viability and sustainability. Key properties include:

- **Porosity:** Porosity measures the proportion of void space within the rock. It determines the amount of hydrocarbons that a reservoir can hold. Porosity is typically measured using core analysis in the laboratory or well logs in the field. High porosity indicates a greater capacity for storing hydrocarbons, making it a crucial parameter in reservoir evaluation. Reservoir properties vary widely depending on rock type and geological history.[1] For instance, high-quality sandstone reservoirs often exhibit porosity between **20% and 35%**, whereas tight reservoirs may have porosity as low as **5%**. Permeability, a key factor in fluid flow, ranges from **less than 0.01 mD in unconventional reservoirs** to **10,000 mD in highly permeable formations**. Understanding these properties allows engineers to estimate hydrocarbon storage capacity and optimize production strategies.

- **Permeability:** Permeability measures a fluid's ability to flow through porous rock. It influences the rate at which hydrocarbons can be produced from the reservoir. Permeability is also measured using core analysis and well tests. High permeability allows for easier fluid flow, which can enhance production rates.
- **Fluid saturation:** Fluid saturation refers to the proportion of pore space occupied by each fluid phase (oil, gas, and water). Determining fluid saturations is crucial for estimating the original oil and gas in place (OOIP/OGIP). Well, logs and core analysis are used to determine fluid saturations. Understanding fluid saturation helps predict recoverable volumes of hydrocarbons.
- **Reservoir pressure and temperature:** Reservoir pressure and temperature influence the behavior of hydrocarbons and affect their flow properties. These parameters are measured using downhole gauges and well tests. Pressure and temperature data are essential for understanding reservoir dynamics and designing production strategies.
- **Reservoir drive mechanisms:** Reservoir drive mechanisms are the natural forces that push hydrocarbons toward the wellbore. Common drive mechanisms include pressure depletion, water drive, gas drive, and gravity drainage. Understanding the drive mechanism is crucial for predicting reservoir performance and designing appropriate production strategies. Each drive mechanism has unique characteristics that impact production rates and recovery factors. Each reservoir drive mechanism influences the ultimate recovery factor. **Solution gas drive reservoirs** typically recover **5–15%** of the original oil in place (OOIP), while **strong water drive reservoirs** can recover up to **50%**. **Gravity drainage**, common in fractured carbonate reservoirs, can yield **10–40%** recovery. Understanding the dominant drive mechanism is critical for selecting the most effective production strategy.

Data Acquisition Methods:

- **Well testing:** Various well tests, such as pressure transient and production tests, are conducted to gather information about reservoir permeability, pressure, and connectivity. Pressure transient tests analyze the reservoir's pressure response to changes in production rates, providing insights into reservoir properties.
- **Core analysis:** Cores are rock samples from the reservoir during drilling. Laboratory analysis of these cores provides detailed information about porosity, permeability, fluid saturations, and other rock properties. Core analysis involves various techniques, including thin-section petrography, X-ray diffraction, and fluid displacement experiments.
- **Seismic surveys:** Seismic surveys provide images of the subsurface, which can be used to infer reservoir geometry, structure, and potential fluid distribution. Seismic data help identify geological features such as faults, folds, and stratigraphic traps that can influence hydrocarbon accumulation.

Advanced Techniques in Reservoir Evaluation:

- **Digital rock physics:** Digital rock physics involves creating digital models of rock samples using high-resolution imaging techniques such as micro-CT scanning. These models simulate fluid flow and mechanical properties, providing detailed insights into reservoir characteristics.
- **Machine learning and artificial intelligence:** Machine learning algorithms are increasingly used to analyze large datasets from well logs, core analysis, and seismic surveys. AI-driven models can identify patterns and correlations that may not be apparent through traditional analysis, improving reservoir characterization and prediction accuracy. Machine learning and artificial intelligence (AI) advancements have transformed reservoir characterization. AI algorithms analyze **large datasets from seismic surveys, well logs, and production history**, enabling engineers to identify **hidden patterns** that may not be apparent in traditional analysis. For example, **deep learning models** can predict permeability trends in heterogeneous reservoirs with **90% accuracy**, reducing **data interpretation time by over 50%**.

> Knowing your rock is half the profit. The other half is knowing your data.
>
> *Schlumberger Reservoir Characterization Division*

Comparison of recovery efficiency across reservoir types, from tight shale to EOR-enhanced systems (Figure 3.2). Highlights the technical limitations and potential of each formation, informing economic modeling and field development strategy.

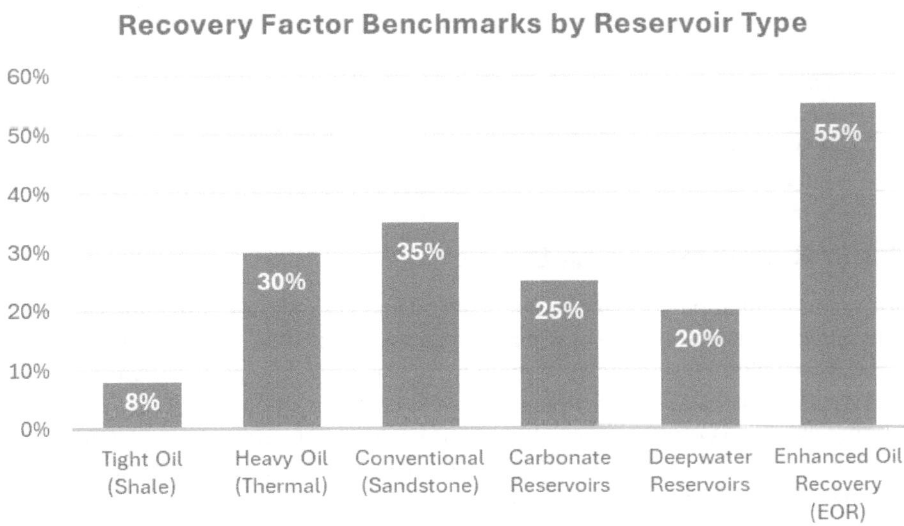

FIGURE 3.2 Oil recovery factors by reservoir type.

3.3 RESERVOIR PERFORMANCE ANALYSIS

Analyzing reservoir performance is crucial for accurately forecasting future production rates and optimizing reservoir management strategies. This process involves a comprehensive assessment of various factors, including pressure transient analysis, production history, and fluid properties. Engineers can identify trends and patterns that influence reservoir behavior by integrating advanced modeling techniques and real-time data. Moreover, understanding the reservoir's rock and fluid characteristics enables the implementation of tailored recovery methods, enhancing overall efficiency and productivity. Ultimately, a thorough analysis not only supports informed. Decision-making also maximizes resource recovery while minimizing operational risks and costs. Common methods include:

- **Decline curve analysis:** Decline curve analysis is a graphical technique used to forecast production rates and estimate ultimate recovery based on historical production data. Different decline curve models (exponential, hyperbolic, harmonic) are used depending on the reservoir characteristics and drive mechanism. Decline curve analysis helps identify trends and predict future production behavior.
- **Material balance calculations:** Material balance is a volumetric method used to estimate OOIP/OGIP based on production data and reservoir pressure changes. Calculations provide insights into reservoir depletion and drive mechanisms.
- **Reservoir simulation:** Reservoir simulation involves building complex computer models of the reservoir to predict its performance under different production scenarios. These models can incorporate detailed geological and engineering data, allowing for the optimization of well placement, production rates, and EOR strategies (Figure 3.3).

"RECOVERY FACTOR BENCHMARKS"

- Primary Recovery: 10–20% of OOIP
- Secondary (Waterflood): 30–50%
- Tertiary (EOR): 40–60%+

3.3.1 Reservoir Simulation Example

Reservoir simulation is a critical tool in the oil and gas industry, providing a dynamic and nuanced understanding of reservoir behavior over time. By integrating diverse datasets—including geological formations, petrophysical properties such as porosity and permeability, and historical production data—simulation models can accurately predict the effects of varying production strategies and operational decisions.

For instance, in a mature oil field undergoing waterflooding, advanced simulation techniques can evaluate optimal water injection rates and strategic well placements, enabling engineers to visualize the reservoir's responsiveness. This process involves analyzing factors like reservoir pressure, fluid dynamics, and saturation levels to maximize efficiency. Moreover, simulations can forecast incremental oil recovery,

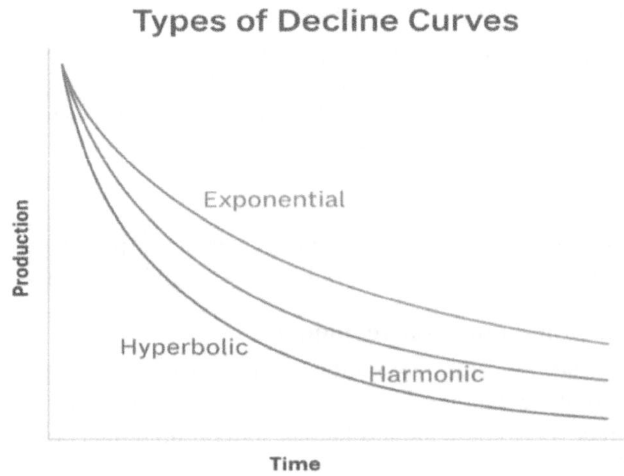

FIGURE 3.3 Types of decline curves in reservoir performance forecasting. Comparison of exponential, hyperbolic, and harmonic decline models used to estimate future production rates. Highlights how reservoir characteristics and drive mechanisms influence decline behavior and recovery forecasting.

helping to identify the most effective methods for enhancing production while minimizing costs and environmental impacts. By leveraging these insights, operators can make informed decisions that significantly improve recovery rates and overall reservoir management.

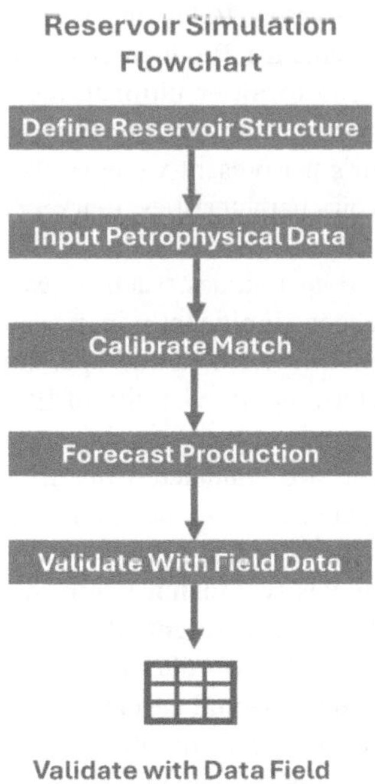

Key Parameters in Reservoir Simulation:

- **Reservoir geometry:** The size, shape, and structure of the reservoir, including the distribution of porosity and permeability.
- **Fluid properties:** The physical and chemical properties of reservoir fluids, including viscosity, density, and phase behavior.
- **Well configuration:** Production and injection wells' number, location, and orientation.
- **Production strategy:** The planned sequence of production activities, including the timing and rates of fluid injection and withdrawal.

Advanced Reservoir Simulation Techniques:

- **EOR simulation:** EOR simulation models evaluate the effectiveness of different EOR techniques, such as CO_2 injection, polymer flooding, and thermal recovery. These models help design optimal injection schemes and predict incremental recovery.
- **Unconventional reservoir simulation:** Unconventional reservoirs, such as shale gas and tight oil formations, require specialized simulation models for hydraulic fracturing and complex fluid flow behavior. Advanced simulation techniques, such as discrete fracture network modeling, simulate production from fractured reservoirs.

3.4 ECONOMIC CONSIDERATIONS IN RESERVOIR MANAGEMENT

Economic considerations play a critical role in reservoir management decisions:

- **Optimizing production rates:** Balancing short-term cash flow with long-term recovery is a key challenge. Producing at high rates can maximize near-term profits but may lead to lower ultimate recovery. Reservoir engineers and economists collaborate to determine the optimal production rates that maximize the reservoir's net present value (NPV). The optimal production rate depends on reservoir characteristics, market conditions, and operational constraints.
- **EOR techniques:** EOR techniques, such as waterflooding, gas injection, and chemical flooding, can significantly increase hydrocarbon recovery but involve substantial capital investment and operating costs. Economic evaluation is essential to determine the viability of EOR projects. Key economic factors include incremental oil recovery, injection costs, and the timing of cash flows. EOR projects are evaluated using discounted cash flow analysis, which considers the time value of money and the risk associated with future cash flows. Reservoir development costs vary significantly based on location and extraction complexity. **Onshore conventional reservoirs** typically cost **$5–15 per barrel of oil equivalent (BOE)** to develop, while **deepwater offshore projects can reach $40–60/BOE** due to high drilling and infrastructure expenses. **Unconventional reservoirs, such as shale and tight oil, range from $20–40/BOE** due to the need for hydraulic fracturing. **EOR projects** can cost **$10–25/BOE**, depending on the method used,[2] with gas

EOR Method	Incremental Recovery (% of OOIP)	Notes
Gas Injection	10%–25%	Includes CO_2, N_2, and hydrocarbon gas. CO_2 is most effective in miscible zones.
Chemical Injection	5%–15%	Depends on polymer, surfactant, or alkaline-polymer mix.
Thermal Recovery	20%–40%	Most effective in heavy oil and bitumen; includes steam flooding and SAGD.
Microbial EOR	3%–10%	Still experimental in many regions; potential varies widely.

FIGURE 3.4 Comparison of enhanced oil recovery (EOR) methods and incremental recovery.

injection techniques generally being more expensive than waterflooding. Understanding these cost variations is crucial for determining project viability and selecting cost-effective recovery strategies.

- **Well placement and completion strategies:** Optimal well placement and completion strategies can minimize drilling costs and maximize production. Economic models evaluate different well locations and completion designs. For example, horizontal wells and multi-stage hydraulic fracturing can enhance production in unconventional reservoirs but involve higher costs and technical complexity.
- **Managing reservoir uncertainty:** Reservoir properties are often uncertain, impacting production forecasts and economic evaluations. Probabilistic methods and sensitivity analysis are used to incorporate uncertainty into economic models and assess the range of potential outcomes. Companies can make more informed decisions by evaluating different scenarios and quantifying risks (Figure 3.4).

Summary of gas injection, chemical flooding, thermal recovery, and microbial EOR techniques, with typical incremental recovery ranges. Highlights how reservoir type and method selection influence recovery efficiency and project economics

3.4.1 Example of Economic Evaluation:

An economic evaluation of an EOR project would involve calculating the expected increase in oil recovery, estimating the costs of implementing the EOR technique, and assessing the project's NPV and IRR. Sensitivity analysis might be conducted to evaluate the impact of different oil price scenarios on project economics. This process helps determine whether the incremental recovery and associated revenue justify the investment in EOR.

Key Economic Metrics:

- **Net present value (NPV):** NPV is the sum of discounted future cash flows minus the initial investment. It reflects the value created by a project and is a key indicator of economic viability.

- **Internal rate of return (IRR):** IRR is the discount rate that makes a project's NPV equal to zero. It represents the project's expected rate of return and is used to compare the profitability of different investment opportunities.
- **Payback period:** The payback period is the time required for a project to recoup its initial investment. It provides a simple measure of project risk and liquidity but does not account for the time value of money.
- **Profitability index (PI):** PI is the ratio of the present value of future cash flows to the initial investment. A PI greater than one indicates a profitable project.

"DID YOU KNOW?"

Ghawar Field's water injection system alone is among the world's largest water processing operations, handling over 9 million barrels per day.

Strategic Considerations in Reservoir Management (continued):

- **Investment timing:** The timing of investments in reservoir development and EOR projects is crucial. Delaying investments may result in missed opportunities, while premature investments may lead to suboptimal returns. Strategic planning involves analyzing market conditions, technological advancements, and reservoir performance to determine the optimal timing for investment.
- **Technology adoption:** Innovative technologies can enhance reservoir performance and reduce costs. However, technological risks and uncertainties must be carefully evaluated. Companies must strike a balance between the potential benefits of innovative technologies and the associated risks and costs.
- **Regulatory compliance:** Compliance with regulatory requirements is essential for reservoir management. Regulations may impact production rates, environmental practices, and reporting standards. Companies must stay informed about regulatory changes and ensure their operations comply with relevant laws and guidelines.
- **Sustainability and environmental considerations:** The growing emphasis on sustainability and environmental stewardship influences reservoir management decisions. Companies are increasingly adopting practices that minimize their environmental impact, such as reducing greenhouse gas emissions, conserving water, and implementing effective waste management practices. Sustainability initiatives enhance a company's reputation and contribute to long-term economic viability.

Comparative analysis of economic thresholds for vertical, horizontal, offshore, and shale wells (Figure 3.5). Highlights how geological complexity and infrastructure requirements shape breakeven prices, informing reservoir development strategy and investment timing.

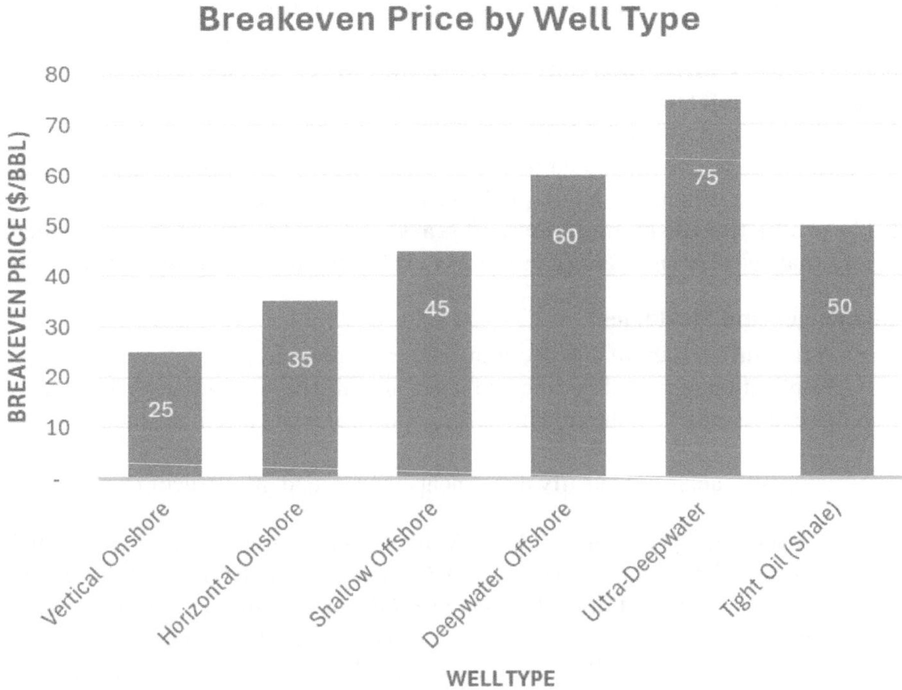

FIGURE 3.5 Breakeven oil prices by well type.

3.5 CASE STUDIES OF RESERVOIR MANAGEMENT

Case Study 1: Prudhoe Bay Field (Alaska, USA): Prudhoe Bay, discovered in **1968**, is the **largest oil field in North America**. Initially estimated to contain **25 billion barrels of oil in place (OOIP)**, the field required advanced reservoir management techniques to maximize recovery. **Waterflooding and gas injection** were implemented, increasing the recovery factor from **30% to over 60%**. The field has produced over **12 billion barrels of oil** and remains an essential energy source despite extreme Arctic conditions. The substantial economic impact has generated **over $250 billion in revenue**. Despite the field's maturity, EOR investments totaling $10 billion have ensured sustained production and economic viability.

Challenges and Strategies:

- Harsh Arctic environment with extreme temperatures and logistical challenges.
- Implement waterflooding and gas injection to maintain reservoir pressure and enhance recovery.
- Economic evaluation of EOR techniques to balance costs and benefits.
- Continuous reservoir monitoring and optimization to adapt to changing conditions.

Case Study 2: Ekofisk Field (North Sea, Norway): In **1969**, the **Ekofisk field** was one of the first commercial offshore discoveries in the North Sea. Initially, the field's **recovery factor was only 17%**, but through the implementation of **waterflooding, gas injection, and horizontal drilling**, it was increased to **over 50%**. The field originally held **3.3 billion barrels of OOIP**, and these enhanced recovery techniques have contributed **over 3 billion barrels of cumulative production**. Continuous investment in **EOR projects exceeding $5 billion** has extended the field's **productive life by more than 30 years**. Today, Ekofisk serves as an example of how **mature offshore fields can remain viable through reservoir optimization**.

Challenges and Strategies:

- Aging infrastructure and declining reservoir pressure.
- Application of waterflooding and horizontal drilling to enhance recovery.
- Use of reservoir simulation models to optimize well placement and production rates.
- Economic analysis to justify investments in EOR and infrastructure upgrades.

Case Study 3: Weyburn Field (Saskatchewan, Canada): The Weyburn field, discovered in **1954**, became a pioneer in **CO_2 injection for EOR** in **2000**. The field originally contained **1.4 billion barrels of OOIP**, with a **natural recovery factor of around 30%**. However, with CO_2 injection, the recovery factor increased by **15–20%**, producing **over 130 million additional barrels**. The project has also been an essential **carbon capture and storage (CCS) initiative**, storing over **40 million tons of CO_2**, reducing greenhouse gas emissions while increasing oil production. The economic feasibility was ensured by **favorable tax incentives and infrastructure investment**. Beyond its oil recovery success, the Weyburn project has generated **millions of dollars in carbon credit benefits**, reinforcing its economic and environmental advantages

Challenges and Strategies:

- Technical and logistical challenges associated with CO_2 capture, transport, and injection.
- Integration of CO_2-EOR with CCS initiatives.
- Economic assessment of CO_2-EOR to ensure financial viability and environmental benefits.
- Collaboration with stakeholders to address regulatory and sustainability considerations.

Case Study 4: Daqing Field (China): Daqing, discovered in **1959**, is **China's largest oil field** and has historically contributed **over 50% of the country's total crude oil production**. With an original oil-in-place (OOIP) of **16 billion barrels**, the field initially operated under **primary and secondary recovery**, yielding recovery factors of **35%**. However, polymer flooding—one of the world's largest EOR applications—increased the recovery factor to **over 50%**, adding **billions of barrels** of recoverable oil. The estimated **investment in polymer flooding exceeded $2–3 billion**, ensuring **sustained production and economic feasibility**.

Challenges and Strategies:

- Declining production rates and reservoir pressure in a mature field.
- Implementation of polymer flooding and chemical flooding to enhance recovery.
- Economic evaluation of EOR techniques to determine the most cost-effective methods.
- Continuous monitoring and optimization to adapt to changing reservoir conditions.

Case Study 5: Cantarell Field (Mexico): The **Cantarell field**, discovered in **1976**, was once **Mexico's largest oil producer**, responsible for over **2 million barrels per day** at its peak in the early 2000s. However, by the late 1990s, production began **declining at an alarming rate** due to falling reservoir pressure. To counteract this, **nitrogen injection** was introduced, making it one of the largest nitrogen injection projects in the world. The **$6 billion investment in nitrogen injection** successfully stabilized production and **extended the field's life by over 15 years**. Despite the initial success, production eventually declined again, illustrating the **challenges of sustaining mature reservoirs over long periods**.

Challenges and Strategies:

- Rapid production declines and declining reservoir pressure.
- Implementation of nitrogen injection to maintain pressure and stabilize production.
- Economic evaluation of nitrogen injection to ensure financial viability.
- Ongoing reservoir monitoring and optimization to adapt to changing conditions.

Case Study 6: Ghawar Field (Saudi Arabia): The **Ghawar field**, discovered in **1948**, is the **largest onshore oil field in the world**, with **an estimated 88 billion barrels of OOIP**. This supergiant field has maintained **production rates of around 5 million barrels per day** for decades. The success of Ghawar's longevity lies in **aggressive water injection programs, smart completions, and advanced reservoir simulation techniques**. Horizontal drilling and **intelligent well technologies** have played a significant role in sustaining high production rates. Over its lifetime, **Ghawar has generated an estimated $1 trillion in revenue**, making it one of the most economically significant oil fields in history.

Challenges and Strategies:

- Managing a massive and highly productive reservoir with complex geology.
- Implementation of water injection and horizontal drilling to enhance recovery.
- Use of advanced reservoir simulation models to optimize production strategies.
- Economic evaluation of reservoir management techniques to ensure long-term profitability.

3.5.1 SUMMARY OF CASE STUDIES

These case studies demonstrate the economic and technical feasibility of various reservoir management strategies. Fields like **Ghawar and Prudhoe Bay** have shown how **long-term investment in EOR** can sustain high production and deliver **trillions of dollars in revenue**. Meanwhile, projects such as **Weyburn's CO_2 injection** highlight the potential for **synergy between carbon capture and oil recovery**, making sustainability a key component of modern reservoir economics. As oil fields mature, integrating **cost-effective enhanced recovery methods** will be essential to maintain production and optimize financial returns.

3.6 CHAPTER SUMMARY

This chapter has explored the critical intersection of reservoir engineering and economics. We discussed the importance of understanding and evaluating reservoir properties, the various methods used to analyze reservoir performance, and the key economic considerations involved in reservoir management decisions. We emphasized the need to strike a balance between technical feasibility and economic viability to maximize the value of hydrocarbon resources.

We demonstrated how advanced reservoir management strategies and economic evaluations contribute to successful field development and production optimization by examining real-world case studies. The next chapter will delve into drilling economics, focusing on the cost analysis, risk assessment, and decision-making processes involved in drilling operations.

NOTES

1. Schlumberger Oilfield Glossary.
2. SPE International, 2022 Enhanced Recovery Report.

4 Drilling Economics

4.1 INTRODUCTION TO DRILLING OPERATIONS

Drilling operations are a critical component of the upstream petroleum sector, representing the direct interface with subsurface hydrocarbon reservoirs. The drilling process involves a complex sequence of operations, starting with site preparation and culminating in well completion, designed to access and produce oil and gas. Drilling activities are capital-intensive, technically challenging,[1] and subject to various geological and operational uncertainties.

QUICK DRILL TYPES RECAP

Type	Purpose	Risk Level
Exploration	Discover new reserves	Very High
Development	Produce from known reservoirs	Moderate
Workover	Enhance/recover production	Low to Medium

Several types of wells are drilled, each serving a specific purpose:

- **Exploration wells:** These wells, also known as wildcats, are drilled to explore new hydrocarbon accumulations in unproven areas. They carry the highest risk but also the potential for significant rewards. Exploration wells help identify new reserves and expand the resource base.
- **Development wells:** Once a discovery is made, development wells are drilled to delineate the reservoir's size and extent and produce oil and gas. Development wells aim to maximize reservoir contact and optimize production.
- **Workover wells:** Workover wells are drilled on existing wells to enhance production, repair equipment, or access bypassed zones. They are performed to maintain or increase production rates and extend the well's life.

Drilling methods: Drilling methods vary depending on the geological conditions, reservoir characteristics, and well objectives. The most common method is rotary drilling, where a rotating drill bit grinds through the rock. Other techniques include:

- **Directional drilling:** This technique involves drilling wells at an angle, allowing access to reservoirs that are not directly below the surface. Directional drilling is valuable for reaching targets beneath obstacles such as cities or environmentally sensitive areas.
- **Horizontal drilling:** Involves drilling a well horizontally through the reservoir to maximize contact with the pay zone and enhance production. Horizontal wells are particularly effective in thin or low-permeability reservoirs.

DOI: 10.1201/9781003750345-5

Site preparation and well planning: Comprehensive planning and site preparation are essential before drilling begins. This includes geological and geophysical studies, environmental impact assessments, permitting, and logistical arrangements. Detailed well planning involves selecting the drilling location, designing the well trajectory, and preparing for potential challenges such as high pressures or unstable formations.

Drilling is where geology meets engineering — and budgets meet reality.

Mark Papa
former CEO, EOG Resources

Illustration of a multi-leg offshore platform with derrick tower, cranes, and auxiliary systems (Figure 4.1). Highlights the engineering complexity and logistical scale of offshore drilling operations, where structural integrity and operational efficiency are paramount.

COST BREAKDOWN—TYPICAL DEEPWATER WELL

- Rig Rental: 35–50%
- Consumables: 20–25%
- Services (cementing, logging, etc.): 20–30%
- Site Prep and Logistics: 10–15%

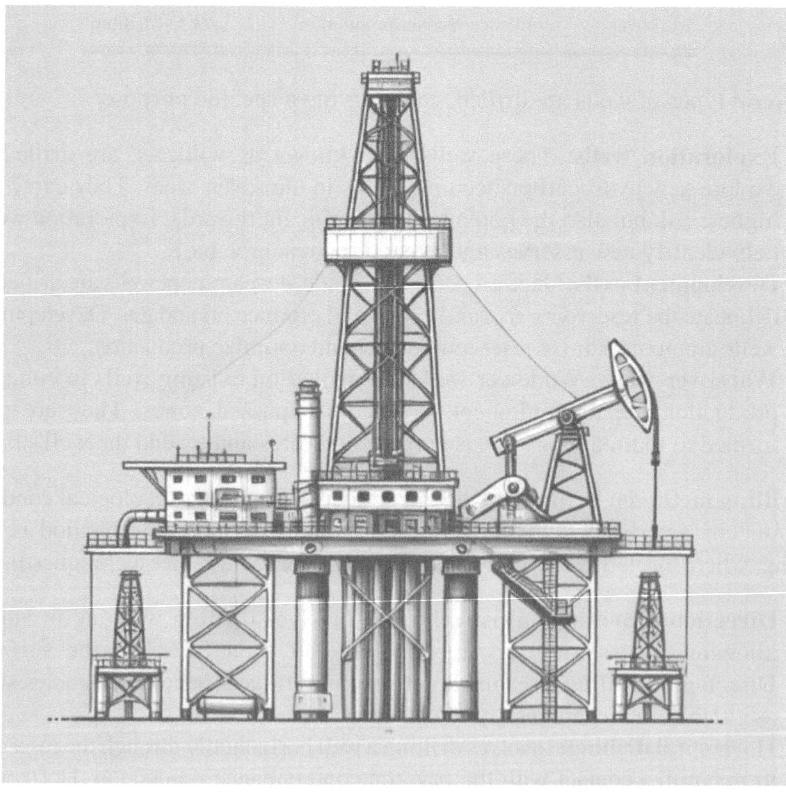

FIGURE 4.1 Offshore drilling platform and surface infrastructure.

4.2 COST STRUCTURES IN DRILLING OPERATIONS

Drilling operations are expensive, and understanding the cost structure is essential for economic evaluation. Key cost components include:

- **Rig costs:** Rig costs account for a massive portion of overall drilling expenses. These costs include rig day rates, mobilization and demobilization fees, crew costs, and maintenance expenses. Rig day rates vary depending on the rig's type, size, and location. High-specification rigs with advanced capabilities command higher day rates.
- **Drilling consumables:** Drilling consumables include drill bits, casing (steel pipe used to line the wellbore), drilling mud (a fluid used to control pressure and carry cuttings), and various chemicals. These costs depend on the well depth, complexity, and drilling conditions. Drilling mud formulations are tailored to specific geological conditions to ensure wellbore stability and efficient cuttings removal.
- **Well services:** Well services encompass a range of specialized operations, such as cementing (pumping cement to secure the casing), logging (running instruments downhole to measure formation properties), perforating (creating holes in the casing to allow fluid flow), and hydraulic fracturing (injecting high-pressure fluid to stimulate production). Each service contributes to well integrity and productivity.
- **Drilling costs**: Drilling costs vary widely depending on well type, depth, and location. **Onshore wells** typically cost **$2–30 million per well**, with deeper wells being more expensive. **Offshore drilling** costs are significantly higher,[2] with **shallow water wells ranging from $30–100 million** and **deepwater wells exceeding $500 million**. **Horizontal and directional drilling** add **$4–15 million** in additional expenses due to complexity. Rig costs also impact drilling economics, with **land rigs averaging $25,000–$100,000 per day**, while **deepwater drillships can cost up to $500,000 per day**. Understanding these cost drivers is crucial for optimizing drilling budgets and maximizing financial returns.
- **Other drilling expenses:** Other expenses include site preparation, transportation of equipment and personnel, waste disposal, supervision, and contingency costs. Site preparation may involve constructing access roads, preparing drilling pads, and ensuring environmental compliance.

Economic analysis of drilling costs: Economic analysis involves estimating total drilling costs and comparing them to the potential revenue from hydrocarbon production. This analysis helps determine the viability of drilling projects and guides investment decisions. Factors such as oil and gas prices, production rates, and cost efficiency are crucial in economic evaluations.

Cost management strategies: Effective cost management is essential for optimizing drilling economics. Strategies include negotiating favorable contracts with service providers, adopting cost-efficient technologies, and implementing best practices for drilling operations. Continuous monitoring and control of drilling costs help identify opportunities for savings and efficiency improvements.

4.3 RIG SELECTION AND DRILLING TECHNIQUES

Selecting the appropriate drilling rig is crucial for efficient and cost-effective drilling operations. Factors influencing rig selection include:

- **Water depth:** Water depth is a primary consideration for offshore drilling. Several types of offshore rigs are used for different water depths, including jack-up rigs (shallow water), semi-submersible rigs (moderate water depth), and drill ships (deep water). Each rig type has specific capabilities and limitations based on water depth and environmental conditions.
- **Location:** The well's location (onshore or offshore, remote or accessible) affects rig selection and transportation costs. Remote locations may require specialized logistics and infrastructure.
- **Well depth:** The well's target depth determines the required rig capacity and drilling equipment. Deep wells demand robust rigs with high horsepower and advanced drilling systems.
- **Well type:** The well type (exploration, development, workover) influences rig selection and drilling techniques. Exploration wells may require rigs with advanced exploration capabilities, while development wells focus on production efficiency.

Various drilling techniques: Various drilling techniques are employed depending on the well objectives and geological conditions:

- **Conventional drilling:** Vertical drilling using a rotary drill bit. Conventional drilling is suitable for simple, vertical wellbores.
- **Directional drilling:** Drilling wells at an angle to access remote targets. Directional drilling is valuable for reaching targets not directly beneath the surface.
- **Horizontal drilling:** Drilling horizontally through the reservoir to maximize production. Horizontal wells provide greater reservoir exposure and enhance recovery in unconventional reservoirs.
- **Managed pressure drilling:** A technique to control formation pressure and prevent wellbore instability. Managed pressure drilling allows for safer and more efficient drilling in challenging formations.

RIG TYPE COMPARISON

Rig Type	Depth Range	Use Case
Jack-up	<500 ft	Shallow offshore
Semi-submersible	500–10,000 ft	Mid-depth offshore
Drillship	>10,000 ft	Ultra-deepwater

The right rig for the right job — that's where you win or lose millions.

Bernard Looney
CEO, BP

Advanced drilling technologies: Advancements in drilling technology have significantly improved drilling efficiency and safety. Technologies such as rotary steerable systems, measurement-while-drilling (MWD) tools, and real-time data monitoring enhance drilling precision and decision-making. Automation and digitalization also transform drilling operations, enabling remote monitoring and control.

Drilling optimization: Optimizing drilling performance is essential for minimizing costs and maximizing efficiency. This involves reducing non-productive time (NPT), improving drilling rates, and utilizing advanced drilling technologies. Drilling optimization strategies include real-time data analysis, predictive maintenance, and continuous improvement initiatives. The adoption of **automation and artificial intelligence (AI) in drilling** is transforming the industry. **AI-driven drilling automation systems** optimize bit pressure, weight, and rotation speed, reducing human error and improving efficiency. **Digital twins**—real-time virtual models of drilling operations—allow engineers to simulate drilling scenarios before executing them. These technologies have led to **20–30% reductions in non-productive time (NPT)**, saving millions in operational costs. With real-time decision-making and predictive maintenance, AI and automation are reshaping modern drilling economics.

4.4 WELL PRODUCTIVITY AND ECONOMIC OPTIMIZATION

Well productivity, the rate at which a well can produce hydrocarbons, is influenced by several factors, including:

- **Reservoir properties:** Porosity, permeability, and fluid saturation. High porosity and permeability enhance fluid flow and production rates.
- **Well design includes well** diameter, completion techniques, and well trajectory. Well design affects the ability to access and produce hydrocarbons from the reservoir.
- **Completion techniques:** Methods used to prepare the well for production, such as perforating and hydraulic fracturing. Completion techniques optimize fluid flow and enhance well performance.

Stimulation techniques: Stimulation techniques, such as hydraulic fracturing and acidizing, enhance well productivity by improving permeability and fluid flow. Hydraulic fracturing involves injecting high-pressure fluid to create fractures in the reservoir, increasing its permeability. Acidizing involves injecting acid to dissolve rock and improve flow pathways. These techniques are particularly effective in tight formations and unconventional reservoirs.

Artificial lift methods: Artificial lift methods, such as pumps and gas lifts, are used to lift fluids to the surface when reservoir pressure is insufficient. Pumps, including electrical submersible pumps (ESPs) and sucker rod pumps, provide mechanical lift to enhance production. Gas lift involves injecting gas into the well to reduce hydrostatic pressure and lift fluids.

Economic optimization: Economic optimization is a critical aspect of drilling operations. Companies must balance drilling costs with expected production and

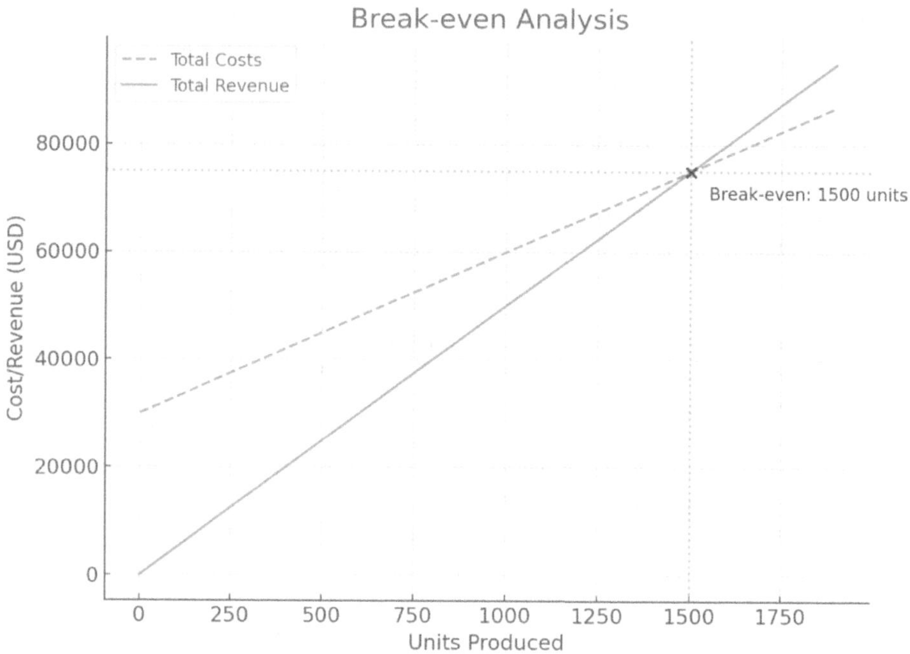

FIGURE 4.2 Break-even analysis of drilling operations.

revenue to maximize the monetary value of their assets. This involves using economic metrics such as net present value (NPV), internal rate of return (IRR), and cost per barrel to evaluate drilling projects and make informed investment decisions. Economic optimization ensures that drilling projects generate positive returns and contribute to overall profitability. Different types of wells have varying break-even prices based on **drilling and production costs**. For example, conventional **onshore oil wells break even at $30–50 per barrel,**[3] while **deepwater wells require oil prices above $50–80 per barrel** to remain profitable. **Unconventional shale and tight oil wells** typically have break-even prices in the **$40–60 per barrel** range, influenced by completion techniques such as hydraulic fracturing. Economic assessments using **sensitivity analysis** allow companies to determine how market fluctuations impact profitability, ensuring drilling investments align with long-term financial objectives (Figure 4.2).

Graph comparing total costs and total revenue as a function of units produced. The intersection point marks the break-even threshold, where cumulative revenue equals cumulative cost—critical for assessing project viability and investment timing.

Economic evaluation example: Consider a drilling project with an initial investment of $10 million and an expected production of 1 million barrels of oil over the well's life. The estimated production rate is 2000 barrels per day, and the operating costs are $10 per barrel. Assuming an oil price of $60 per barrel, the project generates $50 per barrel in net revenue. The NPV and IRR calculations help assess the project's profitability and guide investment decisions.

Risk management: Risk management is an integral part of economic optimization. Drilling projects have various risks, including geological uncertainties, market fluctuations, and operational challenges. Companies use risk assessment and mitigation strategies to address potential issues and enhance project resilience. Sensitivity analysis and scenario planning help evaluate the impact of different risk factors on project economics.

BREAK-EVEN COSTS BY WELL TYPE

Well Type	Breakeven ($/bbl)
Onshore Conventional	$30–50
Shale (Fracked)	$40–60
Deepwater	$50–80

4.5 CASE STUDIES OF DRILLING PROJECTS

Case Study 1: The Mariner Field (North Sea, UK): The **Mariner Field**, discovered in the **early 1980s**, contains **high-viscosity heavy oil**, making its development technically and economically challenging. Statoil (now Equinor) took on the project in **2012**, investing over **$7 billion** in **horizontal drilling and enhanced oil recovery (EOR) techniques** to improve production efficiency. The development required **multilateral wells** to maximize **reservoir contact**, and **waterflooding** was implemented to enhance oil recovery. Despite these complexities, **first oil was achieved in 2019**, and the field is expected to produce for **over 30 years**, with an estimated **300 million barrels of recoverable reserves**.

Challenges and Strategies:

- **Challenges:** Heavy oil viscosity, complex reservoir structure with multiple fault blocks, and high development costs.
- **Drilling techniques used:** Horizontal drilling to maximize **reservoir contact**, multilateral wells to enhance productivity, and **waterflooding** as an **EOR technique**.

Key Economic Factors:

- **Investment:** ~$7 billion
- **Break-even price:** ~$50 per barrel
- **Expected peak production:** ~55,000 barrels per day.
- **Economic justification:** Large recoverable reserves and long production lifespan justify **high upfront costs**.

Outcomes:
The **Mariner Field was successfully developed**, and production began in **2019**. The **combination of advanced drilling techniques and EOR** ensures long-term economic viability, with **sustained production expected for over three decades**. Strategic **economic evaluation and cost optimization** have made Mariner one of the most notable **North Sea heavy oil projects**.

Case Study 2: The Sakhalin-1 Project (Russia): The **Sakhalin-1 project**, operated by **Exxon Neftegas Limited**, is an **offshore development** located off **Sakhalin**

Island in the Russian Far East. The project required **extended-reach drilling (ERD)**, drilling some of the **world's longest wells** to access offshore oil and gas reservoirs from **onshore drilling pads**, significantly reducing infrastructure costs. With an investment of **$17 billion**, the project successfully produced **over 1 billion barrels of oil** since its inception.

Challenges and Strategies:

- **Challenges:** Harsh environmental conditions, remote location, and complex reservoir geology.
- **Drilling techniques used:** Extended-reach drilling (ERD) to access distant reservoirs, advanced drilling technologies for wellbore stability and pressure control.

Key Economic Factors:

- **Investment:** ~$17 billion
- **World record ERD wells: Over 40,000 feet (~12 km) long**
- **Economic strategy:** Use of **onshore drilling** to reduce offshore platform costs

Outcomes

The **Sakhalin-1 project** has set records in **extended-reach drilling**, producing **over 1 billion barrels of oil** while optimizing costs by minimizing offshore infrastructure. The project's success highlights **the economic advantage of ERD in offshore developments**.

> ⚠ **LESSONS FROM MACONDO**
>
> - Overlooked BOP diagnostics
> - No redundancy in cementing verification
> - Regulatory oversight gaps

Case Study 3: The Macondo Prospect (Gulf of Mexico, USA): The **Macondo Prospect**, drilled by **BP in 2010**, became infamous due to the **Deepwater Horizon disaster**. The well, drilled in **ultra-deep water (~5,000 feet depth)**, encountered **high-pressure and high-temperature conditions**. A failure in the **blowout preventer (BOP)** led to an uncontrolled blowout, causing **one of the worst oil spills in history**, with millions of barrels of oil leaking into the Gulf of Mexico. The incident resulted in **over $65 billion** in financial damages, including legal penalties, cleanup costs, and compensation claims.

Challenges and Strategies:

- **Challenges:** Ultra-deep-water drilling risks, well pressure control failures, and regulatory shortcomings.
- **Drilling techniques used:** Deepwater drilling, blowout preventers (BOPs), and advanced well control measures (which failed in this case).

Key Economic Factors:

- **Investment before disaster:** ~$150 million
- **Total financial damage:** Over $65 billion (cleanup, legal, and compensation costs)
- **Regulatory impact:** Led to tighter deepwater drilling safety regulations

Outcomes

The **Macondo disaster** reshaped **deepwater drilling regulations worldwide**, increasing **safety standards and compliance costs**. The economic damage highlighted the **critical need for improved well integrity and risk management** in ultra-deepwater drilling.

> Disaster in the Gulf: BP's Deepwater Horizon Explosion Claims 11 Lives and 65 Billion USD
>
> *New York Times, April 2010*

Case Study 4: The Thunder Horse Field (Gulf of Mexico, USA): The **Thunder Horse Field**, operated by **BP and ExxonMobil**, is one of the **largest deepwater oil discoveries in the Gulf of Mexico**. Discovered in **1999**, the field faced **multiple technical challenges**, including **hurricane damage in 2005**, which delayed production. Despite setbacks, BP invested over **$10 billion** in **dynamically positioned drillships, advanced drilling fluids, and subsea production systems**. Production commenced in **2008**, reaching an output of **250,000 barrels per day**.

Challenges and Strategies:

- **Challenges:** Deepwater drilling exceeding **6000 feet**, high-pressure and high-temperature reservoirs, and subsea infrastructure challenges.
- **Drilling techniques used:** Dynamically positioned drillships, advanced drilling fluids, and subsea production systems.

Key Economic Factors:

- **Investment:** ~$10 billion
- **Production capacity: ~250,000 barrels per day**
- **Economic justification:** Large **recoverable reserves and optimized production systems**

Outcomes

Despite **initial setbacks**, Thunder Horse has become one of the **most productive deepwater fields**, with **advanced technology and continuous production optimization** ensuring long-term economic success.

Case Study 5: The Khurais Megaproject (Saudi Arabia): Saudi Aramco's **Khurais Megaproject** is one of the **largest oil developments in the world**, integrating three fields: **Khurais, Abu Jifan, and Mazalij**. The project aimed to **increase Saudi production capacity by 1.2 million barrels per day**. By utilizing **multilateral horizontal wells, advanced water injection, and real-time well monitoring**, Aramco significantly improved production efficiency and long-term sustainability. The total investment for the project was estimated at **$10 billion**.

Challenges and Strategies:

- **Challenges:** Developing infrastructure in a remote desert location, managing high production rates, and integrating advanced technologies for enhanced recovery.
- **Drilling techniques used:** Multilateral horizontal wells, large-scale water injection, and advanced well monitoring.

Key Economic Factors:

- **Investment:** ~$10 billion
- **Production capacity increase: 1.2 million barrels per day**
- **Economic benefit:** Enabled **Saudi Arabia to meet global supply demands**

Outcomes

The **Khurais Megaproject** successfully increased **Saudi Arabia's oil production capacity**, reinforcing the **economic and strategic importance of large-scale oil field developments**.

4.5.1 Summary of Case Studies

These case studies highlight the **financial and technological challenges** of drilling projects worldwide. They emphasize the **critical role of economic evaluation, risk management, and technological advancements** in ensuring project success. From **deepwater drilling breakthroughs to extended-reach wells and large-scale megaprojects**, these examples showcase how **strategic planning and investment decisions** shape the future of upstream drilling operations.

4.6 CHAPTER SUMMARY

This chapter has provided an overview of drilling economics, covering the drilling process, cost structures, rig selection, drilling techniques, well productivity, and economic optimization. We emphasized the importance of cost-effective drilling operations and balancing costs with expected production and revenue. The case studies highlighted the diverse challenges and solutions in drilling operations, demonstrating the critical role of economic evaluation in ensuring project success.

> The deeper we drill, the more expensive each decision becomes.
>
> **Dave Lesar**
> *former Halliburton CEO*

We demonstrated how advanced drilling techniques and economic optimization contribute to successful field development and production enhancement by examining real-world case studies. Integrating technical expertise with economic analysis is essential for making informed decisions that maximize the value of hydrocarbon resources.

The next chapter will examine production economics, focusing on managing producing fields, optimizing production rates, and maximizing economic returns from hydrocarbon production. We will explore the principles and practices governing production operations and the financial considerations in sustaining and enhancing production levels.

NOTES

1. IADC's 2022 Global Drilling Report.
2. Wood Mackenzie offshore cost benchmarks, 2023.
3. Rystad Energy, 2022.

5 Production Economics

5.1 INTRODUCTION TO PRODUCTION OPERATIONS

Production operations encompass all activities involved in extracting hydrocarbons from subsurface reservoirs and delivering them to processing facilities. This includes managing well performance, optimizing reservoir recovery, and handling the produced fluids (oil, gas, and water). Production operations begin after wells have been drilled and completed and continue throughout the reservoir's productive life.

The production process: The production process involves a combination of surface and subsurface equipment:

- **Subsurface equipment:** This includes downhole pumps (for artificial lift), packers (to isolate production zones), and tubing (to transport fluids to the surface). Downhole pumps, such as electrical submersible pumps (ESPs) and rod pumps, are used to lift fluids to the surface when reservoir pressure is insufficient. Packers are used to isolate different production zones within the wellbore, ensuring that fluids are directed to the desired intervals.
- **Surface equipment:** This includes wellheads, flowlines, separators (to separate oil, gas, and water), storage tanks, and processing facilities. Wellheads control the flow of hydrocarbons from the wellbore to the surface. Separators separate the produced fluids into their individual components—oil, gas, and water. The separated fluids are then stored or transported to processing facilities.

Hydrocarbon recovery stages: Hydrocarbon recovery can be classified into three stages:

- **Primary recovery:** This stage relies on natural reservoir pressure to push hydrocarbons to the wellbore. It typically recovers only a fraction of the original oil and gas in place. Primary recovery methods include natural flow, gas drive, and water drive.
- **Secondary recovery:** Involves injecting fluids (usually water or gas) into the reservoir to maintain pressure and displace additional hydrocarbons toward the production wells. Secondary recovery methods enhance recovery by sustaining reservoir pressure and improving fluid displacement efficiency.
- **Tertiary recovery (enhanced oil recovery—EOR):** This method employs more advanced techniques to alter the reservoir or fluids' physical and chemical properties, further improving hydrocarbon recovery. EOR methods include thermal EOR, chemical EOR, and miscible EOR.

DOI: 10.1201/9781003750345-6

Primary Recovery Techniques:

- **Natural flow:** The reservoir's natural pressure drives oil and gas to the surface without artificial lift.
- **Gas drive:** Natural gas expansion provides the energy to drive hydrocarbons to the wellbore.
- **Water drive:** Natural aquifer pressure pushes oil toward the production wells.

Secondary Recovery Techniques:

- **Water injection:** Injecting water into the reservoir to maintain pressure and displace oil toward the production wells.
- **Gas injection:** Injecting gas (such as natural gas or CO_2) to maintain pressure and enhance oil displacement.

Tertiary Recovery (EOR) Techniques:

- **Thermal EOR:** Involves injecting heat into the reservoir to reduce oil viscosity and improve flow. Examples include steam injection, cyclic steam stimulation, and in-situ combustion.
- **Chemical EOR:** Uses chemicals to improve oil displacement efficiency. Examples include surfactant flooding, polymer flooding, and alkaline flooding.
- **Miscible EOR:** Involves injecting a fluid that mixes with the oil, reducing interfacial tension and improving displacement. Examples include gas injection and CO_2 injection.

5.2 COST ANALYSIS OF PRODUCTION OPERATIONS

Production operations in the upstream oil and gas sector involve a combination of **capital expenditures (CAPEX)** and **operating expenses (OPEX)**. Effective cost management is critical for maintaining profitability, especially in fluctuating oil prices. The key cost components in production operations are categorized into **Exploration, Development, Operation Expenses, and Decommissioning and Abandonment.**

- **Exploration costs:** Exploration costs represent the expenditures incurred in identifying and appraising hydrocarbon reserves. These costs include:
  - **Geological and geophysical (G&G) studies:** Seismic surveys, geological mapping, and other subsurface evaluations.
  - **Seismic surveys:** Advanced imaging techniques to assess reservoir potential.
  - **Exploration licensing and permits:** Regulatory costs associated with acquiring exploration rights.
  - **Exploration drilling:** Drilling of exploratory wells to confirm hydrocarbon presence.
  - **Exploration appraisal drilling:** Follow-up drilling to determine reservoir size and viability.
  - **Exploration others:** Miscellaneous exploration-related costs such as remote sensing and field studies.

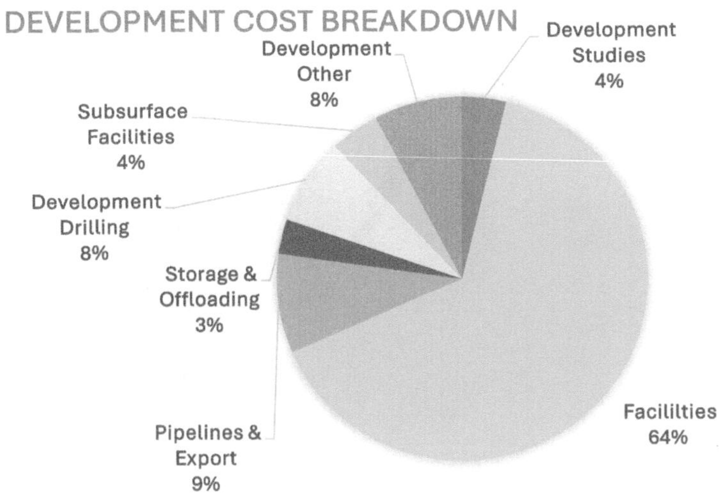

FIGURE 5.1 Development cost breakdown: Distribution of capital expenditures across facilities, drilling, pipelines, and subsurface systems. Highlights how infrastructure dominates development budgets, reinforcing the need for strategic planning and cost control in field execution.

- **Development costs:** Development costs refer to investments made to bring a discovered field into production (Figure 5.1). These include:
  - **Development studies:** Before field development, engineering, feasibility, and reservoir simulation studies.
  - **Facilities:** Construction of production infrastructure, including processing plants and separation units.
  - **Pipelines and export systems:** Installation of transport networks for oil and gas delivery.
  - **Storage and offloading:** Storage facilities, loading terminals, and floating storage units (FPSOs in offshore fields).
  - **Development drilling:** Drilling of production and injection wells.
  - **Subsurface facilities:** Downhole completion equipment, artificial lift systems, and reservoir monitoring tools.
  - **Development other:** Any additional costs not classified under the above categories.
  - **Abandonment:** Preliminary costs related to future field decommissioning.
- **Operating expenses (OPEX):** Operating expenses cover the recurring costs necessary for maintaining day-to-day production (Figure 5.2). These include:
  - **Lifting costs:** Energy and equipment needed to lift hydrocarbons to the surface, including artificial lift mechanisms.
  - **Maintenance and repairs:** Costs associated with maintaining production equipment and infrastructure.
  - **Water disposal costs:** Treatment and disposal of produced water in compliance with environmental regulations.

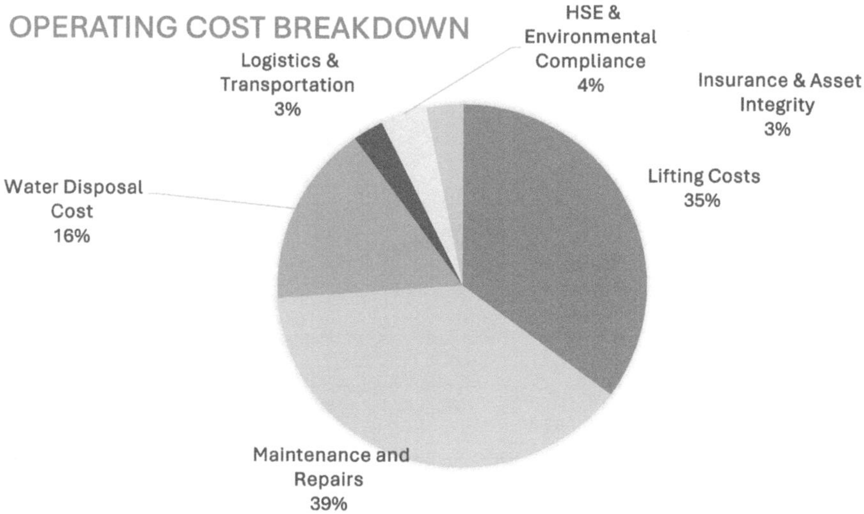

FIGURE 5.2 Operating cost breakdown: Distribution of recurring production expenses across lifting, maintenance, water disposal, logistics, insurance, and HSE compliance. Highlights how equipment upkeep and artificial lift dominate operating budgets, reinforcing the need for preventive maintenance and efficiency strategies.

  - **Logistics and transportation:** Transporting personnel, equipment, and hydrocarbon costs.
  - **HSE and environmental compliance:** Expenditures on health, safety, and environmental protection measures.
  - **Insurance and asset integrity:** Coverage for production facilities, wells, and operational risks.
- **Decommissioning and abandonment costs:** As oil and gas fields reach the end of their productive life, decommissioning and abandonment costs become significant (Figure 5.3). These costs include:
  - **Well plugging and abandonment (P&A):** Permanent sealing of wells to prevent fluid migration.
  - **Facilities removal and cleanup:** Dismantling production infrastructure and restoring the site.
  - **Long-term monitoring:** Environmental and regulatory monitoring after decommissioning.

5.2.1 Economic Analysis of Production Costs

Understanding the financial impact of these cost components is crucial for assessing project profitability. Key financial metrics include:

- **Cost per barrel:** The total production cost divided by the number of barrels produced.
- **Netback:** Revenue per barrel after deducting OPEX, royalties, and taxes.

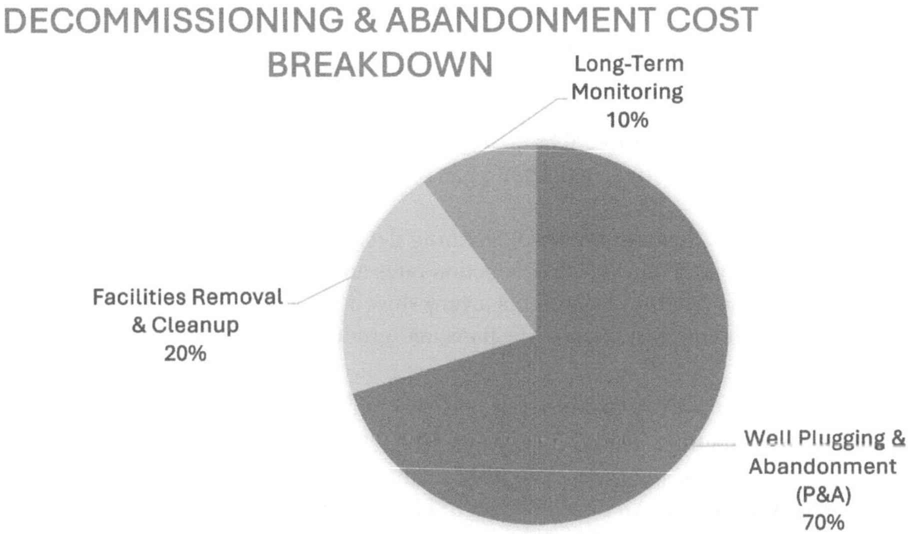

FIGURE 5.3 Decommissioning and abandonment cost breakdown: Distribution of end-of-life expenditures across well plugging, infrastructure removal, and long-term monitoring. Highlights how well abandonment dominates decommissioning budgets, reinforcing the need for early planning and regulatory compliance.

5.2.2 Cost Management Strategies

To optimize production economics, companies implement the following strategies:

- **Operational efficiency:** Implementing best practices to maximize output while minimizing costs.
- **Preventive maintenance:** Reducing unplanned downtime through scheduled equipment servicing.
- **Economies of scale:** Scaling up production to lower per-unit costs.
- **Supply chain optimization:** Streamlining procurement and logistics to reduce operational costs.

Following a structured cost management approach, upstream operators can ensure sustainable production while maintaining profitability across different oil price scenarios.

5.3 PRODUCTION FORECASTING AND DECLINE CURVE ANALYSIS

Forecasting future production rates and estimating ultimate recovery are crucial for economic planning and reservoir management. Decline curve analysis is a commonly used technique for this purpose.

Decline curve analysis: Decline curve analysis involves analyzing historical production data to identify trends and project future production. Common decline curve models include:

- **Exponential decline:** Characterized by a constant percentage decline rate. It is commonly used for reservoirs with stable production decline rates.
- **Hyperbolic decline:** Shows a declining decline rate over time. It is suitable for reservoirs with varying production rates and recovery mechanisms.
- **Harmonic decline:** Represents a very slow decline rate. It is used for reservoirs with long production lifespans and gradual declines.

Steps in Decline Curve Analysis:

1. **Data collection:** Gather historical production data, including production rates, reservoir pressures, and well performance metrics.
2. **Trend analysis:** Analyze the data to identify production trends and decline patterns.
3. **Model selection:** Select the appropriate decline curve model based on reservoir characteristics and historical data.
4. **Forecasting:** Use the selected model to project future production rates and estimate ultimate recovery.

Limitations of Decline Curve Analysis:

- Decline curve analysis relies on historical data and may not accurately predict future production if reservoir conditions change.
- The introduction of secondary or tertiary recovery methods can alter production trends, requiring adjustments to the decline curve model.

Advanced Forecasting Techniques:

- **Reservoir simulation:** This involves using detailed reservoir models to simulate production scenarios and forecast future performance. It incorporates geological, petrophysical, and production data to provide a comprehensive understanding of reservoir behavior.
- **Machine learning:** Applying machine learning algorithms to analyze production data and improve forecasting accuracy. Machine learning models can identify complex patterns and correlations in production data that traditional methods may overlook.

Production forecasting example: Consider a field with historical production data showing a gradual decline in production rates. By applying decline curve analysis, the operator can identify the declining trend and project future production rates. If the field is expected to implement water injection in the future, the operator can adjust the decline curve model to account for the impact of secondary recovery on production (Figure 5.4).

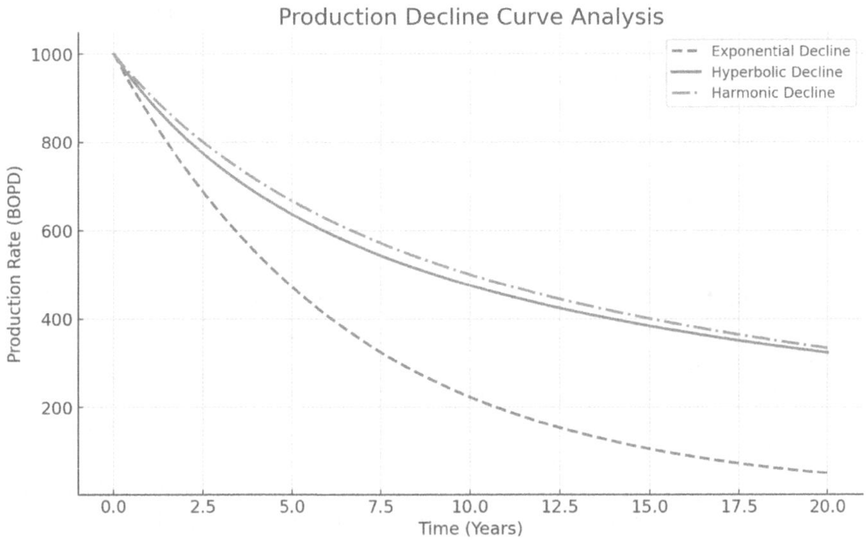

FIGURE 5.4 Decline curve analysis: Graph comparing exponential, hyperbolic, and harmonic decline models over a 20-year production horizon. Highlights how reservoir behavior and recovery strategy influence decline rates, with implications for forecasting accuracy and economic planning.

5.4 ENHANCED OIL RECOVERY (EOR) ECONOMICS

EOR methods are used to increase hydrocarbon recovery beyond what is achievable with primary and secondary recovery. EOR techniques can be broadly classified into:

- **Thermal EOR:** Involves injecting heat into the reservoir to reduce oil viscosity and improve flow. Examples include steam injection, cyclic steam stimulation, and in-situ combustion.
  - **Steam injection:** Steam is injected into the reservoir to heat the oil, reduce its viscosity, and improve flow. Steam flooding is a continuous process where steam is injected into injection wells, and oil is produced from production wells.
  - **Cyclic steam stimulation:** Also known as "huff and puff," this process involves injecting steam into a well, allowing it to soak, and then producing oil from the same well.
  - **In-situ combustion:** Air or oxygen is injected into the reservoir to ignite a portion of the oil, generating heat and producing lighter hydrocarbons that are easier to produce.

FAST FACT—SHALE VS. CONVENTIONAL DECLINE

Shale wells typically experience production declines of 70–90% in the first three years, while conventional reservoirs may sustain output for decades. This sharp contrast shapes economic and operational strategies in upstream development.

- **Chemical EOR:** Uses chemicals to improve oil displacement efficiency. Examples include surfactant flooding, polymer flooding, and alkaline flooding.
    - **Surfactant flooding:** Surfactants are injected to reduce interfacial tension between oil and water, improving oil mobility.
    - **Polymer flooding:** Polymers are injected to increase the viscosity of the injected water, improving sweep efficiency and displacing more oil.
    - **Alkaline flooding:** Alkali chemicals are injected to react with acidic components in the oil, creating surfactants in situ and reducing interfacial tension.
- **Miscible EOR:** Involves injecting a fluid that mixes with the oil, reducing interfacial tension and improving displacement. Examples include gas injection and CO_2 injection.
    - **Gas injection:** Hydrocarbon gases or nitrogen are injected to mix with the oil and improve flow. Gas injection can be miscible or immiscible, depending on reservoir conditions.
    - **CO_2 injection:** Carbon dioxide is injected to mix with the oil, reducing its viscosity and improving displacement. CO_2 injection is widely used due to its effectiveness and the availability of CO_2 from industrial sources.

Cumulative oil recovery and CO_2 injection volumes over a 24-year span (Figure 5.5). Highlights the correlation between sustained CO_2 injection and incremental recovery, reinforcing the economic and environmental viability of integrated EOR-CCS strategies.

Economic evaluation of EOR projects[1]: EOR projects require significant capital investment and operating costs. Economic evaluation is essential to determine the viability of EOR projects. Key economic factors include:

- **Incremental oil recovery:** The additional volume of oil recovered due to EOR methods. Incremental recovery must be sufficient to justify the additional costs.
- **Injection fluid costs:** Costs associated with purchasing or producing the injection fluid (e.g., steam, chemicals, CO_2). The cost of injection fluids can significantly impact project economics.
- **Operating costs:** Costs for energy, labor, and maintenance required to implement EOR techniques. Operating costs must be managed to ensure project profitability.

Weyburn Project – Timeline of Oil Recovery and CO_2 Injection

Year	Cumulative Oil Recovered (Million Barrels)	Cumulative CO_2 Injected (Million Tonnes)
2000	180	0.5
2005	210	6
2010	235	13
2015	255	18.5
2020	270	23
2024*	282 (est.)	25.5 (est.)

FIGURE 5.5 Weyburn project: Timeline of oil recovery and CO_2 injection (2000–2024).

CAPEX vs. OPEX Bar Chart for EOR Projects

EOR Method	CAPEX (Million USD)	OPEX (USD/Barrel)	Notes
Gas Injection	150–300	6–10	High compression and recycle costs
Chemical Injection	250–400	10–15	Polymers/surfactants require continuous dosing
Thermal Recovery (Steam)	300–600	15–25	High steam generation and fuel costs
Microbial EOR	80–120	4–6	Lower CAPEX but uncertain long-term performance
CO_2-EOR (with Capture)	400–800	8–12	Includes CO_2 capture, transport, and recycle infrastructure

FIGURE 5.6 CAPEX and OPEX comparison across EOR methods. Summary of capital and operating expenditures for gas injection, chemical flooding, thermal recovery, microbial EOR, and CO_2-EOR. Highlights how method selection impacts upfront investment and per-barrel operating costs, guiding strategic decisions in mature field redevelopment.

- **Capital costs:** Costs for equipment, facilities, and infrastructure needed for EOR operations. High capital costs can be a barrier to EOR project implementation.
- **Profitability metrics:** Metrics such as net present value (NPV) and internal rate of return (IRR) to assess project profitability. Positive NPV and IRR values indicate economically viable projects.

Government incentives for EOR: Governments often provide fiscal incentives, such as tax credits or royalty reductions, to encourage EOR projects. These incentives help offset the excessive costs of EOR and make projects more economically attractive. For example, in the United States, the federal government provides tax credits for CO_2-EOR projects to promote carbon capture and storage (CCS) and reduce greenhouse gas emissions (Figure 5.6).

EOR case study: The Weyburn field (Canada): The Weyburn Field is a prominent example of successful EOR implementation using CO_2 injection. Since 2000, the project has injected CO_2 captured from industrial sources to enhance oil recovery and reduce greenhouse gas emissions. The project has demonstrated the economic and environmental benefits of CO_2-EOR, with significant incremental oil recovery and carbon storage.

- **Challenges:** Implementing CO_2-EOR in a mature field, managing CO_2 supply and infrastructure, and addressing environmental concerns.
- **Production strategies:** Continuous injection of CO_2 to maintain reservoir pressure and improve oil displacement, integration of carbon capture and storage (CCS) initiatives, and regular monitoring and optimization of injection and production operations.
- **Outcomes:** Increased oil recovery and extended field life, successful integration of CO_2-EOR with CCS, and positive economic returns. The Weyburn project has sequestered millions of tons of CO_2, contributing to environmental sustainability.[2]

> CO_2-EOR is one of the few proven technologies that delivers both energy security and emissions reductions.
>
> *IEA, Energy Technology Perspectives, 2023*

EOR Method	CO_2 Utilization (MCF/Barrel)	Recovery Efficiency (%)	Notes
Continuous CO_2 Injection	5,000–6,000	7%–15%	Common in conventional reservoirs; simple but may result in early gas breakthrough.
Water-Alternating-Gas (WAG)	3,000–4,500	10%–18%	Improves sweep efficiency; widely adopted in the Permian Basin.
Tapered WAG	2,000–3,500	12%–20%	Adjusts gas/water ratio over time; better for heterogeneous reservoirs.
Simultaneous WAG	1,500–2,500	15%–22%	Injects water and gas together; enhances miscibility and reduces gas override.
Advanced Mobility Control	< 2,000	18%–25%	Uses foams or gels with CO_2; best in fractured reservoirs.

FIGURE 5.7 CO_2-EOR method comparison: Utilization and recovery efficiency.

One of the critical economic decisions in field management is whether to **extend the life of an existing reservoir using EOR or abandon it and invest in drilling new wells**. The choice depends on factors such as **recovery potential, cost per barrel, oil prices, and infrastructure availability**. In **conventional heavy oil fields**, operators often favor **EOR techniques like steam flooding or CO_2 injection** over new drilling, as they provide a more cost-effective method of recovering additional oil. For instance, **heavy oil fields in Canada** rely heavily on steam-assisted gravity drainage (SAGD) to improve production rather than continuously drilling new wells. Conversely, **tight oil and shale plays**, such as those in the **Permian Basin and Bakken**, prioritize **continuous drilling and hydraulic fracturing** because production declines rapidly, making new wells the most efficient way to sustain output. The economic viability of EOR vs. new wells is also influenced by fiscal incentives, regulatory frameworks, and technological advancements that enhance **reservoir monitoring and recovery efficiency**. Analyzing these trade-offs allows companies to **optimize their capital allocation and maximize the economic potential of their assets**.

Comparison of five CO_2-based EOR techniques by gas utilization and recovery performance (Figure 5.7). Highlights how tapered and simultaneous WAG methods improve sweep efficiency, while advanced mobility control delivers the highest recovery in fractured reservoirs.

5.5 PRODUCTION OPTIMIZATION AND MANAGEMENT

Optimizing production operations is essential for maximizing economic returns. Key strategies include:

- **Well performance optimization:** Monitoring well performance, identifying production bottlenecks, and implementing interventions to improve flow rates. Techniques include well stimulation, artificial lift optimization, and adjusting production rates.
  - **Well stimulation:** Techniques such as hydraulic fracturing, acidizing, and matrix stimulation to enhance well productivity by improving permeability and fluid flow.
  - **Artificial lift optimization:** Adjusting the settings and operation of artificial lift systems (e.g., pumps, gas lift) to maximize production and minimize energy consumption.
  - **Production rate management:** Adjusting production rates to balance short-term cash flow with long-term reservoir management and recovery optimization.

- **Reservoir management strategies:** Optimizing production rates to balance short-term cash flow with long-term recovery. Implementing water injection or gas injection programs to maintain reservoir pressure and improve sweep efficiency.
  - **Water injection:** Continuous injection of water to maintain reservoir pressure and displace oil toward production wells. Water injection helps sustain production rates and improve recovery.
  - **Gas injection:** Injection of gas (such as natural gas, nitrogen, or CO_2) to maintain pressure and enhance oil displacement. Gas injection is particularly effective in miscible and immiscible displacement processes.
  - **Reservoir surveillance:** Continuous monitoring of reservoir performance using downhole sensors, production data analysis, and reservoir simulation models to optimize injection and production strategies.
- **Production allocation:** Distributing production among different wells and reservoirs to maximize overall field production and economic value. This involves prioritizing high-performing wells and managing production constraints.
  - **Production allocation models:** Using data-driven models to allocate production based on well performance, reservoir characteristics, and economic considerations.
  - **Optimization algorithms:** Applying optimization algorithms to determine the optimal production allocation that maximizes field-wide recovery and economic returns.

Advanced Production Management Techniques:

- **Digital oilfield technologies:** Using real-time data monitoring and analytics to optimize production operations and improve decision-making. Digital oilfield technologies enable remote monitoring, predictive maintenance, and data-driven optimization.
- **Smart wells:** Implementing intelligent well-completion systems that enable remote monitoring and control of production parameters. Smart wells use downhole sensors and control valves to optimize production in real-time.
- **Integrated asset management:** Coordinating reservoir management, well operations, and surface facilities to optimize production and reduce costs. Integrated asset management involves collaboration across disciplines to achieve holistic optimization.

Case Study 1: The Prudhoe Bay Field (Alaska, USA):

- **Background:** Discovered in 1968, Prudhoe Bay is the largest oil field in North America, with **25 billion barrels of original oil in place (OOIP)**.
- **Challenges:** Harsh Arctic conditions, complex infrastructure logistics, and declining reservoir pressure.
- **Production strategy:**
  - Waterflooding and gas injection increased the recovery factor from **30% to 60%**.
  - Over **12 billion barrels produced**, with continued production through enhanced recovery techniques.
  - Investment in **EOR exceeded $10 billion**, ensuring sustained production.

- **Economic impact:**
  - **Revenue generated:** Over **$250 billion** since production began.
  - **Current production: ~280,000 bpd**, down from a peak of 2 million bpd in the 1980s.
  - Economic feasibility driven by **high oil prices and long-term field optimization**.

Case Study 2: The Weyburn Field (Saskatchewan, Canada):

- **Background:** The Weyburn Field, discovered in **1954**, became a global leader in **CO_2-EOR** starting in **2000**.
- **Challenges:** Mature reservoir with declining pressure; need for sustainable oil recovery.
- **Production strategy:**
  - Injecting **3 million tons of CO_2 annually** from a coal gasification plant in North Dakota.
  - Increased recovery factor by **15–20%**, producing over **130 million additional barrels**.
  - Integrated **carbon capture and storage (CCS)**, reducing **40 million tons of CO_2 emissions**.
- **Economic impact:**
  - **Incremental recovery:** ~130 million barrels added.
  - **Investment:** ~$2 billion in CO_2-EOR infrastructure.
  - **Long-term viability:** CO_2-EOR extended the field's life **by over 25 years**.

Case Study 3: The Cantarell Field (Mexico):

- **Background:** Discovered in **1976**, Cantarell was Mexico's largest oil field, producing over **2 million bpd at peak** in the early 2000s.
- **Challenges:** Rapid production decline by the late 1990s due to falling reservoir pressure.
- **Production strategy:**
  - Implemented **one of the world's largest nitrogen injection projects**.
  - Investment in nitrogen injection: **~$6 billion**.
  - The injection **stabilized production for 15 years**, delaying decline.
- **Economic impact:**
  - **Peak production:** 2.1 million bpd (2004).
  - **Current production:** ~150,000 bpd (2023), highlighting the limitations of nitrogen injection for long-term sustainability.
  - The case underscores **the risks of overproducing fields too quickly**.

Case Study 4: The Ghawar Field (Saudi Arabia):

- **Background:** Ghawar, discovered in **1948**, remains the world's **largest onshore oil field**, with **88 billion barrels of OOIP**.
- **Challenges:** Managing reservoir pressure, maximizing ultimate recovery, and maintaining high output.

- **Production strategy:**
  - **Massive water injection system** sustaining reservoir pressure.
  - **Horizontal drilling and smart well technology** for production optimization.
  - **Integration of digital oilfield technologies** for real-time monitoring.
- **Economic impact:**
  - **Peak production:** Over **5 million bpd**.
  - **Revenue generated:** Estimated at **$1 trillion+**.
  - **Ongoing investment in technology** ensures Ghawar remains a top producer for decades.

Case Study 5: The Thunder Horse Field (Gulf of Mexico, USA):

- **Background:** Thunder Horse is one of the largest deepwater oil discoveries in the Gulf of Mexico, operated by BP and ExxonMobil.
- **Challenges:** Deepwater drilling complexity, high development costs, and reservoir heterogeneity.
- **Production strategy:**
  - Advanced deepwater drilling and subsea production systems.
  - Use of dynamically positioned drillships for extreme depths.
  - Real-time monitoring and predictive maintenance to prevent downtime.
- **Economic impact:**
  - **Development cost:** ~$4 billion.
  - **Peak production:** ~250,000 bpd.
  - Despite early setbacks (hurricane damage and platform tilt in 2005), Thunder Horse remains one of BP's **most profitable offshore assets**.

5.5.1 Summary of Case Studies

The case studies in this chapter highlight how advanced production techniques, EOR, and economic planning impact field performance and long-term viability.

- **Prudhoe Bay Field (USA)**: Waterflooding and gas injection increased recovery from 30% to over 60%, producing 12 billion barrels and generating $250 billion in revenue.
- **Weyburn Field (Canada): A leading CO_2-EOR project, adding 130 million barrels and storing 40 million tons of CO_2, supported by $2 billion in infrastructure investments.**
- **Cantarell Field (Mexico)**: Nitrogen injection stabilized production at 2 million bpd for 15 years but declined sharply, illustrating the limitations of large-scale pressure maintenance.
- **Ghawar Field (Saudi Arabia)**: The world's largest oil field (88 billion barrels OOIP) maintained high production for over 70 years using water injection, horizontal drilling, and digital monitoring, generating over $1 trillion in revenue.
- **Thunder Horse (USA)**: A deepwater success story, despite $4 billion in costs, reaching peak production of 250,000 bpd, demonstrating the profitability of high-risk offshore projects.

Key Insights:

- **EOR techniques** significantly improve recovery rates and field longevity.
- **Economic feasibility** depends on oil prices, production costs, and government incentives.
- **Digital monitoring and advanced drilling** are critical for maximizing production efficiency.
- **Balanced depletion strategies** prevent rapid declines and maximize asset value.

5.6 DECLINE MANAGEMENT IN PRODUCTION OPERATIONS

Production decline is an inevitable phase in the lifecycle of oil and gas wells, but companies employ various strategies to mitigate and manage this decline. Common techniques include workovers, artificial lift enhancements, well stimulations, and EOR expansions. Workovers involve replacing downhole equipment, re-perforating productive zones, or implementing chemical treatments to remove blockages. Artificial lift improvements, such as optimizing gas lift systems, ESPs, or rod pumps, can help maintain production rates by assisting fluid flow. Well stimulations, including hydraulic fracturing and acidizing, improve permeability, allowing hydrocarbons to flow more efficiently. In mature fields, companies also expand EOR efforts, such as polymer flooding, CO_2 injection, and thermal recovery, to extract additional reserves. For example, shale oil wells typically experience steep production declines of 60–90% within the first three years, necessitating continuous drilling of new wells or innovative recovery strategies to sustain output. Effective decline management strategies are crucial to ensuring long-term field profitability while maximizing resource recovery.

5.7 CHAPTER SUMMARY

This chapter has provided an overview of production economics, covering the production process, cost structures, production forecasting, EOR economics, and production optimization. We emphasized the importance of efficient and cost-effective production operations to maximize the economic value of hydrocarbon assets. The case studies highlighted the diverse challenges and solutions in production operations, demonstrating the critical role of economic evaluation in ensuring project success.

We demonstrated how advanced production techniques and economic optimization contribute to successful field development and production enhancement by examining real-world case studies. Integrating technical expertise with economic analysis is essential for making informed decisions that maximize the value of hydrocarbon resources.

The next chapter will discuss field development economics, focusing on the planning, designing, and executing field development projects. We will explore the principles and practices that govern field development and the economic considerations involved in optimizing field performance.

5.8 ROLE OF DIGITAL TECHNOLOGY

The integration of digital technologies, data analytics, and artificial intelligence (AI) is transforming production economics by improving forecasting accuracy, optimizing field performance, and reducing operational costs. AI-driven predictive models analyze historical production data to detect patterns, anticipate equipment failures, and recommend proactive maintenance, minimizing downtime and improving efficiency. Advanced machine learning algorithms enhance reservoir simulations, decline curve analysis, and real-time production monitoring, enabling operators to make data-driven decisions. Automated well surveillance uses real-time sensors and digital twins to track well performance, allowing for immediate adjustments to artificial lift parameters, injection rates, or production schedules. Additionally, cloud-based collaboration platforms improve coordination between field engineers and data scientists, enhancing decision-making and operational response times. By leveraging digital transformation, oil and gas companies can achieve higher recovery rates, improved asset reliability, and significant cost savings, making technology a vital component of modern production economics.

NOTES

1. International Energy Agency (IEA), Enhanced Oil Recovery – Technologies, Potentials, and Costs, 2022.
2. Weyburn is one of the world's longest-running CCS projects, storing over 40 million tons of CO_2 since 2000.

6 Field Development Economics

6.1 INTRODUCTION TO FIELD DEVELOPMENT PLANNING

Field development planning is a **multi-stage** process that translates a discovered hydrocarbon resource into a **commercially viable** producing asset. It involves a series of **technical, economic, and strategic** decisions aimed at maximizing **hydrocarbon recovery**, optimizing capital expenditures (CAPEX), and ensuring long-term profitability while managing risks and uncertainties. This process begins after **a discovery well confirms hydrocarbon presence**, followed by **appraisal drilling** to estimate reserves and determine reservoir properties. Before making a **final investment decision (FID)**, companies conduct **detailed reservoir modeling, economic feasibility studies, and infrastructure planning** to assess the project's commercial viability. A well-executed field development plan (FDP) ensures that capital is deployed efficiently, and that production is sustained over the asset's life cycle.

6.1.1 Key Considerations in Field Development Planning

Effective field development planning requires an integrated approach, balancing reservoir engineering, production optimization, and cost management. Several key factors influence the selection of development strategies:

- **Reservoir characteristics:** The field's size, shape, porosity, permeability, fluid properties, and pressure regime dictate the recovery potential and influence well placement. Fields with high permeability (>100 mD) often support natural flow production, while low-permeability reservoirs (<1 mD) may require hydraulic fracturing or enhanced oil recovery (EOR). Reservoir complexity, such as fractured carbonate formations (e.g., Middle East reservoirs) or deepwater turbidite systems (e.g., Gulf of Mexico fields), affects development feasibility.
- **Production capacity:** The desired production rate and plateau duration must align with reservoir deliverability and market demand. High-rate fields (e.g., Ghawar, producing ~5 million barrels per day) can sustain long production plateaus. In contrast, unconventional fields, such as shale plays (e.g., Permian Basin, with steep decline rates of ~60% in the first year), require continuous well drilling to maintain output.
- **Infrastructure requirements:** Development plans must account for wells, pipelines, processing facilities, transportation, and storage systems. Offshore projects, such as floating production storage and offloading (FPSO) units in Brazil's deepwater pre-salt fields, require specialized infrastructure due to remote locations and harsh environments. Onshore

DOI: 10.1201/9781003750345-7

fields with high associated gas production may require gas processing and reinjection facilities, as seen in Algeria's Hassi R'Mel gas field.

- **Economic factors:** Oil and gas prices, CAPEX, OPEX, tax regimes, royalties, and discount rates dictate the project's viability. Fields with low break-even costs (<$20/bbl, e.g., Middle East onshore fields[1]) are more resilient to price fluctuations than deepwater projects, which typically require >$50/bbl to be profitable.[2] Developers use sensitivity analysis to test project economics against price swings, drilling costs, and fiscal changes.

A phased development approach is often adopted, especially for large or complex fields. Phased development allows companies to manage risk, optimize capital spending, and incorporate learnings from earlier phases into subsequent development stages. However, it can also lead to delays and potentially forgo some early production revenue (Figure 6.1).

FIGURE 6.1 Field development planning within a multi-energy infrastructure context isometric schematic showing offshore platforms, renewables, and energy systems. Highlights the complexity of planning hydrocarbon field development amid evolving infrastructure demands and energy diversification strategies.

6.1.2 Phased Development Approach

To optimize capital efficiency and reduce geological and financial risks, many field developments are executed in phases. This approach ensures early cash flow generation while incorporating lessons learned into later stages. The three main phases of development are:

- **Phase 1: Pilot and Initial Development**: The first phase includes appraisal drilling, pilot production, and early-stage infrastructure. This stage is critical for confirming reservoir continuity, assessing well productivity, and gathering fluid properties. In offshore projects, early production may be facilitated using tiebacks to existing platforms or a leased FPSO unit instead of full-scale infrastructure investment. For instance, Guyana's Stabroek Block (Liza Phase 1) initiated production in 2019 using an FPSO, generating early revenue while refining development plans for later phases.
- **Phase 2: Full-Scale Development and Production Plateau**: Once commercial viability is confirmed, full-field development proceeds with additional wells, full infrastructure deployment, and production system expansion. This phase often includes constructing processing plants, pipelines, export terminals, and storage facilities. The Johan Sverdrup Field (Norway), one of the most significant North Sea discoveries in recent decades, expanded from an initial Phase 1 production capacity of 440,000 bpd to Phase 2 output of 755,000 bpd through additional drilling and infrastructure expansion.
- **Phase 3: Enhanced Recovery and Late-Life Optimization**: As reservoirs mature, natural pressure declines, leading to production drops. Companies implement secondary recovery (water/gas injection) and tertiary recovery (CO_2, steam, or polymer flooding) to sustain output and maximize ultimate recovery. For example, the Weyburn Field (Canada) extended its life by 25+ years using CO_2 injection, increasing recovery from 30% to ~45%. Late-life strategies also include well workovers, artificial lift enhancements, and facility upgrades to maintain economic production.

> Efficiency in field development is not about cutting costs—it's about maximizing every dollar's value.
>
> **Amin Nasser**
> *CEO, Saudi Aramco*

6.1.3 Project Economics and Investment Analysis

A successful field development strategy **must balance technical feasibility with economic viability.** Given the capital-intensive nature of oil and gas projects, investment decisions rely on **robust financial modeling, sensitivity analysis, and risk assessment.**

Key Economic Components of Field Development:

- **Capital expenditures (CAPEX):**
  - **CAPEX constitutes 60-80% of total field development costs** and covers **well drilling, production facilities, pipelines, and processing plants.**

    - Offshore deepwater projects have higher CAPEX than onshore fields due to complex infrastructure.
    - Example: The **Kashagan Field (Kazakhstan)** required **$55 billion** in CAPEX due to **high-pressure drilling and extreme Arctic conditions.**
- **Operating expenditures (OPEX):**
    - OPEX includes **well interventions, labor, power costs, and facility maintenance.**
    - Low OPEX fields maximize profitability, especially during oil price downturns.
    - Example: The **Tengiz Field (Kazakhstan)** optimized its OPEX by integrating **sulfur management strategies**, reducing long-term handling costs.
- **Revenue projections:**
    - Based on **estimated production profiles, crude quality, and market conditions.**
    - **Brent vs. WTI price differentials** affect revenue streams, especially for landlocked fields.
    - Example: **The Johan Sverdrup Field (Norway)** benefits from **low breakeven costs of $20/bbl**, ensuring profitability in low-price environments (Figure 6.2).
- **Taxes, royalties and fiscal terms:**
    - Government fiscal policies **directly influence profitability.** Royalty rates range from **5% (Norway) to 50%+ (Middle East NOCs).**
    - Countries with **Production Sharing Agreements (PSAs)** allow cost recovery before profit-sharing, improving project economics.
    - Example: The **Guyana Stabroek Block PSA** permits **75% cost recovery** before splitting profit oil, making deepwater investment attractive.

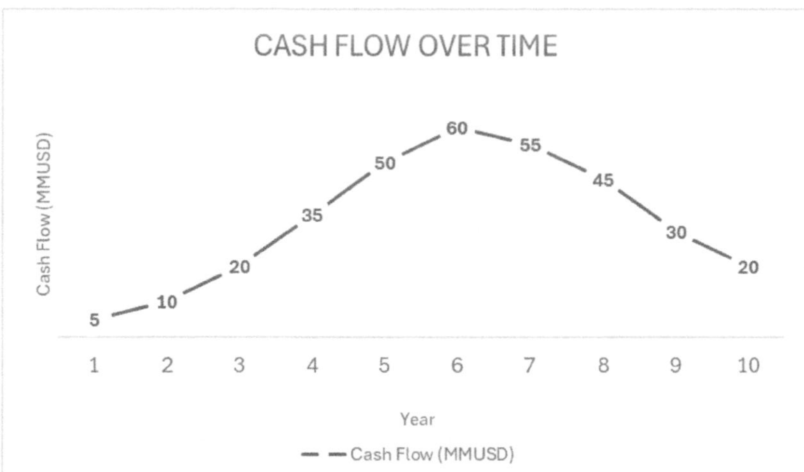

FIGURE 6.2 Cash flow profile over a 10-Year field development horizon: Illustrates annual cash flow progression from initial investment through production plateau and decline. Highlights peak earnings in years 5–6, followed by tapering returns—Critical for evaluating payback period, NPV, and IRR in field development decisions.

Revenue	Income generated from oil and gas sales is influenced by production levels and market prices.
Capital Expenditures (CAPEX)	Initial and ongoing investments in exploration, drilling, infrastructure, and facility development.
Operating Expenditures (OPEX)	Recurring costs for production operations, maintenance, transportation, and compliance.
Taxes & Royalties	Government-imposed financial obligations, including royalties and corporate income taxes.
Cash Flow	Total revenue minus CAPEX, OPEX, and taxes, representing the actual earnings from operations.
Net Present Value (NPV)	The present value of expected future cash flows is discounted to reflect investment risk.
Internal Rate of Return (IRR)	The discount rate at which NPV equals zero indicates the project's profitability threshold.
Breakeven Price	The minimum oil price is required to cover costs and sustain profitable operations.
Sensitivity Analysis	Evaluates how financial outcomes change with variations in oil prices, costs, and production rate
Scenario Analysis	Analyzes multiple financial scenarios (e.g., base case, high-cost, low-cost) for decision-making.

FIGURE 6.3 Key economic metrics in field development modeling.

Definitions of core financial and operational terms—NPV, IRR, breakeven price, sensitivity analysis—used in evaluating field development strategies (Figure 6.3). Serves as a foundational reference for scenario planning, investment screening, and production optimization.

Economic Metrics for Field Development Decisions:

- **Net present value (NPV):** This measure measures the **total project value discounted to present dollars.** A positive NPV signals a viable project.
- **Internal rate of return (IRR):** The discount rate where **NPV = 0.** If IRR exceeds the required return (hurdle rate), the project is economically sound.
- **Payback period:** Time required to **recover initial investment** via project cash flow. Shorter payback periods reduce financial risk.
- **Breakeven oil price:** The minimum oil price needed for the **project cash flow to reach zero.** Lower breakeven = higher resilience to price volatility.

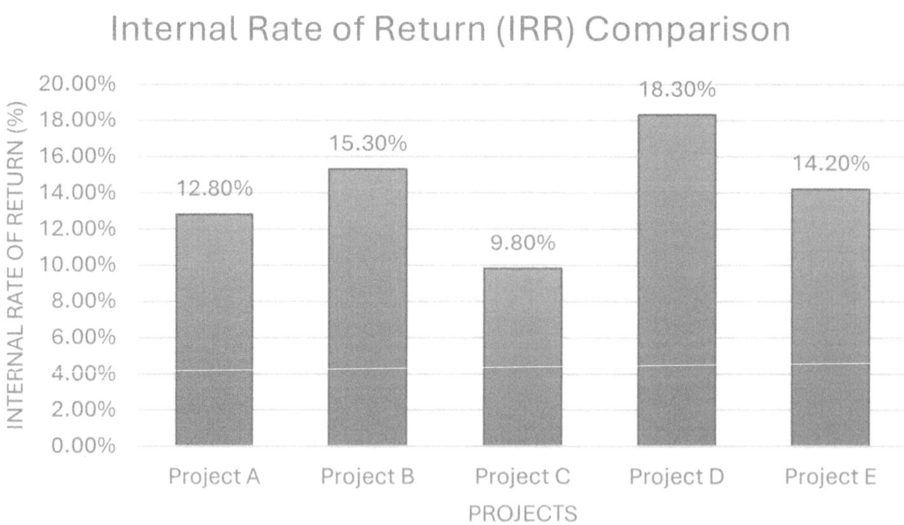

FIGURE 6.4 Internal rate of return (IRR) comparison across field development projects.

Bar chart comparing IRR values for five hypothetical projects (Figure 6.4). Highlights Project D as the most financially attractive, with an IRR of 18.3%, guiding capital allocation and prioritization in portfolio management.

6.2 SENSITIVITY ANALYSIS AND SCENARIO PLANNING

Field economics depend on **oil prices, production rates, and costs.** Sensitivity analysis quantifies how changes in these factors affect project profitability.

- **Example:** A deepwater project with a **$40/bbl breakeven** becomes unprofitable if oil prices fall to **$35/bbl**, requiring cost reduction strategies.
- **Monte Carlo simulations** are used to model price fluctuations and reservoir uncertainties, providing **probabilistic economic forecasts.**
- **Real-world case:** The **Kashagan Field's initial development** suffered **$20 billion in cost overruns** due to unexpected high-pressure zones, underscoring the importance of **robust risk assessment.**

6.3 FIELD DEVELOPMENT PLANNING AND OPTIMIZATION

Strategic field development planning is crucial for **maximizing hydrocarbon recovery, optimizing costs, and ensuring long-term profitability.** A well-designed development plan integrates **reservoir characterization, well placement, facility design, and production optimization** to create **a technically and economically sound project.**

1. **Production Profile Optimization**
   Determining the optimal production rate over a field's lifespan requires balancing:
   - **Short-term cash flow needs vs. long-term reservoir recovery**
   - **Reservoir pressure maintenance to avoid premature decline**
   - **Infrastructure capacity to handle peak production rates**

   Example: **The Statfjord Field (Norway)** adjusted its production strategy by integrating **gas reinjection and waterflooding,** increasing its recovery factor from **42% to over 65%** while extending field life beyond 40 years.
2. **Well Placement and Completion Optimization**
   Well placement decisions directly impact **field productivity, drilling efficiency, and operational costs.** Techniques include:
   - **Reservoir simulation models:** Predict optimal well spacing and reservoir drainage patterns.
   - **Horizontal and multilateral wells:** Increase reservoir exposure and reduce the number of wells needed.
   - **Smart completion technology:** Enables zonal control, allowing operators to **dynamically adjust production** from different layers of the reservoir.

   Example: **The Tengiz Field (Kazakhstan)** optimized well placement using **horizontal drilling and extended-reach wells,** boosting oil recovery without additional surface infrastructure.

3. **Infrastructure Design and Optimization**
   Field infrastructure includes:
   - **Flowlines and gathering systems:** Transport produced hydrocarbons to processing facilities.
   - **Separation and processing units:** Handle oil, gas, and water separation.
   - **Storage and export terminals:** Facilitate product delivery.

   Optimizing infrastructure reduces **CAPEX and OPEX** while improving production efficiency.

   Example: **The Johan Sverdrup Field** (**Norway**) optimized its offshore facilities to **minimize energy consumption**, reducing carbon emissions by **90% compared to similar projects.**
4. **Integration of Reservoir Management and Production Operations**
   Effective field development requires **continuous reservoir monitoring and adaptive production strategies.** Key techniques:
   - **Water and gas injection programs:** Maintain reservoir pressure and improve sweep efficiency.
   - **Real-time data analytics:** AI-driven insights optimize injection rates and fluid recovery.
   - **Digital twin technology:** Simulates reservoir behavior and infrastructure performance for proactive optimization.

   Example: **The Ghawar Field (Saudi Arabia)** successfully integrated **smart field monitoring technologies** to **sustain production above 5 million barrels daily for decades.**

Advanced Field Development Technologies:

- **3D seismic imaging:** High-resolution seismic imaging provides detailed subsurface images for accurate reservoir characterization and well placement.
- **Horizontal drilling and multilateral wells:** Advanced drilling techniques enhance reservoir contact and improve recovery.
- **Digital twins:** Virtual models of physical assets that enable real-time monitoring, simulation, and optimization of field operations.
- **Artificial intelligence (AI) and machine learning (ML):** AI and ML algorithms analyze large datasets to identify patterns, predict reservoir behavior, and optimize production strategies.

6.4 INFRASTRUCTURE DEVELOPMENT AND COST CONSIDERATIONS

Infrastructure is a major cost driver in field development, with expenses covering drilling and completion, production facilities, pipelines, and transportation systems. Efficient infrastructure design optimizes production efficiency while minimizing CAPEX and OPEX.

FAST FACT: CAPEX VS. OPEX IN OFFSHORE FIELDS

In offshore developments, CAPEX can account for up to 80% of total project costs due to platform construction, subsea installations, and pipeline networks. OPEX in deepwater fields often includes logistics costs such as helicopter transport and remote maintenance.

1. **Key Components of Field Infrastructure**
   Onshore fields:
   - **Well pads and gathering systems**: Connect production wells to central processing facilities.
   - **Storage facilities**: Oil storage tanks, underground gas storage, and export terminals.
   - **Pipeline networks**: Transport hydrocarbons to refineries, liquefaction plants, or export hubs.

   Offshore fields:
   - **Fixed platforms:** Used for shallow water developments (e.g., **North Sea's Brent Field).**
   - **Floating production storage and offloading (FPSO):** Deployed in deepwater fields with no pipeline infrastructure (e.g., Brazil's Lula Field).
   - **Subsea production systems: Utilize remotely operated equipment to optimize deepwater production (e.g., Thunder Horse Field, Gulf of Mexico).**
2. **Cost Estimation and Budgeting**
   Accurate cost estimation is critical for investment decisions and financial planning. Cost drivers include:
   - **Reservoir complexity**: Deep, high-pressure fields (e.g., Kashagan Field, Kazakhstan) require specialized drilling and high-cost infrastructure.
   - **Project location:** Remote locations (e.g., Arctic fields) have higher logistics costs and require winterized facilities.
   - **Regulatory compliance:** Stringent environmental laws increase CAPEX for emissions management and decommissioning obligations.
3. **Infrastructure Sharing and Cost Reduction Strategies**
   Operators often collaborate to share infrastructure and reduce costs:
   - **Pipeline and processing hubs:** Multiple companies use shared facilities to lower per-barrel processing costs (e.g., Johan Sverdrup's electrified platform reducing OPEX).
   - **Joint FPSO deployment:** Smaller deepwater fields co-develop floating production units to optimize costs.
   - **Enhanced logistics coordination:** Shared supply bases, helicopter transport, and supply vessels reduce offshore operation expenses.
4. **Case Example: Johan Sverdrup (Norway): Efficient Infrastructure Development**
   The Johan Sverdrup Field exemplifies cost-efficient offshore development.
   - **Electrified platform:** 90% lower CO_2 emissions than conventional projects.
   - **Phased infrastructure rollout:** Spread out investment risk while optimizing CAPEX.
   - **Shared export pipelines:** Reduced costs and improved economic efficiency across North Sea fields.

Equinor Electrifies Offshore Platforms, Cutting 90% of CO_2 Emissions

The Johan Sverdrup Field sets new standards in decarbonized offshore development, thanks to full electrification powered by hydroelectric grids.

6.5 RISK MANAGEMENT IN FIELD DEVELOPMENT

Field development projects are high-risk, capital-intensive investments affected by geological uncertainty, technical challenges, market volatility, and regulatory constraints. Effective risk management strategies enhance project viability and protect financial returns.

1. **Categories of Risk in Field Development**
   Summary of technical, economic, regulatory, and operational risks with real-world examples (Figure 6.5). Highlights how projects like Kashagan, Johan Sverdrup, and Gulf of Mexico fields illustrate the diverse challenges that must be mitigated through strategic planning and technology adoption.
   - **Technical risks:** Reservoir uncertainty, drilling challenges, production problems, and equipment failures. Technical risks can impact production rates, recovery factors, and operational efficiency.
   - **Economic risks:** Commodity price volatility, cost overruns, regulatory changes, and political instability. Economic risks can affect project profitability, cash flow, and financial stability.
2. **Risk Mitigation Strategies**
   i. **Diversification of Assets**
      - Investing in **a portfolio of projects** (onshore/offshore, conventional/unconventional) minimizes exposure to any single risk.
      - **Example: BP's global asset diversification** helps balance losses from high-cost projects with low-cost assets.

   ii. **Advanced Reservoir Modeling and Seismic Imaging**
      - High-resolution **3D and 4D seismic imaging** improves reservoir characterization, reducing drilling and production risks.
      - **Example: Thunder Horse Field (Gulf of Mexico)** used 4D seismic to refine well locations and avoid dry holes.

   iii. **Contingency Planning and Scenario Analysis**
      - Sensitivity analysis evaluates **oil price impacts on field economics.**
      - **Example:** Kashagan Field's delays prompted contingency budgeting, **raising project costs by $20B** but ensuring long-term viability.

Risk Type	Description	Example Cases
Technical Risks	Uncertainties in reservoir performance, drilling challenges, and equipment failures.	**Kashagan Field, Kazakhstan:** Delays due to **high-pressure reservoirs and sour gas corrosion.**
Economic Risks	Oil price fluctuations, cost overruns, fiscal instability, funding challenges.	**Brazil's Pre-Salt Fields:** Government subsidies are required due to high deepwater development costs.
Regulatory Risks	Environmental restrictions, local laws, production quotas, and tax changes.	**Norway's Johan Sverdrup:** Stringent emissions targets influenced platform electrification.
Operational Risks	Logistics, workforce safety, extreme weather events, supply chain disruptions.	**Gulf of Mexico Fields:** The hurricane-prone region demands **hurricane-resistant platforms** and emergency response planning.

FIGURE 6.5 Categorization of risks in field development projects.

iv. **Integrated Supply Chain Management**
   - **Bulk purchasing and long-term contracts** lower CAPEX/OPEX volatility.
   - **Example:** Johan Sverdrup leveraged Norway's shared logistics network to minimize offshore costs.

v. **Technology Adoption for Risk Reduction**
   - **Digital oilfields and real-time monitoring** detect anomalies before failures occur.
   - **Example: Ghawar Field (Saudi Arabia)** implemented AI-driven predictive maintenance to prevent unplanned shutdowns.

vi. **Risk Transfer via Insurance and Hedging**
   - **Political risk insurance,** cost overrun protection, and price hedging protect against financial losses.
   - **Example:** Deepwater projects use hedging strategies to ensure stable cash flow despite oil price swings.

SMART FIELDS: THE RISE OF DIGITAL OILFIELDS

Saudi Aramco's Ghawar Field uses smart sensors and AI to monitor more than 5,000 wells in real time, reducing intervention costs by 30% and boosting productivity.

Case Study: Risk Management in the Kashagan Field (Kazakhstan): The Kashagan Field, located in the Caspian Sea, is one of the largest offshore oil discoveries in recent decades. The field's development faced significant technical and economic risks, including high reservoir pressure, harsh environmental conditions, and complex regulatory requirements. Risk mitigation strategies included advanced drilling techniques, robust project management, and strong stakeholder engagement. Despite initial delays and challenges, the field commenced production in 2016, demonstrating the importance of comprehensive risk management in field development.

6.6 CASE STUDIES OF FIELD DEVELOPMENT PROJECTS

6.6.1 Case Study 1: The Johan Sverdrup Field (Norway)

Background: Lundin Petroleum discovered the Johan Sverdrup Field in 2010, making it one of the largest oil discoveries in the North Sea in recent decades. With estimated recoverable reserves of **2.7 billion barrels of oil equivalent (boe)**, the project was developed in phases to optimize costs and minimize environmental impact.

Challenges:

- Harsh offshore environment with high waves and extreme weather conditions.
- Stringent regulatory requirements in Norway emphasizing carbon emissions reduction.
- High initial CAPEX required for large-scale infrastructure.

Development Plan and Strategies:

- **Phased Development:** The project was split into two phases, reducing upfront CAPEX and incorporating early learnings into later stages.
- **Electrification Strategy:** The field is fully powered by **renewable electricity from onshore**, reducing CO_2 emissions by **over 90% compared to similar projects.**
- **Digital Oilfield Technologies:** Real-time monitoring systems improve operational efficiency and reduce downtime.

Outcomes and Economic Impact:

- **CAPEX Investment:** Over **$11 billion** spent across both phases.
- **Production Capacity:** Peaked at **755,000 barrels per day (bpd)**, making it Norway's most productive offshore field.
- **Economic Performance:** With a breakeven oil price of **$20–30 per barrel**, the field is highly profitable, contributing billions in tax revenue to the Norwegian government.

6.6.2 CASE STUDY 2: THE KASHAGAN FIELD (KAZAKHSTAN)

Project Background:

- One of the world's largest offshore discoveries with 13 billion barrels of recoverable oil.
- High-pressure, sour gas reservoir required special metallurgy and complex drilling.

Key Risks and Challenges:

1. **Extreme reservoir conditions:** 800-bar pressure and high hydrogen sulfide (H_2S) content led to severe pipeline corrosion and leaks.
2. **Regulatory and environmental issues:** Ice-covered waters and Caspian Sea ecosystem required strict compliance with emissions limits.
3. **Cost overruns:** The initial budget of $57B ballooned to over $120B due to delays and technical failures.

Risk Mitigation Strategies and Outcomes:

- **Use of corrosion-resistant alloys** → Reduced equipment failure risks.
- **Implementation of gas reinjection** → Minimized H_2S emissions and improved recovery factor.
- **Revised project timeline and cost control measures** → Despite delays, Kashagan now produces over 400,000 bpd.

6.6.3 Case Study 3: The Thunder Horse Field (Gulf of Mexico, USA)

Background: Thunder Horse, discovered in 1999 by BP and ExxonMobil, is one of the largest deepwater oil discoveries in the Gulf of Mexico. Located at a depth of **6,200 feet (1,890 meters),** it represents a breakthrough in deepwater drilling and production technology.

Challenges:

- **Hurricane damage:** The Thunder Horse platform suffered structural damage from **Hurricane Dennis in 2005**, delaying production by **three years**.
- **Complex reservoir management:** High-pressure and multi-layered reservoirs required sophisticated reservoir modeling.
- **High CAPEX and operating costs:** Deepwater drilling is significantly more expensive than onshore or shallow-water operations.

Development Plan and Strategies:

- **Subsea production systems:** Advanced subsea wells and remotely operated vehicles (ROVs) are used to minimize surface infrastructure.
- **Automated pressure management:** Intelligent well monitoring ensures optimal production and prevents well blowouts.
- **Production capacity optimization:** Initial production infrastructure was expanded to handle increased volumes.

Outcomes and Economic Impact:

- **CAPEX investment: $4 billion** spent on development, including rebuilding costs after hurricane damage.
- **Peak production:** Over **250,000 bpd**, making it one of BP's most valuable assets.
- **Long-term economic impact:** Expected to generate **over $50 billion in total revenue** by 2040.

6.6.4 Case Study 4: The Tengiz Field (Kazakhstan)

Background: The Tengiz Field, discovered in 1979, is one of Kazakhstan's most productive onshore oil fields, containing **26 billion barrels of original oil in place (OOIP)**. It was one of the first fields to implement large-scale **gas injection for EOR.**

Challenges:

- **High sulfur content:** Required special refining and gas reinjection technology to handle **hydrogen sulfide (H_2S)**.
- **Pressure maintenance issues:** Needed aggressive reservoir pressure support to sustain high production rates.
- **Infrastructure development:** Required **a dedicated export pipeline** to deliver crude to global markets.

Development Plan and Strategies:

- **Full-scale gas injection:** Over **8 billion cubic meters of gas reinjected annually** to maintain reservoir pressure.
- **Advanced water handling facilities:** Installed **dedicated water injection plants** to prevent formation damage.
- **Pipeline expansion:** Linked to the **Caspian Pipeline Consortium (CPC)** to improve export capacity.

Outcomes and Economic Impact:

- **Production rates:** Currently producing **700,000–800,000 bpd**, with future expansion aiming for **1 million bpd**.
- **Economic returns:** Over **$200 billion in revenue generated** since production began.
- **Future expansion investment: $37 billion** allocated for further development to sustain output until 2045.

6.6.5 Case Study 5: The Statfjord Field (North Sea, Norway)

Background: The Statfjord Field, discovered in **1974**, is one of the largest oil fields in the North Sea. Located in Norwegian waters, the field has been producing oil and gas for over **four decades** and is considered a prime example of successful **EOR implementation.** The field originally had **3.6 billion barrels of oil in place (OOIP)** and has undergone multiple recovery optimization phases.

Challenges:

- **Aging infrastructure:** Many facilities required extensive maintenance and upgrades due to long-term use.
- **Declining reservoir pressure:** The field needed a pressure support mechanism to sustain production rates.
- **Harsh offshore conditions:** Located in the North Sea, the field faces strong currents, storms, and cold temperatures that challenge operations.

Development Plan and Strategies:

- **Water injection for pressure maintenance:** The project implemented **one of the largest waterflooding operations in the North Sea**, improving recovery rates.
- **Gas injection for enhanced recovery: Natural gas reinjection** was deployed to sustain pressure and optimize oil displacement.
- **Continuous infrastructure upgrades:** Several **platform modifications and pipeline expansions** were made to extend the field's lifespan.
- **Advanced digital monitoring:** Real-time reservoir monitoring systems were installed to enhance production efficiency.

Outcomes and Economic Impact:

- **CAPEX investment:** Over **$6 billion** spent on EOR and facility upgrades since the 1980s.
- **Recovery factor increase:** Initially projected to recover only **40% of OOIP**, but with EOR, the recovery factor reached **67%**, unlocking **an additional 1 billion barrels.**
- **Production rates:** Peaked at **700,000 bpd** in the 1980s; currently producing around **50,000–70,000 bpd** as the field nears the end of its life.
- **Total revenue generated:** The field has contributed **over $230 billion in oil revenues** to Norway's economy.

Lessons Learned:

- **EOR techniques can significantly extend a field's life and increase profitability.**
- **Proactive infrastructure upgrades prevent costly downtime and maximize long-term returns.**
- **Investment in digital technologies can optimize late-stage production and delay decommissioning.**

6.6.6 Summary of Case Studies

The case studies presented in this chapter illustrate how various field development strategies, economic evaluations, and advanced technologies impact project success. These real-world examples showcase **a variety of geological, technical, and economic challenges** and the strategies used to overcome them.

6.6.7 Key Takeaways from Each Case Study

Overview of challenges, strategies, and economic outcomes across five major oil fields (Figure 6.6). Highlights how phased development, advanced recovery techniques, and infrastructure innovation drive long-term profitability and resilience in diverse geological and regulatory environments.

Key Insights from the Case Studies:

✓ Phased Development Mitigates Risk

- Johan Sverdrup and Kashagan demonstrated that phased field development reduces risk exposure, allows companies to refine strategies before full-scale investment, and improves capital efficiency.

Case Study	Challenges	Strategies Implemented	Economic Impact
Johan Sverdrup Field (Norway)	High development costs, environmental regulations, complex reservoir geology	Phased development, advanced seismic imaging, water injection, digital oilfield integration	Initial investment of **$10 billion**, expected total revenue **$200+ billion**, field life **50+ years**, breakeven price **$20–30/bbl**
Kashagan Field (Kazakhstan)	High-pressure, high-temperature reservoir, remote offshore conditions, regulatory complexities	Artificial islands, advanced HPHT drilling, extensive infrastructure investment	CAPEX **$50 billion**, production target **450,000 bpd**, suffered delays but remains a key asset in Kazakhstan's oil sector
Thunder Horse Field (Gulf of Mexico, USA)	Deepwater drilling complexity, high development costs, hurricane-related setbacks	Subsea production systems, dynamic-positioning drillships, real-time monitoring	CAPEX **$4 billion**, peak production **250,000 bpd**, among the most profitable Gulf of Mexico assets
Tengiz Field (Kazakhstan)	High reservoir pressure and temperature, sulfur management issues, remote location	Sour gas injection, advanced drilling, infrastructure expansion	Over **$50 billion** in investments, production **700,000+ bpd**, extended field life beyond initial estimates
Statfjord Field (North Sea, Norway)	Aging infrastructure, declining pressure, offshore weather challenges	Waterflooding, gas injection, continuous facility upgrades	Recovery factor **67%**, additional **1 billion barrels recovered**, total revenue **$230 billion**, peak production **700,000 bpd**

FIGURE 6.6 Comparative summary of field development case studies.

✓ Advanced Recovery Techniques Extend Field Life and Profitability
  - Statfjord and Tengiz showcased how waterflooding, gas injection, and sour gas reinjection can extend production by decades and recover additional reserves.

✓ Deepwater and Harsh Environments Require High-Tech Solutions
  - Thunder Horse and Kashagan proved that challenging environments necessitate high CAPEX investment, but with proper technology and execution, projects remain highly profitable.

✓ Economic Viability Hinges on Oil Prices and Breakeven Costs[3]
  - Johan Sverdrup ($20/bbl breakeven) vs. Kashagan ($50/bbl breakeven) shows that fields with lower breakeven prices remain competitive even during market downturns.

✓ Digitalization and Infrastructure Upgrades Prevent Premature Decline
  - Statfjord and Thunder Horse highlighted the role of real-time monitoring, predictive analytics, and continuous facility upgrades in prolonging field productivity and reducing operational disruptions.

6.6.8 Final Thoughts on Field Development Economics

Field development is a complex, high-stakes process that requires a balance of technical expertise, economic analysis, and risk mitigation. The best-performing projects:

- Integrate advanced recovery techniques to maximize ultimate recovery.
- Adopt digital monitoring and automation to optimize real-time production efficiency.
- Ensure infrastructure scalability to handle future production expansions.
- Evaluate fiscal policies and incentives to maintain economic resilience.

By learning from these case studies, companies can refine their development strategies, improve cost efficiency, and make data-driven investment decisions that maximize hydrocarbon asset value.

6.7 CHAPTER SUMMARY

This chapter has provided an overview of field development economics, covering the field development process, project economics, planning and optimization, infrastructure considerations, and risk management. We emphasized the importance of comprehensive economic evaluation and strategic planning to maximize the value of oil and gas field developments. The case studies highlighted the diverse challenges and solutions in field development projects, demonstrating the critical role of economic evaluation in ensuring project success.

We demonstrated how advanced field development techniques and economic optimization contribute to successful field development and production enhancement by examining real-world case studies. Integrating technical expertise with economic analysis is essential for making informed decisions that maximize the value of hydrocarbon resources.

The next chapter will discuss risk management in upstream operations, focusing on identifying, assessing, and mitigating risks associated with exploration, drilling, production, and field development activities. We will explore the principles and practices that govern risk management, as well as the tools and techniques used to enhance operational resilience.

6.7.1 Final Thoughts on Field Development Economics

Field development economics is a **multi-disciplinary process that integrates engineering, financial analysis, risk assessment, and strategic decision-making** to maximize hydrocarbon asset value.

- **Successful projects** balance **technical feasibility with economic efficiency** while maintaining flexibility to adapt to uncertainties.
- **Investment timing, infrastructure scalability, and regulatory frameworks** play key roles in long-term project sustainability.
- **Oil price volatility and breakeven cost management** dictate which projects remain competitive during market downturns.

As the energy industry continues evolving, **emerging technologies, environmental considerations, and improved reservoir management techniques** will shape the next generation of field development strategies.

NOTES

1. International Energy Agency (IEA), *World Energy Investment 2023*.
2. U.S. EIA—Deepwater Cost Recovery Comparisons.
3. BP Statistical Review of World Energy (2023).

Part II

Risk, Regulation, and Sustainability

7 Risk Management in Upstream Operations

7.1 INTRODUCTION TO RISK IN UPSTREAM OPERATIONS

Risk is an inherent and unavoidable aspect of the upstream oil and gas sector. It refers to the **uncertainty of outcomes** that can negatively impact a project's financial, technical, environmental, or operational performance. Given the capital-intensive nature of upstream operations, companies must **identify, analyze, and manage** risks throughout the exploration, development, and production phases.

> If you don't actively manage risk, it will actively manage you.
>
> **Peter Bernstein**
> *Risk Management Expert*

Two key components typically characterize risk:

- **Probability:** The likelihood that an event will occur.
- **Consequence:** The magnitude of the negative impact if the event occurs.

In upstream operations, risks are **interconnected** and can affect project viability in multiple ways. For example, a **technical failure** in well design can lead to **environmental spills**, which in turn trigger **legal and financial penalties**.

Upstream companies face a wide range of risks that can be classified into the following categories:

1. **Technical Risks**
   Technical risks stem from **uncertainties in reservoir characteristics**, drilling complexities, equipment failures, and production challenges. These risks often arise due to:
   - Poorly understood subsurface geology, leading to **dry wells** or unexpected reservoir behavior.
   - **Unconventional reservoir challenges**, such as high-pressure/high-temperature (HPHT) wells or ultra-deepwater drilling.
   - **Equipment failure**, including **blowout preventers (BOPs)** or subsea wellhead malfunctions.
   - **Unexpected reservoir depletion**, requiring enhanced recovery techniques (EOR).

   Example: The **Kashagan Field** in Kazakhstan faced severe technical risks due to its high-pressure reservoir and high hydrogen sulfide (H_2S) content. Early production attempts led to pipeline corrosion, delaying the project for years.[1]

DOI: 10.1201/9781003750345-9

2. **Economic Risks**
   Economic risks result from fluctuations in **oil and gas prices, operational costs, and market demand**. These risks are influenced by:
   - **Oil price volatility**, which can make high-cost projects (e.g., deepwater or Arctic drilling) uneconomical.
   - **Project cost overruns**, especially in megaprojects, where delays lead to budget escalations.
   - **Exchange rate fluctuations** for international projects, impacting CAPEX and OPEX.
   - **Inflation and interest rates** affecting project financing.

   Example: The **deepwater horizon disaster (BP, 2010)** led to over **$65 billion** in legal costs, penalties, and settlements, proving how technical failures can trigger **massive financial losses**.
3. **Political and Regulatory Risks**
   Upstream operations are highly regulated, and political instability can pose **significant threats**. Political risks include:
   - **Changes in government policies**, such as tax hikes or production restrictions.
   - **Resource nationalism**, where governments seek greater control over oil and gas assets.
   - **Sanctions and trade restrictions**, limiting access to key markets or technologies.
   - **Permitting and environmental regulations**, delaying or halting projects.

   Example: In **Venezuela**, the government nationalized oil assets, forcing companies like ExxonMobil and ConocoPhillips to **exit the country**, leading to billions in legal disputes.
4. **Environmental Risks**
   Environmental risks are among the most **high-profile challenges** in the industry. Key concerns include:
   - **Oil spills and gas leaks**, which can lead to **catastrophic environmental damage**.
   - **Carbon footprint and climate change regulations**, increasing compliance costs.
   - **Water contamination**, especially in hydraulic fracturing and offshore operations.
   - **Seismic activity from drilling and injection wells**, leading to regulatory shutdowns.

   Example: Shell's **Prelude FLNG project** in Australia faced **delays due to environmental permitting issues**, as concerns over emissions and marine life protection required significant project redesign.
5. **Social and Reputational Risks**
   The upstream sector operates in many **sensitive environments** where public perception and local community engagement play a critical role. Social risks include:
   - **Community opposition**, leading to protests or project shutdowns.
   - **Workforce health and safety incidents** impact labor relations and productivity.

- **Corporate reputation damage**, affecting investor confidence and stock performance.

  Example: Shell faced strong **local opposition in Nigeria**, where oil spills in the Niger Delta led to long-standing conflicts with local communities, lawsuits, and reputational damage.

7.1.1 Why Effective Risk Management Is Critical

The impact of upstream risks can range from **minor delays to full project abandonment**. A **comprehensive risk management strategy** ensures that companies:

- **Identify risks early** and develop proactive mitigation strategies.
- **Improve financial resilience** through hedging and cost control measures.
- **Enhance safety and environmental compliance**, reducing liability and reputational damage.
- **Adapt to regulatory changes**, ensuring long-term operational stability.

> BP's $65 Billion Spill: A Decade Later, Industry Still Grappling with Risk Accountability
>
> *Financial Times, April 2020*

7.2 RISK ASSESSMENT METHODOLOGIES

Risk assessment is a **systematic process** of identifying, analyzing, and evaluating risks in upstream operations. It helps companies **prioritize risks** based on their likelihood and impact, allowing for targeted mitigation strategies.

There are **two primary approaches** to risk assessment:

1. **Qualitative risk assessment**: Uses descriptive categories to assess risks (e.g., high, medium, low).
2. **Quantitative risk assessment**: Uses numerical data and probabilistic models to measure risks.

7.2.1 Qualitative Risk Assessment

A **qualitative approach** is often used when data is limited or when assessing risks at a **high level**. It categorizes risks based on **expert judgment, past experiences, and structured frameworks** such as:

- **Risk registers**: A centralized document listing all identified risks, their probability, impact, and mitigation plans.
- **Risk matrices**: A **visual tool** that plots risks based on **likelihood and severity**, helping prioritize high-impact risks.

WHAT WENT WRONG AT MACONDO

In 2010, the Macondo well experienced a catastrophic blowout due to multiple failures in cementing, well integrity, and emergency response systems. Despite known risks, mitigation measures were under-prioritized. The result was the worst offshore oil spill in U.S. history.

Likelihood ↓ / Impact →	Low Impact	Medium Impact	High Impact
Rare (1-10%)	Low	Low	Medium
Unlikely (11-30%)	Low	Medium	High
Possible (31-60%)	Medium	High	High
Likely (61-90%)	Medium	High	Critical
Almost Certain (91-100%)	High	Critical	Critical

FIGURE 7.1 Risk assessment matrix for upstream operations.

7.2.2 Example of a Risk Matrix

Matrix categorizing risks by likelihood and impact, ranging from low to critical (Figures 7.1 and 7.2). Highlights how high-impact events with moderate-to-high probability—such as equipment failure or regulatory shutdowns—require immediate mitigation in upstream planning.

7.2.2.1 Practical Application

A risk categorized as critical requires **immediate mitigation**, whereas a **low-risk event** may only require **monitoring**.

Case Example: BP Macondo Well (Deepwater Horizon, 2010):

- **Risk identified:** Well integrity issues, BOP failure, and cementing concerns.
- **Risk rating (before incident):** Medium.
- **Actual outcome: Blowout resulted in the largest offshore spill in U.S. history.**
- **Lesson learned:** Failure to properly classify and escalate **technical risks** can lead to **catastrophic consequences**.

7.2.3 Quantitative Risk Assessment

Quantitative risk assessment uses **data-driven models** to measure risks **objectively**. It helps in **financial risk estimation, decision-making, and scenario analysis.**

Likelihood ↓ / Impact →	Low Impact	Medium Impact	High Impact
Rare (1-10%)	Low	Low	Medium
Unlikely (11-30%)	Low	Medium	High
Possible (31-60%)	Medium	High	High
Likely (61-90%)	Medium	High	Critical
Almost Certain (91-100%)	High	Critical	Critical

FIGURE 7.2 Risk assessment matrix for upstream operations.

Key Quantitative Methods:

1. **Monte Carlo Simulation**
   - A computational model that runs **thousands of simulations** using probability distributions.
   - Helps assess **uncertain variables** (e.g., oil prices, drilling success rates).
   - Provides **expected monetary value (EMV)** for different scenarios.

   Example:
   - A company wants to estimate the **NPV of a deepwater project**.
   - Oil prices, CAPEX, and production rates are uncertain.
   - Monte Carlo runs **10,000 simulations**, generating a **range of possible outcomes** (Figure 7.3).

   Frequency distribution of simulated outcomes from a probabilistic model. Highlights the mean, 5th percentile, and 95th percentile boundaries, illustrating the range of potential financial or operational results under uncertainty—critical for scenario planning and contingency budgeting.
2. **Decision Tree Analysis**
   - A graphical method to analyze **various decision paths** and their possible outcomes.
   - Used in **field development planning** (e.g., drill a high-risk well vs. invest in a lower-risk asset).

   Example:
   - **Option 1:** Drill an exploratory well (**70% success, 30% dry hole**).

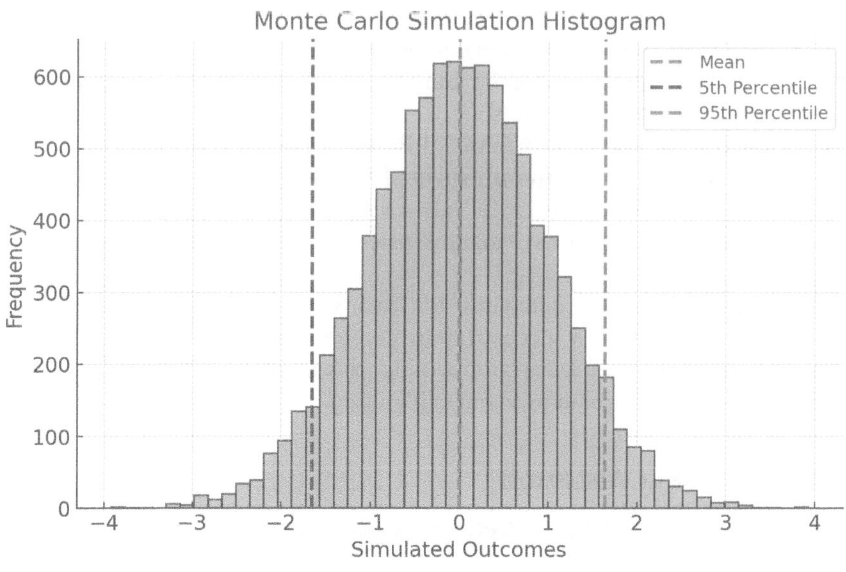

FIGURE 7.3 Monte Carlo simulation histogram for upstream project risk.

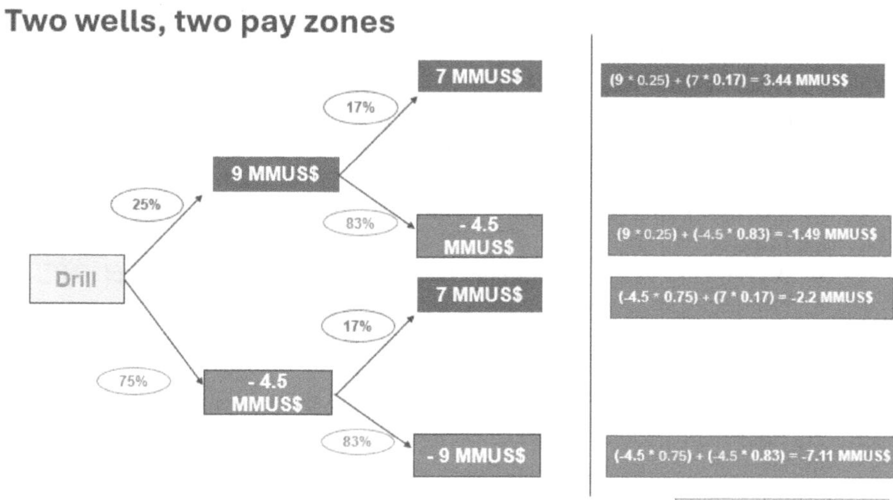

Prospect	COS	EMV
2 Success	4.25%	3.44
P1 S + P2 F	20.75%	(1.49)
P1 F + P2 S	12.75%	(2.20)
2 Failures	62.25%	(7.11)
EMV		(7.36)

FIGURE 7.4 Decision tree analysis for drilling outcomes and expected monetary value (EMV).

- **Option 2:** Invest in a smaller, known-producing field (**lower risk, lower reward**).
- The decision tree calculates the **expected value** of each option.

3. **Failure Mode and Effects Analysis (FMEA)**
   - Identifies **failure points** in equipment or operations.
   - Used in **pipeline integrity management, well control systems, and BOP reliability analysis**.
4. **Value at Risk (VaR) for Financial Risk**
   - Measures the **maximum expected loss** over a given time period, at a confidence level (e.g., 95% VaR = $50M loss).
   - Used by oil companies to **manage price fluctuations and investment risk**.

Illustrates probabilistic outcomes for two wells across multiple pay zones (Figure 7.4). Highlights how success/failure combinations yield varying EMVs, reinforcing the importance of scenario modeling, risk-adjusted valuation, and strategic decision-making in upstream investment planning.

7.2.4 Steps in Risk Assessment

Regardless of the method used, **effective risk assessment** follows these five key steps:

a. **Risk Identification**
   - Techniques: Expert brainstorming, historical analysis, and regulatory reviews.

- Tools: **SWOT analysis (Strengths, Weaknesses, Opportunities, Threats)** and **PESTLE (Political, Economic, Social, Technological, Legal, Environmental)**.
- *Example:* Identifying potential drilling risks in **HPHT wells** before project execution.

b. **Risk Analysis**
- Assess the probability and impact of each risk.
- Determine **cause-effect relationships** and worst-case scenarios.
- *Example:* Using Monte Carlo Simulation to analyze **CAPEX overruns** in a deepwater development.

c. **Risk Evaluation**
- Compare risks to the company's **risk tolerance** and investment criteria.
- Prioritize risks that need **urgent mitigation**.
- *Example:* Evaluating whether an **ultra-deepwater drilling project** is too risky given market conditions.

d. **Risk Treatment**
- Choose a **response strategy**:
  - ✓ Avoid (do not proceed with the risk).
  - ✓ Mitigate (reduce impact through technology or redundancy).
  - ✓ Transfer (use insurance or joint ventures).
  - ✓ Accept (if the risk is within company tolerance).
- *Example:* Shell's Prelude FLNG **transferred risk** using joint venture partnerships.

e. **Monitoring and Review**
- Continuously monitor risk performance.
- Adapt risk strategies based on **new data, regulatory changes, or operational learnings**.

Real-World Case Studies of Risk Assessment:

a. **Case Study: Total's Egina Deepwater Project (Nigeria)**
- **Challenge:** Harsh deepwater environment (1,600 m depth), supply chain disruptions, and political risks.
- **Approach:**
  - Monte Carlo analysis for **cost and schedule risks**.
  - Hedging strategies to mitigate **currency exchange risks**.
  - **Scenario analysis** for potential tax changes in Nigeria.
- **Outcome:** Project delivered **on time and within budget** due to **proactive risk management**.

b. **Case Study: Shell's Arctic Drilling (Alaska)**
- **Challenge:** Extreme weather conditions, environmental regulations, and logistical constraints.
- **Approach:**
  - **Risk matrix assessment** to classify operational hazards.
  - Decision tree analysis comparing Arctic drilling vs. investing in other regions.
  - **Probability modeling** of ice sheet movement to avoid rig damage.

- **Outcome:** Due to **high environmental risk and regulatory hurdles**, Shell **abandoned Arctic exploration in 2015**, avoiding potential billions in losses.

Key Takeaways from 7.2: Risk Assessment Methodologies:

✓ **Qualitative assessments (risk matrices, registers)** help prioritize quick risk.

✓ **Quantitative assessments (Monte Carlo, decision trees, VaR)** are crucial for financial and technical **decision-making**.

✓ **A structured risk assessment process (identify → analyze → evaluate → treat → monitor)** ensures risks are properly managed.

✓ **Case studies (BP Macondo, Shell Arctic, Total Egina)** highlight how **proper risk assessment** can make or break an upstream project.

Grid mapping risk severity based on likelihood and impact, with scores ranging from 1 (low risk) to 25 (critical risk) (Figure 7.5). Highlights how rare but severe events—such as blowouts or regulatory shutdowns—demand immediate mitigation, while frequent low-impact issues may be monitored.

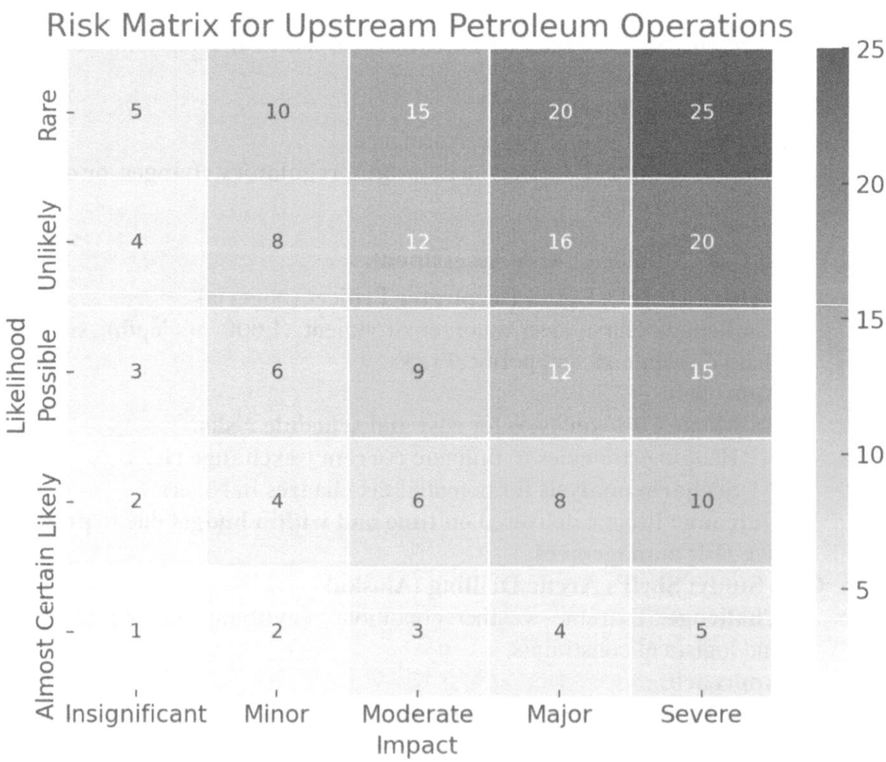

FIGURE 7.5 Numerical risk matrix for upstream petroleum operations.

Key Takeaways from 7.2: Risk Assessment Methodologies:

- **Qualitative assessments** (risk matrices, registers) help in quick risk prioritization.
- **Quantitative assessments** (Monte Carlo, decision trees, VaR) are crucial for financial and technical decision-making.
- **A structured risk assessment** process (identify → analyze → evaluate → treat → monitor) ensures risks are properly managed.
- **Case studies** (BP Macondo, Shell Arctic, Total Egina) highlight how proper risk assessment can make or break an upstream project.

7.3 RISK MITIGATION STRATEGIES IN EXPLORATION AND PRODUCTION

Once risks have been **identified, analyzed, and evaluated**, the next step is to develop **mitigation strategies** to reduce their **probability and impact**.

Upstream oil and gas projects face **multiple technical, economic, political, environmental, and social risks**. Effective **risk mitigation** strategies involve a **combination of operational best practices, financial tools, technological innovations, and strategic planning**.

7.3.1 Technical Risk Mitigation Strategies

Technical risks arise from **reservoir uncertainty, drilling challenges, equipment failure, and operational inefficiencies**. These can lead to **cost overruns, production delays, and even well failures**.

Key Strategies for Technical Risk Reduction:

a. **Reservoir Characterization and Seismic Imaging**
   - **3D and 4D seismic surveys** help in precise reservoir mapping and well placement.
   - **AI-driven seismic interpretation** enhances accuracy in detecting oil/gas-bearing formations.
   - *Example:* The **Johan Sverdrup Field (Norway)** used advanced seismic imaging to optimize well placement, reducing drilling failures.

b. **Well Engineering and Drilling Best Practices**
   - Implement **managed pressure drilling (MPD)** for **HPHT wells**.
   - **Cement bond log (CBL) analysis** ensures proper casing integrity.
   - **Real-time well monitoring** prevents kick/blowout events.
   - *Example:* BP's **Mad Dog Field (Gulf of Mexico)** used **dual-gradient drilling (DGD)** to manage deepwater wellbore pressure.[2]

c. **Production Optimization and Artificial Lift Systems**
   - Use **intelligent completions** to adjust flow rates in real-time.
   - **Gas lift and ESP (Electrical Submersible Pumps)** enhance production from depleting reservoirs.
   - *Example:* **Saudi Aramco's Ghawar Field** optimized artificial lift strategies, maintaining production above 5 million bpd.

d. **Equipment Reliability and Maintenance**
   - Implement **predictive maintenance** using IoT sensors.
   - **FMEA** to anticipate system breakdowns.
   - *Example:* Shell's **Prelude FLNG** used predictive analytics to reduce equipment failures in extreme offshore conditions.

7.3.2 Economic and Financial Risk Mitigation Strategies

Economic risks include **oil price volatility, cost overruns, funding shortages, and revenue fluctuations**.

Key Strategies for Financial Risk Reduction:

a. **Oil Price Hedging and Derivatives**
   - Use **futures contracts, swaps, and options** to lock in prices.
   - *Example:* Chevron hedged part of its **LNG production** to stabilize cash flow amid price swings.

b. **Cost Control and Lean Project Execution**
   - Apply **stage-gate processes** to review project feasibility before full-scale investment.
   - Use **modular construction** to reduce offshore project costs.
   - *Example:* ExxonMobil's **Liza FPSO (Guyana)** used modular construction to cut CAPEX.

c. **Investment Diversification**
   a. Develop **multiple assets** across **different geographies** to spread risk.
   b. *Example:* **TotalEnergies** balances **low-cost Middle East fields with high-risk offshore Africa** projects.

d. **Dynamic Budgeting and Scenario Planning**
   - Run **Monte Carlo simulations** to predict CAPEX/OPEX variations.
   - Maintain a **contingency budget** of **10–20%** for megaprojects.
   - *Example:* The **Tengiz Expansion Project (Kazakhstan)** incorporated **cost buffers** to manage unpredictable inflationary pressures.

7.3.3 Political and Regulatory Risk Mitigation Strategies

Political risks include **policy changes, nationalization, expropriation, and unstable governments**.

Key Strategies for Political Risk Reduction:

a. **Geographical Risk Diversification**
   - Reduce exposure by investing in **low-risk jurisdictions** while limiting exposure in volatile regions.
   - *Example:* BP diversified **from Russia to U.S. shale plays** to mitigate political risk.

b. **Production Sharing Agreements (PSAs) and Joint Ventures**
   - Partner with **national oil companies (NOCs)** to share risks.
   - Negotiate **fiscal stability clauses** in contracts.
   - *Example:* ExxonMobil's **joint venture with Qatar Energy** secured stable fiscal terms for LNG exports.

c. **Political Risk Insurance (PRI)**
   - Cover risks related to **expropriation, civil unrest, and regulatory changes**.
   - *Example:* Total used **PRI in Venezuela** to hedge against nationalization risks.

d. **Strong Stakeholder Engagement and Government Relations**
   - Engage in early discussions with governments and **local authorities**.
   - *Example:* **Equinor (Norway)** collaborated with the government to ensure smooth operations in the North Sea.

7.3.4 Environmental and Social Risk Mitigation Strategies

Environmental and social risks include **oil spills, carbon emissions, ESG concerns, and local community conflicts**.

Key Strategies for Environmental and Social Risk Reduction:

a. **Environmental Impact Assessments (EIA) and Carbon Management**
   - Conduct **EIAs** before project execution.
   - Use **carbon capture and storage (CCS)** to minimize emissions.
   - *Example:* BP's **Net Zero Initiative** integrates CCS in major oil fields.

b. **Spill Prevention and Emergency Response**
   - Install **double-hulled tankers, leak detection systems, and BOPs**.
   - Train emergency teams for **quick spill containment**.
   - *Example:* The **Macondo (Deepwater Horizon) spill** exposed poor emergency preparedness.

c. **Community Engagement and CSR Initiatives**
   - Provide **local employment opportunities**.
   - Invest in **healthcare, education, and infrastructure**.
   - *Example:* Shell's **CSR projects in Nigeria** support local job creation and community development.

d. **ESG Compliance and Green Investments**
   - Develop **low-carbon portfolios** (e.g., hydrogen, renewables).
   - Follow **IFRS and SEC climate disclosure frameworks**.
   - *Example:* **Equinor invests in offshore wind energy** as part of its energy transition strategy.

Comparison of CO_2 emissions per barrel across five extraction techniques (Figure 7.6). Highlights the environmental intensity of coal-to-liquids and oil sands, reinforcing the need for ESG screening and carbon mitigation strategies in upstream project selection.

7.3.5 Case Studies in Risk Mitigation

7.3.5.1 Case Study 1: Shell's Prelude FLNG (Australia)

Challenges:

- **Harsh offshore conditions, high development costs, and new FLNG technology risks.**

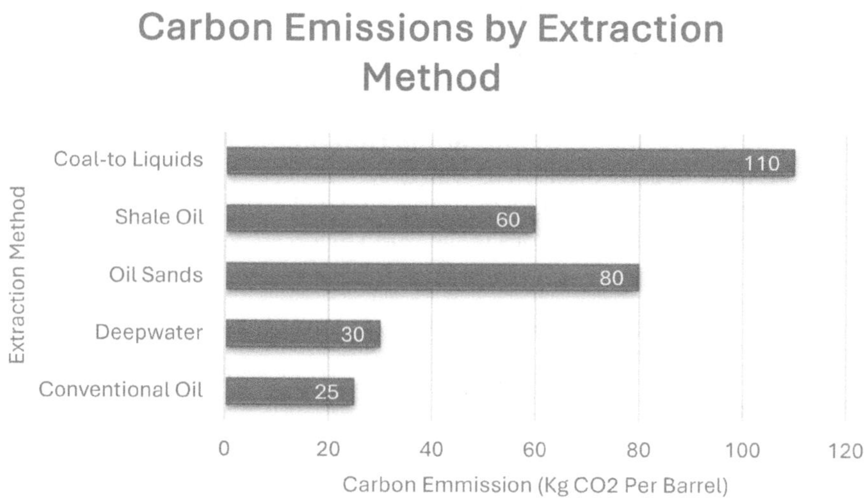

FIGURE 7.6 Carbon emissions by oil extraction method.

Risk Mitigation:

✓ Deployed **AI-driven predictive maintenance** to prevent equipment failures.
✓ Used **phased construction** to manage cost overruns.
✓ Secured **long-term LNG contracts** to hedge price risks.

Outcome:

✓ Successfully delivered the **first FLNG gas production** despite technical and financial challenges.

7.3.5.2 Case Study 2: BP's Mad Dog Phase 2 (Gulf of Mexico)

Challenges:

- **Deepwater drilling risks**, **hurricane exposure**, and **complex subsea architecture**.

Risk Mitigation:

✓ Used **DGD** for better pressure control.
✓ Designed a **hurricane-resistant floating production system**.
✓ Applied **scenario-based budgeting** to control unexpected costs.

Outcome:

✓ BP successfully **reduced the project breakeven** to **below $50 per barrel**.

7.3.5.3 Case Study 3: ExxonMobil's Liza FPSO (Guyana)

Challenges:

- **First deepwater project in Guyana**, **logistical challenges**, and **regulatory hurdles**.

Risk Mitigation:

✓ Used the **FPSO leasing model** to lower initial CAPEX.
✓ Partnered with **Guyana's government** to ensure stable fiscal terms.
✓ Applied **3D seismic data** for optimal well placement.

Outcome:

✓ Production reached **380,000 bpd**, making Guyana a **major oil exporter**.

7.4 INSURANCE AND HEDGING MECHANISMS

Financial risk is one of the most significant challenges in upstream oil and gas projects. **Commodity price volatility, geopolitical instability, project cost overruns, and unexpected events** can impact profitability and long-term viability.

Companies use insurance policies and hedging strategies as risk management tools to mitigate financial exposure.

> **REAL-WORLD HEDGING: OCCIDENTAL VS. UNHEDGED FIRMS (2020)**
>
> When oil crashed in 2020, Occidental Petroleum cushioned losses using swaps and collars, while other producers without hedging saw their revenue plummet. This case emphasized the importance of risk-adjusted financial planning in volatile markets.

7.4.1 Insurance in Upstream Oil and Gas

Insurance protects against unforeseen risks, reducing the economic burden of accidents, equipment failures, and liability claims.

7.4.1.1 Types of Insurance in Upstream Operations

Overview of key insurance categories—property, well control, liability, business interruption, and political risk—used to mitigate operational, financial, and geopolitical exposures (Figure 7.7). Highlights how tailored coverage supports resilience across exploration, production, and transport phases.

Insurance Type	Coverage Scope
Property Insurance	Covers **physical damage** to drilling rigs, production platforms, and equipment from **fire, explosions, storms, and earthquakes.**
Well Control Insurance	Covers costs associated with **blowouts, well failures, and emergency response operations.**
Business Interruption Insurance	Protects against **income loss** due to unexpected shutdowns or disasters.
Liability Insurance	Covers legal costs and damages for **injuries, spills, and environmental damage claims.**
Political Risk Insurance (PRI)	Protects against **expropriation, nationalization, civil unrest, and contract breaches** in politically unstable countries.
Cargo & Marine Insurance	Covers risks related to **oil and gas transportation via tankers and pipelines.**
Environmental Liability Insurance	Covers costs associated with **oil spills, pollution cleanup, and regulatory fines.**

FIGURE 7.7 Insurance types and coverage scope in upstream oil and gas operations.

Case Study: BP Deepwater Horizon and Well Control Insurance:

- The **2010 Deepwater Horizon disaster** resulted in **$65 billion in financial losses**.
- BP had **well control insurance**, but it **only covered a fraction** of the total cost.
- **Lesson:** Companies must ensure **adequate coverage** for catastrophic failures.

Challenges and Limitations of Insurance:

- **High premium costs:** Offshore projects in extreme environments face **higher insurance rates** (e.g., Arctic drilling).
- **Exclusions and coverage gaps:** Some policies **do not cover all risks** (e.g., regulatory changes, long-term environmental damage).
- **Claim delays and litigation risks:** Insurance claims can take **years to settle** (e.g., Exxon Valdez oil spill claim disputes).

7.4.2 Hedging Strategies in Upstream Operations

Commodity price volatility is a major risk in upstream operations. Oil prices fluctuate due to:

- **Global supply-demand imbalances**
- **Geopolitical tensions**
- **Regulatory policies and OPEC+ decisions**
- **Economic downturns (e.g., 2020 oil price crash)**

Hedging allows companies to **stabilize cash flows** and protect against price fluctuations.

7.4.2.1 Types of Hedging Instruments

Overview of futures, options, swaps, collars, and forward contracts (Figure 7.8). Highlights how each strategy stabilizes revenue, protects against downside risk, and supports financial planning under volatile market conditions

Example: Hedging in Action:

- In 2014, **oil prices crashed from $100 to $40 per barrel**.
- Companies like **Occidental Petroleum and Shell** had hedged **part of their production** using swaps and options.[3]
- Result: **Less financial impact** compared to unhedged firms.

Hedging Strategy	How It Works
Futures Contracts	Agreement to buy/sell crude oil at a fixed price on a future date. Used to lock in prices and reduce uncertainty.
Options Contracts	Gives the right (but not obligation) to buy/sell oil at a pre-agreed price. Provides flexibility to benefit from favorable price movements.
Swaps	Agreement to exchange floating oil price payments for fixed-price payments. Used by companies to **stabilize revenues**.
Collars	Combines a **put and call option** to create a price range (floor & ceiling) for oil sales. Ensures minimum cash flow protection.
Forward Contracts	Customized contracts between two parties for future crude oil delivery at an agreed price.

FIGURE 7.8 Hedging instruments used in upstream oil price risk management.

Challenges of Hedging:

- **High costs:** Hedging contracts require **upfront payments**, which can be expensive.
- **Missed market gains:** If oil prices rise above the hedged price, the company **loses potential profits**.
- **Liquidity risks:** Hedging may require **margin deposits**, tying up company funds.

7.4.3 Insurance vs. Hedging—When to Use What?

Hedging is best for:

✓ Managing **short-term price volatility**
✓ Ensuring **stable cash flow for CAPEX and OPEX planning**
✓ Protecting against **market downturns**

Insurance is best for:

✓ Protecting against **catastrophic losses (blowouts, oil spills, lawsuits)**
✓ Managing **operational and liability risks**
✓ Ensuring **financial recovery after major disasters**
❖ **Best Practice: Many companies use both**—insurance for **operational risks** and hedging for **financial risks**.

7.4.3.1 Case Study 1: Shell's Oil Price Hedging Strategy

Background:

- In 2020, oil prices collapsed due to the COVID-19 pandemic.
- Shell **hedged a significant portion** of its production at ~$50 per barrel.

Risk Mitigation Strategy:

- Used a **mix of swaps and futures** to stabilize cash flow.
- Reduced exposure to **short-term price fluctuations**.
- Avoided **major financial losses** faced by unhedged competitors.

Outcome:

- Despite market volatility, Shell maintained a **stable cash flow** to fund new projects.

7.4.3.2 Case Study 2: Petrobras and Political Risk Insurance (Brazil)

Background:

- **Petrobras operates in Brazil**, a country with **political instability and high taxation risks**.
- The company faced **expropriation threats and corruption scandals** in 2014.

Risk Mitigation Strategy:

✓ Purchased **PRI** to protect against government intervention.
✓ Diversified investments into **offshore fields** with more stable regulatory frameworks.
✓ Partnered with **foreign investors** (e.g., TotalEnergies) to share risks.

Outcome:

✓ Ensured financial protection against **government policy shifts**.
✓ Maintained stable operations despite **domestic political uncertainty**.

Key Takeaways from 7.4: Insurance and Hedging Mechanisms:

✓ **Insurance** protects against **physical damage, liability, and catastrophic losses**.
✓ **Hedging** provides **financial stability against oil price volatility**.
✓ **Futures, swaps, and options** are commonly used hedging instruments.
✓ **PRI** is crucial for **operations in unstable regions**.
✓ **Best practice**: Companies use a **mix of insurance and hedging** to reduce risk exposure.

7.5 RISK MANAGEMENT FRAMEWORKS AND BEST PRACTICES

For effective risk management, upstream companies must implement **structured risk frameworks** that integrate risk assessment, mitigation, and continuous monitoring into their operations. A well-defined framework ensures that **risks are proactively managed**, reducing financial losses, operational disruptions, and environmental liabilities.

7.5.1 Key Components of a Risk Management Framework

A robust **Risk Management Framework (RMF)** includes policies, strategies, and tools for assessing, mitigating, and monitoring risks across all oil and gas project phases (Figure 7.9).

Outlines the five foundational elements of risk management: identification, assessment, mitigation, monitoring, and compliance. Highlights how each component contributes to proactive decision-making, regulatory alignment, and long-term operational resilience.

Key Objective:

✓ **Minimize uncertainty and financial exposure**
✓ **Ensure regulatory compliance**
✓ **Improve decision-making efficiency**

Component	Purpose & Description
Risk Identification	The systematic process of recognizing risks across technical, economic, environmental, and geopolitical categories.
Risk Assessment	Evaluating the likelihood and severity of each risk using qualitative or quantitative methods.
Risk Mitigation	Implementing strategies to **reduce, transfer, accept, or avoid risks.**
Monitoring & Review	Ongoing evaluation of risk exposure, adjusting mitigation strategies based on **real-time data and lessons learned.**
Compliance & Reporting	Ensuring regulatory compliance and reporting key risks to **investors, government agencies, and stakeholders.**

FIGURE 7.9 Core components of a risk management process in upstream operations.

7.5.2 Risk Governance and Leadership Commitment

A **top-down approach** to risk management ensures that risk-related decisions align with corporate objectives. **Senior leadership and board-level oversight** are crucial in fostering a risk-aware culture.

Best Practices for Risk Governance:

✓ Establish a **Chief Risk Officer (CRO)** or dedicated risk management team.
✓ Define a **Risk Appetite Statement (RAS)** to guide risk-taking activities.
✓ Conduct **regular risk audits** to evaluate exposure across projects.
✓ Align risk frameworks with **corporate sustainability and ESG (Environmental, Social, Governance) goals**.

Example: ExxonMobil's Risk Governance Approach:

- ExxonMobil integrates **risk-based decision-making** across **all major projects**.
- Uses **internal risk committees** to evaluate project feasibility based on technical and economic risks.
- Risk frameworks align with **long-term investment cycles** to reduce uncertainty in volatile markets.[4]

7.5.3 Risk Identification and Categorization

Risk categorization is a fundamental step in risk management. By grouping risks into key categories, companies can develop targeted mitigation strategies.

7.5.3.1 Common Risk Categories in Upstream Operations

Summary of technical, economic, geopolitical, environmental, and operational risks (Figure 7.10). Highlights how each category presents distinct challenges—from reservoir uncertainty to regulatory shifts—requiring tailored mitigation strategies within a unified risk framework.

Risk Identification Methods:

- **SWOT analysis**—Identifies **Strengths, Weaknesses, Opportunities, and Threats** related to upstream projects.
- **PESTLE analysis**—Assesses **Political, Economic, Social, Technological, Legal, and Environmental** risks.
- **Scenario planning**—Evaluate **best-case, worst-case, and baseline scenarios** for decision-making.

Category	Examples
Technical Risks	Reservoir uncertainty, drilling failures, and well integrity issues.
Economic Risks	Commodity price fluctuations, cost overruns, and CAPEX inflation.
Geopolitical Risks	Resource nationalism, regulatory changes, tax instability.
Environmental Risks	Oil spills, emissions regulations, extreme weather events.
Operational Risks	Equipment failures, logistics disruptions, labor strikes.

FIGURE 7.10 Categorization of risk types in upstream oil and gas operations.

Approach	Description
Risk Avoidance	Eliminating risk by choosing **less risky projects or locations.**
Risk Reduction	Implementing **technical controls, redundancies, and monitoring systems.**
Risk Transfer	Using **insurance, joint ventures, and hedging** to shift financial exposure.
Risk Acceptance	Acknowledging risk when mitigation is **too costly or impractical.**

FIGURE 7.11 Risk treatment approaches in upstream risk management.

7.5.4 Risk Mitigation and Response Planning

Mitigation strategies vary based on **risk type, severity, and controllability**.

7.5.4.1 Four Risk Treatment Approaches

Summary of four core strategies—avoidance, reduction, transfer, and acceptance—used to manage upstream risks (Figure 7.11). Highlights how companies tailor responses based on risk severity, controllability, and cost-effectiveness across technical, financial, and geopolitical domains.

Example: Shell's Risk Mitigation in Arctic Drilling:

- **Avoidance:** Suspended Arctic drilling due to **regulatory and environmental risks**.
- **Reduction:** Used **advanced ice monitoring and subsea BOPs** for safety.
- **Transfer:** Partnered with **local companies** to share risk and expertise.

7.5.5 Continuous Monitoring and Risk Audits

Effective risk management requires **continuous monitoring** to identify emerging threats.

Best Practices for Risk Monitoring:

- Implement **real-time risk dashboards** using AI-driven analytics.
- Conduct **quarterly risk audits** to reassess project vulnerabilities.
- Use **Predictive Maintenance Systems** to detect equipment failures **before they occur**.

Example: Ghawar Field (Saudi Arabia):

- **Largest oil field in the world** (over 5 million bpd production).
- Uses **digital twin technology and real-time sensors** to monitor well integrity.
- Predictive analytics **prevents unplanned shutdowns**, increasing operational efficiency.

> The future of resilience in oil and gas lies in digitization, not just detection.
>
> *Wood Mackenzie Analyst, 2023*

7.5.6 Compliance and Risk Reporting

Adhering to **industry regulations** and **best practices** enhances corporate credibility.

Regulatory Standard	Applicable Region	Purpose
OGP Risk Frameworks	Global	Best practices for oil & gas risk management.
ISO 31000	Global	International standard for risk assessment methodologies.
SEC Reporting Rules	USA	Requires public disclosure of material risks to investors.
HSE (Health, Safety, Environment) Guidelines	Global	Defines safety and environmental risk protocols.

FIGURE 7.12 Regulatory standards governing risk management in upstream operations.

7.5.6.1 Key Compliance Standards in Upstream Operations

Summary of global and U.S. frameworks—including ISO 31000, SEC rules, and HSE guidelines—that shape risk governance, disclosure, and operational safety (Figure 7.12). Highlights how upstream companies align risk practices with international standards and investor expectations.

- ❖ **Example:** After the **Deepwater Horizon disaster**, the **Bureau of Safety and Environmental Enforcement (BSEE)** imposed **stricter offshore drilling regulations**.

7.6 CASE STUDIES OF RISK MANAGEMENT

7.6.1 Case Study 1: BP Deepwater Horizon (2010)

Risk category: Technical, Environmental, Financial

Challenge: Blowout preventer (BOP) failure led to the **largest offshore oil spill in history**.

Response:

- BP invested **$500 million in oil spill prevention R&D**.
- Developed **Stronger BOPs and emergency response systems**.
- **Lesson:** Regulatory oversight is **critical** in high-risk offshore drilling.

7.6.2 Case Study 2: Shell's Prelude FLNG (Australia)

Risk category: Technical, Economic

Challenge: The World's **first floating LNG plant** faced engineering and cost overrun risks.

Response:

- Deployed **digital twin simulations** to optimize design.
- Used **modular construction** to reduce project delays.
- **Lesson:** Innovation and digital monitoring enhance risk mitigation.

7.6.3 Case Study 3: Petrobras' Tupi Field (Brazil)

Risk category: Economic, Geopolitical
Challenge: Deepwater pre-salt drilling with high CAPEX.

Response:

- Used **joint venture partnerships** to share costs (e.g., with TotalEnergies).
- **Lesson: Risk-sharing models improve project feasibility.**

7.7 CHAPTER SUMMARY

- **Risk frameworks** integrate assessment, mitigation, and continuous monitoring.
- **Leadership commitment and risk governance** are critical for long-term success.
- **Risk categorization (technical, economic, geopolitical, environmental, operational)** helps tailor mitigation strategies.
- **Advanced analytics, digital twins, and real-time monitoring** enhance risk detection.
- **Best practice:** Combining **insurance, hedging, and regulatory compliance** ensures sustainable risk management.

NOTES

1. Kashagan Field delays due to H_2S corrosion were documented in Kazakhstan's Ministry of Energy Report, 2018.
2. BP Investor Presentation, 2021, outlined dual-gradient drilling success and CAPEX savings in Mad Dog Phase 2.
3. Shell Annual Report, 2020: Explained swap/option contracts used during COVID-19 price collapse.
4. ExxonMobil Corporate Citizenship Report, 2022—Describes risk governance and sustainability integration.

8 Regulatory and Environmental Considerations

8.1 INTRODUCTION TO REGULATORY FRAMEWORKS

The upstream petroleum sector operates within a complex web of government regulations designed to ensure responsible resource development, protect the environment, and safeguard public health and safety. These regulations govern various aspects of upstream operations, from exploration and drilling to production and decommissioning. Regulatory frameworks vary across regions based on **their geopolitical priorities—for example, Norway prioritizes sustainability, while some OPEC nations focus on maximizing resources**.

8.1.1 Key Types of Regulations

8.1.1.1 Licensing and Permitting

Upstream companies must obtain **exploration licenses** to assess reserves before applying for a **production license**. These permits specify conditions related to environmental protection, safety, and operational standards.

- **Exploration vs. production licensing:** Exploration licenses allow seismic studies and test drilling, while production licenses require field development plans and long-term environmental monitoring.
- **Offshore vs. onshore licensing:** Offshore drilling requires additional environmental assessments (e.g., U.S. BOEM mandates extensive risk studies before granting Gulf of Mexico drilling leases).
- **Example:** Brazil's National Agency of Petroleum (ANP) requires a detailed field development plan before issuing a production license.

8.1.1.2 Environmental Regulations

These regulations aim to minimize the environmental impact of upstream operations, covering **air pollution, water management, waste disposal, and habitat protection**.

- **Air pollution:** Methane emissions regulations include U.S. **EPA's Methane Rule** and the EU's **Zero Pollution Action Plan**.

DOI: 10.1201/9781003750345-10

- **Water management:** Produced water disposal must comply with **reinjection standards** or **treatment before discharge** (e.g., U.S. Clean Water Act standards).
- **Stricter compliance for arctic drilling:** The **Sakhalin-1 project** in Russia required **special ice-resistant oil spill response plans** due to extreme environmental sensitivity.

8.1.1.3 Safety Regulations

These focus on **worker safety, well integrity, pipeline operations, and emergency response protocols**.

- **Pipeline safety act (USA):** Mandates **integrity management programs** to prevent oil and gas pipeline leaks.
- **Workforce safety requirements:** Offshore drilling workers must comply with **fire-resistant PPE requirements**, **blowout prevention training**, and **helicopter transport safety protocols** (Figures 8.1 and 8.2).

8.1.2 Regulatory Agencies by Region

Region	Regulatory Agency	Key Focus Areas
USA	EPA, BOEM, OSHA	Environmental impact, offshore safety, worker protection
Norway	Norwegian Petroleum Directorate (NPD)	Sustainability, carbon capture incentives
EU	European Environment Agency (EEA)	Emission reduction, renewable energy transition
Australia	NOPSEMA	Offshore drilling safety & environmental compliance
Saudi Arabia	Ministry of Energy	Maximizing resource extraction efficiency

FIGURE 8.1 Key regulatory agencies by region. Overview of major upstream regulatory bodies and their focus areas, illustrating regional priorities in environmental protection, safety, and resource management.

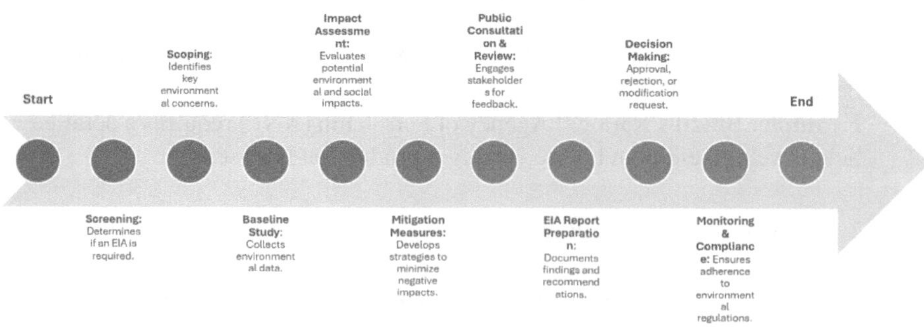

FIGURE 8.2 Environmental impact assessment (EIA) process flowchart. Step-by-step overview from screening to compliance, illustrating how upstream projects integrate environmental safeguards into planning and execution.

8.2 ENVIRONMENTAL IMPACT ASSESSMENT (EIA)

Environmental Impact Assessment (EIA) is a process used to identify and evaluate the potential environmental impacts of proposed upstream projects. It ensures that environmental considerations are integrated into project planning and decision-making.

The EIA Process Typically Involves:

- **Screening**: Determines whether a project requires a full EIA. Projects with minimal environmental impact may be exempt, while high-risk projects (e.g., offshore drilling) undergo a detailed review.
- **Scoping**: This step identifies key environmental concerns and sets the boundaries for the assessment. It involves consultations with regulators, scientists, and stakeholders.
- **Baseline study**: This study collects data on existing environmental conditions, including air and water quality, biodiversity, and local community health.
- **Impact assessment**: Evaluate how the project will affect land, water, air, wildlife, and communities. This includes predicting both direct and indirect environmental consequences.
- **Mitigation measures**: Develop strategies to prevent, reduce, or offset negative environmental impacts. Common mitigation actions include pollution control, habitat restoration, and spill prevention measures.
- **Public consultation**: Engages local communities and stakeholders to gather concerns and ensure transparency.
- **Reporting and approval**: The EIA report is submitted to regulatory authorities for review. If the project meets environmental standards, approval is granted.
- **Monitoring and compliance**: Continuous monitoring ensures the project adheres to mitigation measures and regulatory conditions (Figure 8.3).

EIA Process—Flowchart

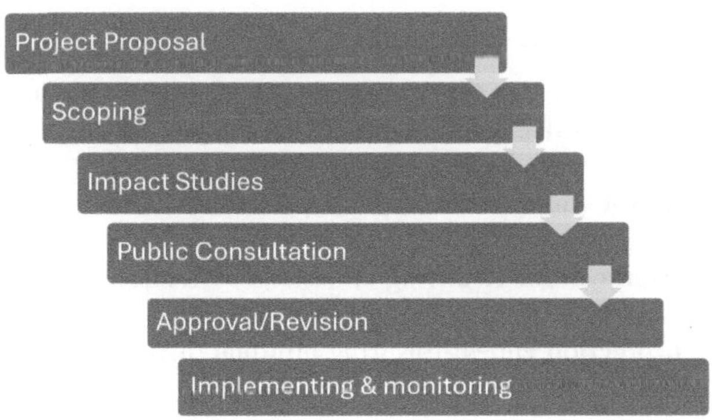

FIGURE 8.3 Project development stages in the EIA process: Sequential steps from proposal to monitoring, illustrating how upstream projects progress through environmental planning and regulatory approval.

Key Environmental Concerns Related to Upstream Operations Include:

- **Air pollution:** Emissions of greenhouse gases, volatile organic compounds, and other pollutants. Air pollution can result from flaring, venting, and combustion engine operation.
- **Water pollution:** Drilling fluid discharges produce water and other contaminants. Water pollution can affect surface water, groundwater, and marine environments.
- **Land disturbance:** Clearing land for drilling sites, pipelines, and other infrastructure. Land disturbance can lead to soil erosion, habitat loss, and landscape alteration.
- **Noise pollution:** Noise from drilling operations, construction, and production facilities can impact wildlife and local communities.
- **Biodiversity loss:** Land clearing for infrastructure and offshore disturbances can disrupt wildlife habitats.

Mitigation measures: Mitigation measures can include pollution control technologies, waste management practices, site restoration and remediation, and habitat conservation. Examples include:

- **Air pollution control:** Using low-emission equipment, implementing flaring reduction programs, and adopting best practices for combustion processes.
- **Water management:** Treating and recycling produced water, using closed-loop drilling systems, and implementing spill prevention measures.
- **Land restoration:** Rehabilitating disturbed land, planting native vegetation, and restoring natural habitats.
- **Noise reduction:** Using noise barriers, implementing quiet drilling technologies, and scheduling noisy activities to minimize impact on local communities.

8.2.1 Case Example: The Kashagan Field (Kazakhstan)

The Kashagan offshore field in the Caspian Sea faced strict EIA requirements due to its environmentally sensitive location. The project implemented:

- **Zero discharge policy:** Preventing wastewater and drilling fluid disposal into the sea.
- **Sour gas reinjection:** Reducing emissions by reinjecting high-sulfur gas into the reservoir.
- **Wildlife protection zones:** Ensuring minimal disruption to Caspian seals and migratory birds.

8.3 COMPLIANCE COSTS AND SOCIAL RESPONSIBILITY

Regulatory compliance in the upstream petroleum sector comes with significant costs, but it also plays a vital role in ensuring environmental protection, worker safety, and social responsibility. Compliance costs impact project economics but are necessary for maintaining a company's social license to operate.

8.3.1 Compliance Costs in Upstream Operations

Regulatory compliance costs vary depending on the project's location, environmental sensitivity, and regulatory framework. These costs include:

1. **Licensing and Permitting Fees:**
   - Companies must pay fees for exploration, drilling, and production permits.
   - Example: In the U.S., offshore operators pay lease fees to the Bureau of Ocean Energy Management (BOEM) and must comply with the Bureau of Safety and Environmental Enforcement (BSEE) regulations.
2. **Environmental Monitoring and Reporting:**
   - Regular air, water, and soil quality assessments are required to ensure compliance with environmental standards.
   - Example: The European Union's Emissions Trading System (ETS) requires companies to report and offset their greenhouse gas emissions.
3. **Pollution Control and Waste Management:**
   - Costs include installing emission reduction systems, wastewater treatment plants, and hazardous waste disposal facilities.
   - Example: Many offshore platforms use **Zero Discharge Systems**, meaning no waste is dumped into the ocean.
4. **Workplace Safety Compliance:**
   - Includes training workers, maintaining emergency response systems, and implementing safety technologies.
   - Example: In Norway, oil companies must follow strict HSE (Health, Safety, and Environment) requirements under the Petroleum Safety Authority (PSA).
5. **Fines and Penalties for Non-Compliance:**
   - Non-compliance can result in financial penalties, project shutdowns, or legal action.
   - Example: BP's **Deepwater Horizon oil spill (2010)** led to over $60 billion in fines, settlements, and cleanup costs.

FAST FACT—ENVIRONMENTAL PENALTIES

- BP paid $60+ billion after Deepwater Horizon.
- Equinor faced $1.5B in carbon tax since 2017.
- Shell's flaring penalties in Nigeria: $1 million annually (estimated).

If you think compliance is expensive, try non-compliance.

Paul McNulty
Former U.S. Deputy Attorney General

8.3.2 Corporate Social Responsibility (CSR) in Upstream Operations

Beyond legal compliance, companies engage in **Corporate Social Responsibility (CSR)** to foster positive relationships with stakeholders, enhance community well-being, and mitigate social risks.

Key Areas of CSR in Upstream Operations:

1. **Community Engagement and Development:**
   - Companies consult local communities, address concerns, and support economic development.
   - Example: Shell's **Nigeria Community Development Program** funds education, healthcare, and infrastructure projects in oil-producing regions.
2. **Social Investment Initiatives:**
   - Investments in education, vocational training, and local businesses create long-term benefits.
   - Example: ExxonMobil sponsors STEM (Science, Technology, Engineering, and Mathematics) programs in developing countries to create job opportunities.
3. **Environmental Conservation Programs:**
   - Oil companies often invest in reforestation, wildlife protection, and carbon offset programs.
   - Example: Chevron's **Gorgon LNG Project in Australia** includes a **Carbon Capture and Storage (CCS)** facility, injecting 3.4 million metric tons of CO_2 underground annually.
4. **Indigenous Rights and Cultural Preservation:**
   - Respecting land rights and engaging with Indigenous groups ensures ethical resource development.
   - Example: Canada's oil sands operators collaborate with Indigenous communities to create employment and business opportunities.

8.3.3 Case Example: TotalEnergies in Uganda—The Tilenga and EACOP Project

TotalEnergies faced backlash for its **East African Crude Oil Pipeline (EACOP) project**, which crosses Uganda and Tanzania. Environmental groups opposed the project due to concerns over wildlife impact and community displacement. In response, the company:

- Conducted extensive stakeholder consultations.
- Established compensation programs for displaced communities.
- Implemented biodiversity conservation initiatives, including reforestation efforts.

This case illustrates how balancing compliance costs with CSR initiatives can improve project sustainability and social acceptance (Figure 8.4).

Compliance Category	Estimated Cost per Barrel (USD)	Notes
Permits & Licensing	$0.35	Includes exploration, drilling, environmental, and emissions permits
Environmental Monitoring	$0.25	Covers continuous air, water, and soil quality monitoring
Emissions Control	$0.50	Costs related to capturing or reducing CO_2, methane, and VOC emissions
Safety Training & HSE	$0.20	Safety programs, emergency drills, and mandatory HSE certifications
Corporate Social Responsibility (CSR)	$0.30	Community engagement, local hiring, and infrastructure development
Total Estimated Regulatory Cost: **$1.60 per barrel**		

FIGURE 8.4 Estimated regulatory compliance cost per barrel. Breakdown of upstream regulatory costs by category, illustrating the financial impact of permits, monitoring, emissions control, safety, and CSR initiatives.

8.4 INTERNATIONAL REGULATORY FRAMEWORKS AND STANDARDS

The oil and gas industry operates across multiple jurisdictions, each with unique regulations. However, international agreements, organizations, and industry standards play a crucial role in setting global best practices for environmental protection, safety, and sustainability. These frameworks ensure consistency across regions and help companies manage regulatory risks in a globally integrated industry.

8.4.1 Key International Environmental Agreements

Several global agreements shape environmental regulations for the upstream sector:

1. **The Paris Agreement (2015)**
    - A legally binding treaty adopted by 196 countries to limit global warming to below **2°C** (preferably **1.5°C**).
    - Upstream companies must monitor and reduce greenhouse gas emissions (GHG), adopt low-carbon technologies, and align with **Nationally Determined Contributions (NDCs)**.
    - Example: Norway's oil sector integrates **carbon pricing** and emissions reduction targets in line with Paris commitments.
2. **The Kyoto Protocol (1997) and Carbon Market Mechanisms**
    - Established **carbon trading** systems and emissions reduction targets.
    - Countries and companies participate in **carbon offset programs** to balance emissions.
    - Example: Shell and BP invest in **carbon credits** to offset emissions from oil operations.
3. **The Convention on Biological Diversity (CBD, 1992)**
    - Focuses on biodiversity conservation and sustainable resource use.
    - Oil companies operating near sensitive ecosystems must conduct **EIAs** and implement mitigation strategies.
    - Example: **Guyana's offshore Stabroek Block** development includes measures to protect marine biodiversity.
4. **The Basel Convention on Hazardous Waste (1989)**
    - Regulates the transboundary movement of hazardous waste, including **oilfield chemicals and drilling waste**.
    - Example: Companies must properly manage **produced water, drilling mud, and chemical waste** when decommissioning offshore wells.

8.4.2 Major International Regulatory Bodies and Industry Organizations

Several organizations establish environmental, health, and safety (EHS) standards in upstream oil and gas:

1. **The International Maritime Organization (IMO)**

- Regulates offshore and marine petroleum activities through conventions such as:
  - **MARPOL 73/78**: Prevents pollution from ships, including oil spills.
  - **London Protocol (1996)**: Regulates subsea waste disposal and decommissioning of offshore infrastructure.

2. **The International Energy Agency (IEA)**
   - Provides policy recommendations, promotes energy efficiency, and monitors global emissions trends.
   - Example: The IEA's **Net Zero by 2050** roadmap influences corporate sustainability targets.
3. **The United Nations Environment Programme (UNEP)**
   - Monitors environmental performance in oil and gas, promoting best practices for emissions reduction and ecosystem protection.
4. **The Extractive Industries Transparency Initiative (EITI)**
   - Focuses on **financial transparency** in resource-rich countries, ensuring that oil revenues benefit national development.

FRAMEWORK SNAPSHOT—ISO 14001

- Focus: Environmental management systems
- Global adoption: Over 300,000 certificates
- Key benefit: Tracks emissions, waste, and energy use

8.4.3 Key Industry Standards and Certifications

In addition to government regulations, oil and gas companies adhere to industry standards to ensure compliance and best practices:

1. **ISO 14001: Environmental Management Systems**
   - The global benchmark for environmental management in industrial operations.
   - Helps companies track emissions, waste, and resource efficiency.
   - Example: Equinor (Norway) integrates **ISO 14001-certified environmental systems** across its offshore platforms.
2. **IFC Performance Standards (International Finance Corporation)**
   - Guidelines for environmental and social risk management in energy projects.
   - Used by banks and investors to assess **sustainable financing eligibility**.
   - Example: Many upstream projects must align with **IFC standards** to secure funding.
3. **The Oil and Gas Climate Initiative (OGCI)**
   - A coalition of major oil companies working to reduce emissions and develop carbon capture technologies.
   - Example: OGCI members (ExxonMobil, Shell, Chevron, etc.) invest in **low-carbon R&D**.

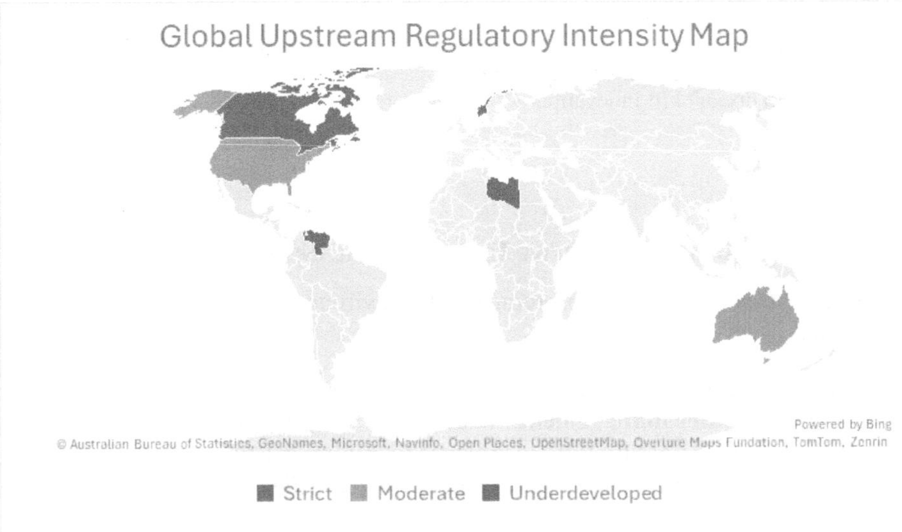

FIGURE 8.5 Global upstream regulatory intensity map Visual comparison of regulatory environments across regions, highlighting countries with strict, moderate, and underdeveloped upstream oversight.

4. **API Standards (American Petroleum Institute)**
   - Technical and operational standards for drilling, well integrity, and equipment safety.
   - Example: API 14C governs **offshore safety systems** and platform shutdown procedures (Figure 8.5).

8.4.4 Case Study: Carbon Pricing in Canada

Canada implemented **carbon pricing** regulations requiring oil companies to **pay per ton of CO_2 emissions**. Companies must either:

- Reduce emissions using **clean technology** (e.g., Carbon Capture and Storage—CCS), or
- Purchase **carbon credits** to offset emissions.

Result: **Oil sands producers like Suncor and Cenovus invested in CCS projects**, reducing their carbon tax liability while improving environmental performance.

8.5 ENVIRONMENTAL REGULATIONS AND TECHNOLOGICAL INNOVATION

Environmental regulations drive innovation in the upstream sector as companies seek cost-effective solutions to reduce emissions, minimize waste, and improve operational efficiency. Regulatory pressure has accelerated the adoption of **clean technologies**, digital monitoring, and advanced engineering solutions.

> Regulation doesn't kill innovation—it channels it.
>
> **Fatih Birol**
> *Executive Director, IEA*

8.5.1 How Regulations Drive Technological Innovation

Governments worldwide are enforcing stricter environmental laws, compelling companies to:

- Reduce **greenhouse gas (GHG) emissions** (e.g., CO_2, methane).
- Improve **water management** (reduce usage, recycle wastewater).
- Adopt **low-impact drilling** techniques to minimize land disturbance.
- Enhance **spill prevention and response capabilities**.

Example: The European Union's **Emission Trading System (ETS)** sets a carbon price, incentivizing oil companies to adopt **Carbon Capture and Storage (CCS)** technology.

8.5.2 Key Environmental Technologies in Upstream Operations

1. **Carbon Capture, Utilization, and Storage (CCUS)**
    - Captures CO_2 emissions from oil production and either **stores it underground** or **reuses it for Enhanced Oil Recovery (EOR)**.
    - Example: **Norway's Sleipner Project** has stored over **20 million tons of CO_2** since 1996.
2. **Methane Leak Detection and Repair (LDAR) Technologies**
    - Uses **drones, infrared cameras, and AI** to detect and seal methane leaks.
    - Reduces methane emissions by up to **80%**.
    - Example: **ExxonMobil and Shell** use AI-driven **satellite monitoring** for methane leak detection.
3. **Advanced Water Management Systems**
    - Includes **water recycling**, **membrane filtration**, and **zero-liquid discharge systems** to minimize freshwater use.
    - Example: **Chevron's Permian Basin operations** recycle **98% of produced water**, reducing freshwater dependency.
4. **Low-Impact Drilling and Multi-Well Pads**
    - Reduces **surface disturbance** by drilling multiple wells from a single location.
    - Example: **Equinor's Johan Sverdrup Field** in Norway uses **horizontal drilling** to limit land use.
5. **Flare Gas Recovery Systems (FGRS)**
    - Captures and reuses associated gas instead of flaring it, reducing emissions.
    - Example: **Nigeria's "Zero Routine Flaring" Initiative** has cut gas flaring by **70%** since 2010.

6. **Blockchain for Regulatory Compliance**
   - Enables **transparent tracking** of emissions, water usage, and waste disposal.
   - Example: BP and Shell use **blockchain-based tracking** for **carbon credit trading**.

8.5.3 Government Incentives for Environmental Innovation

To encourage investment in clean technology, governments provide:

- **Carbon credits:** Companies can trade excess emissions reductions for financial gain.
- **Subsidies and tax credits:** Reduce the cost of implementing **low-emission technologies**.
- **R&D grants:** Fund new solutions like **biodegradable drilling fluids** and **subsea CO_2 storage**.

Example: The U.S. **45Q Tax Credit** provides **\$50 per ton of CO_2 captured**, making CCS projects more viable.

8.5.4 Case Study: Saudi Aramco's Green Drilling Initiative

- Developed **nano-based drilling fluids** to **reduce water use** by **40%**.
- Implemented **solar-powered oilfield operations** in remote desert locations.
- Reduced **flaring emissions by 60%** using **flare gas recovery technology**.

8.6 Case Studies of Regulatory and Environmental Challenges

This section highlights **real-world examples** where upstream petroleum projects encountered **regulatory and environmental challenges** and the strategies used to **overcome them**.

8.6.1 Case Study 1: Chevron's Gorgon LNG Project (Australia)

- **Challenges:**
  - Located on **Barrow Island**, a Class A nature reserve with **strict environmental regulations**.
  - High **CO_2 content in natural gas** (14% CO_2), requiring carbon capture.
  - **Cost overruns and delays** due to regulatory compliance requirements.
- **Solutions Implemented:**
  - Conducted **comprehensive EIA**.
  - Implemented a **Carbon Capture and Storage (CCS) system** to store 4 million tons of CO_2 per year.
  - Engaged with **Indigenous communities** and **environmental NGOs** to ensure compliance.

- **Outcomes:**
  - One of the **largest CCS projects** globally, capturing over **6 million tons of CO_2** by 2023.
  - Established a **model for large-scale LNG projects** balancing economic and environmental goals.
- **Lessons Learned:**
  - **Early stakeholder engagement** prevents legal delays.
  - **Advanced CCS technology** helps meet emissions targets.
  - **Regulatory compliance adds costs**, but **proactive planning minimizes overruns**.

8.6.2 Case Study 2: BP's Thunder Horse Field (Gulf of Mexico, USA)

- **Challenges:**
  - **Deepwater drilling risks**, requiring compliance with **stricter offshore safety laws** post-Deepwater Horizon.
  - Exposure to **hurricanes** and extreme weather, affecting **safety planning**.
  - Regulatory scrutiny on **subsea well control systems**.
- **Solutions Implemented:**
  - Strengthened **well blowout preventers (BOPs)** and **real-time well monitoring** systems.
  - **Reinforced offshore platforms** to withstand extreme weather.
  - Adopted **digital twin technology** to model and predict **equipment failures**.
- **Outcomes:**
  - Despite early setbacks (platform tilt in 2005), **Thunder Horse became one of BP's most productive fields**, peaking at **250,000 bpd**.
  - Set new **safety benchmarks** for deepwater operations.
- **Lessons Learned:**
  - **Regulatory compliance improves safety**, despite increasing costs.
  - **Real-time monitoring** reduces **well integrity risks**.
  - **Extreme weather planning** is essential for deepwater operations.

8.6.3 Case Study 3: Shell's Prelude FLNG Project (Australia)

- **Challenges:**
  - **First-of-its-kind floating LNG (FLNG) facility**, leading to regulatory uncertainty.
  - High **development costs** (~$17 billion) and **delays due to safety and environmental compliance**.
  - Complex **marine environmental approvals** in Australian waters.
- **Solutions Implemented:**
  - Close coordination with **Australian regulatory agencies** to set **FLNG-specific safety standards**.

- Designed **low-emission LNG operations** to comply with **marine pollution controls**.
- Adopted **autonomous monitoring systems** for **real-time environmental compliance tracking**.
- **Outcomes:**
  - Achieved **first LNG production in 2018**, proving the feasibility of **FLNG technology**.
  - Met Australia's strict **marine biodiversity regulations**, setting a **benchmark for offshore gas projects**.
- **Lessons Learned:**
  - **Pioneering projects require close regulatory coordination**.
  - **Automation improves compliance monitoring**.
  - **High CAPEX makes early feasibility studies essential**.

8.6.4 Case Study 4: Equinor's Johan Sverdrup Field (Norway)

- **Challenges:**
  - Strict **Norwegian environmental regulations** requiring **low-carbon** production methods.
  - Need to integrate **offshore electrification** to cut emissions.
  - Managing stakeholder concerns regarding **seabed impact and fisheries**.
- **Solutions Implemented:**
  - Developed **fully electrified platforms**, reducing **CO_2 emissions by 90%**.
  - Used **advanced 4D seismic imaging** to optimize well placement and **minimize seabed disturbance**.
  - Established **stakeholder engagement programs** to maintain local support.
- **Outcomes:**
  - Norway's **most sustainable offshore oil project**, with **a breakeven price of $20/bbl**.
  - Became a **global model** for **low-carbon offshore production**.
- **Lessons Learned:**
  - **Early investment in electrification reduces long-term emissions costs**.
  - **Stakeholder engagement minimizes legal disputes**.
  - **Data-driven well planning optimizes environmental footprint**.

8.6.5 Case Study 5: Nigeria's Zero Routine Flaring Initiative

- **Challenges:**
  - Nigeria was one of the **top 10 gas-flaring countries**, causing **GHG emissions and resource loss**.
  - Flaring reduction faced **infrastructure gaps**, requiring **policy incentives**.
  - Oil companies needed **financial justification** to invest in flare gas recovery.

- **Solutions Implemented:**
  - Enforced **gas flare reduction policies** with **strict penalties** for violators.
  - Provided **tax incentives** for companies investing in **flare gas recovery**.
  - Supported **public-private partnerships (PPPs)** to build gas processing infrastructure.
- **Outcomes:**
  - Gas flaring reduced by **70% since 2010**, lowering emissions and **boosting Nigeria's gas exports**.
  - Created **new economic opportunities** in natural gas utilization.
- **Lessons Learned:**
  - **Strong regulatory enforcement leads to measurable impact**.
  - **Financial incentives encourage compliance**.
  - **Infrastructure investment is critical to sustainable solutions** (Figure 8.6).

8.6.6 Summary of Case Studies

See Figure 8.6.

8.7 CHAPTER SUMMARY

This chapter explored the **regulatory and environmental considerations** in upstream petroleum operations, highlighting the importance of compliance, risk management, and sustainable development. The discussion covered key **regulatory frameworks, EIA, compliance costs, social responsibility**, and **technological innovations** in environmental management.

Case Study	Challenges	Solutions Implemented	Key Outcomes	Lessons Learned
Gorgon LNG (Australia)	CO_2-heavy gas, nature reserve restrictions, high CAPEX	CCS system, early EIAs, community engagement	First large-scale CCS in LNG, LNG production success	Early stakeholder engagement, CCS improves compliance
Thunder Horse (USA)	Deepwater risks, hurricane exposure, regulatory safety overhaul	BOP upgrades, real-time monitoring, reinforced structures	One of BP's most productive fields	Stronger regulations improve safety, weather risk planning is key
Prelude FLNG (Australia)	First FLNG, unclear regulations, high development costs	Set FLNG standards, low-emission design, digital monitoring	Proved FLNG feasibility, met marine safety laws	Regulatory collaboration needed for pioneering tech
Johan Sverdrup (Norway)	Strict CO_2 laws, seabed impact concerns	Electrified platforms, 4D seismic, community programs	90% CO_2 reduction, breakeven price $20/bbl	Early investment in sustainability pays off
Nigeria's Zero Flaring Initiative	High gas flaring, lack of infrastructure	Flaring penalties, tax incentives, PPP investment	70% flaring reduction, increased gas exports	Strong enforcement and financial incentives work

FIGURE 8.6 Summary of regulatory and environmental case studies. Comparison of challenges, solutions, outcomes, and lessons learned across five upstream projects, illustrating best practices in compliance and environmental management.

8.7.1 Key Takeaways

1. **Regulatory Frameworks Shape Upstream Operations**
   - Governments enforce **licensing, environmental, and safety regulations** to ensure responsible resource extraction.
   - Agencies such as the **EPA (USA), NOPSEMA (Australia), and EEA (Europe)** play a crucial role in **monitoring compliance** and enforcing regulations.
   - Strict **regulatory oversight** ensures projects meet environmental and safety standards while preventing negative societal impacts.
2. **EIA are Essential for Project Approval**
   - The **EIA process** identifies potential environmental risks before project execution, ensuring **early mitigation strategies** are in place.
   - Effective EIAs reduce the **likelihood of regulatory penalties** and community resistance, streamlining project approvals.
   - Upstream companies integrate **pollution control, habitat conservation, and land restoration** into EIAs to enhance long-term sustainability.
3. **Compliance Costs vs. Social Responsibility**
   - **Regulatory compliance** entails significant costs (permits, environmental monitoring, emissions control), but **avoiding fines and delays makes it a necessary investment**.
   - **Social responsibility (CSR initiatives)** helps companies build trust, gain local support, and minimize operational risks.
   - Companies investing in **local infrastructure, environmental conservation, and fair labor practices** benefit from a **stronger social license to operate**.
4. **Global and Industry Standards Drive Sustainability**
   - **International agreements (Paris Agreement, Convention on Biological Diversity)** set global environmental standards.
   - **Industry standards (ISO 14001, IEA guidelines)** help oil companies implement best practices for **environmental risk management**.
   - Aligning with these frameworks improves **corporate reputation, regulatory compliance, and investor confidence**.
5. **Technology Plays a Key Role in Environmental Compliance**
   - Innovations in **carbon capture and storage (CCS), methane detection, and flaring reduction** are helping upstream companies meet **strict environmental standards**.
   - **Digital twins, real-time monitoring, and AI-driven predictive maintenance** optimize environmental performance and ensure early detection of compliance risks.
   - Government **tax credits and subsidies** encourage the adoption of clean technologies, reducing the **cost burden on companies**.
6. **Case Studies Demonstrate Best Practices in Compliance and Environmental Management**
   - **Gorgon LNG (Australia)**: Showcased **carbon capture success** in LNG production, proving CCS viability.

- **Thunder Horse (USA)**: Reinforced **deepwater safety measures** post-deepwater Horizon disaster, improving regulatory trust.
- **Prelude FLNG (Australia)**: Set **new regulatory benchmarks** for floating LNG operations through collaborative compliance efforts.
- **Johan Sverdrup (Norway)**: Proved that **early investment in sustainability (electrified platforms) reduces long-term CO_2 emissions**.
- **Nigeria's Zero Routine Flaring Initiative**: Demonstrated that **strict flaring penalties and financial incentives drive meaningful environmental improvements**.

8.7.2 Final Thoughts

As global pressure for **environmental responsibility increases**, the oil and gas sector must **adapt to stricter regulations, invest in cleaner technologies, and prioritize sustainable development**.

- Companies that **proactively manage environmental risks** will **reduce costs, avoid legal challenges, and maintain long-term profitability**.
- **Regulatory compliance is not just an obligation—it is a competitive advantage** that secures investment, minimizes disruptions, and fosters public trust.

Part III

Innovation and Strategy

9 Technology and Innovation

9.1 INTRODUCTION TO TECHNOLOGY IN UPSTREAM OPERATIONS

Technology and innovation are indispensable drivers of progress in the upstream oil and gas sector. They are crucial in enhancing efficiency, reducing costs, improving hydrocarbon recovery rates, and minimizing environmental impact. As energy demand rises and reservoirs become more complex, adopting cutting-edge technologies has become essential for sustaining production, optimizing operations, and meeting regulatory and environmental expectations.

Key Areas of Technological Innovation:

1. **Exploration:** Advances in seismic imaging, data analytics, and machine learning enhance subsurface imaging, improving the accuracy of hydrocarbon discoveries. Technologies such as 4D seismic[1] and artificial intelligence (AI)-driven data interpretation to reduce exploration risks and improve drilling success rates.
2. **Drilling:** Innovations in drilling techniques, automation, and drill bit technology improve efficiency, reduce costs, and enable access to deeper and more geologically complex reservoirs. Managed pressure drilling (MPD) and rotary steerable systems (RSS) allow precise well placement while minimizing drilling risks.
3. **Production:** Enhanced Oil Recovery (EOR) techniques, subsea production systems, and advanced artificial lift technologies optimize hydrocarbon recovery and production efficiency. Innovations like gas lift optimization, smart completions, and downhole sensors enable real-time production monitoring and adjustments.
4. **Reservoir management:** Advanced reservoir simulation models, smart well technologies, and real-time data analytics improve reservoir performance, extending field life and maximizing recovery. Digital rock

PRECISION IN MOTION: ROTARY STEERABLE SYSTEMS (RSS)

Rotary Steerable Systems (RSS) allow real-time directional drilling without stopping rotation—enabling smoother well paths, reduced drag, and improved rate of penetration. RSS has become essential in complex reservoirs where precise trajectory control determines well productivity and cost efficiency.

Example: In tight offshore formations, RSS enables drilling long laterals with high accuracy, minimizing reservoir damage and maximizing contact.

DOI: 10.1201/9781003750345-12

physics and AI-based decline curve analysis help optimize production strategies.

5. **Digitalization:** Integrating big data, cloud computing, AI, and digital twins transforms decision-making, predictive maintenance, and asset management. Digital oilfields improve efficiency by leveraging real-time monitoring and automated control systems.

These advancements are driven by a combination of factors, including:

- There is a need to meet global **energy demand** and ensure supply security.
- Cost pressures require operators to **optimize capital and operational efficiency**.
- Increasing **environmental concerns and regulatory requirements** drive investment in cleaner, more efficient technologies.
- The depletion of conventional reserves necessitates **innovative recovery techniques and reservoir management strategies**.

Technology continues to be a **key enabler** of sustainable and economically viable upstream operations, shaping the future of oil and gas exploration, development, and production.

9.2 EXPLORATION TECHNOLOGIES

Exploration technologies continually evolve to enhance the likelihood of discovering new hydrocarbon reserves while minimizing exploration risks and costs. Advances in seismic imaging, data analytics, and unconventional exploration methods enable more accurate identification of potential reservoirs, leading to higher exploration success rates.

Key Advancements in Exploration Technologies:

1. **Advanced Seismic Imaging:**
    - 3D and 4D seismic surveys provide high-resolution images of subsurface structures, enabling geoscientists to identify hydrocarbon traps better and reduce dry well risks.
    - Full Waveform Inversion (FWI) and seismic attribute analysis enhance seismic data interpretation, improving the accuracy of subsurface models.
    - **Example:** 3D seismic imaging has played a crucial role in deepwater discoveries in the Gulf of Mexico,[2] where complex geological formations require precise imaging to identify viable drilling locations.
2. **Data Analytics and Machine Learning:**
    - Machine learning algorithms analyze vast amounts of geological, geophysical, and well-log data to detect patterns, predict reservoir properties, and optimize exploration strategies.
    - AI models integrate historical exploration data with real-time information to improve prospect evaluation and reduce uncertainty.

- **Example:** In the North Sea, machine learning techniques have helped predict reservoir quality and improve exploration efficiency by identifying overlooked hydrocarbon-bearing formations.[3]

3. **Unconventional Exploration Techniques:**
   - Electromagnetic (EM) and gravity surveys are used alongside seismic imaging to enhance subsurface characterization and detect hydrocarbon accumulations.
   - Remote sensing and satellite-based methods analyze surface anomalies that may indicate the presence of hydrocarbons, especially in remote or frontier exploration areas.
   - **Example:** Controlled-source electromagnetic (CSEM) surveys have been successfully applied in offshore deepwater environments to differentiate between water-filled and hydrocarbon-filled reservoirs.

Innovations in Seismic Imaging:

- **Full waveform inversion (FWI):** A cutting-edge technique that uses the entire seismic wavefield to generate high-resolution velocity models, improving the accuracy of subsurface interpretations.
- **Seismic attribute analysis:** This involves analyzing seismic attributes such as amplitude, frequency, and phase variations to identify lithology changes, fluid content, and potential reservoir characteristics.
- **Ocean bottom nodes (OBN):** Deploy seismic sensors directly on the seafloor to acquire high-quality seismic data in deepwater and complex geological settings, particularly subsalt imaging in offshore basins.

Advancements in exploration technologies continue to improve discovery success rates, reduce geological uncertainties, and enhance decision-making in frontier and mature basins. These innovations are crucial for unlocking new reserves and maintaining a sustainable energy supply.

9.3 DRILLING TECHNOLOGIES

Advancements in drilling technologies have revolutionized the upstream oil and gas sector by improving efficiency, reducing costs, and enabling access to previously inaccessible reservoirs. Modern drilling techniques, automation, and innovative drill bit designs contribute to safer and more cost-effective well construction, enhancing overall hydrocarbon recovery.

Key Innovations in Drilling Technologies:

1. **Advanced Drilling Techniques:**
   - **Directional and horizontal drilling:** These techniques allow wells to be steered toward specific reservoir targets, increasing reservoir exposure and maximizing hydrocarbon extraction.
   - **Managed pressure drilling (MPD):** A drilling method that precisely controls wellbore pressure, reducing formation damage and mitigating the risk of blowouts.

- **Underbalanced drilling (UBD):** A technique that maintains wellbore pressure below formation pressure to minimize formation damage and improve drilling efficiency.
- **Example:** Horizontal drilling has significantly EOR in tight shale formations like the **Bakken** and **Eagle Ford**. By drilling laterally through the reservoir, operators increase wellbore exposure to productive zones, leading to higher recovery rates.

2. **Drilling Automation and Robotics:**
   - Automated drilling rigs integrate sensors, real-time data analytics, and AI to optimize drilling parameters, reducing human intervention and improving accuracy.
   - Robotic systems are used for **pipe handling, well intervention, and pressure monitoring**, minimizing worker exposure to hazardous conditions.
   - **Example:** Automated drilling rigs equipped with robotic pipe handlers in the **Permian Basin** have improved drilling efficiency, reducing trip times and enhancing safety in high-risk operations.
3. **Drill Bit Technology:**
   - **Polycrystalline diamond compact (PDC) bits:** Made with synthetic diamond material, these drill bits offer superior durability, faster penetration rates, and longer lifespans in challenging drilling environments.
   - **Hybrid drill bits:** Combining roller cone and PDC technology, these bits improve performance in complex lithologies by optimizing cutting efficiency.
   - **Example:** PDC bits have been widely adopted for drilling **hard rock formations** in deepwater and unconventional reservoirs, increasing penetration rates while reducing bit wear and downtime.

Illustration of a vertical-to-horizontal well trajectory within a subsurface reservoir (Figure 9.1). Highlights how horizontal drilling maximizes contact with hydrocarbon-bearing zones, improves recovery efficiency, and reduces surface footprint—especially in tight formations and offshore developments

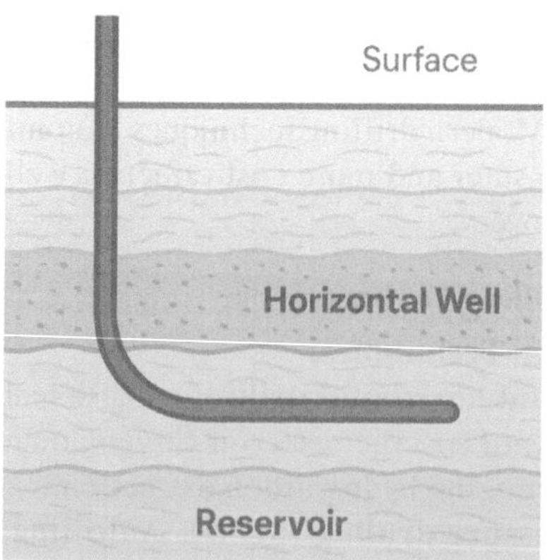

FIGURE 9.1 Horizontal drilling technique for reservoir access.

Innovations in Drilling Techniques:

- **Extended Reach Drilling (ERD):**
  - Enables wells to be drilled over long horizontal distances from a single drilling location, reducing surface footprint and improving access to remote reservoirs.
  - **Example:** The **Sakhalin-1 project in Russia** set a record for ERD with a well extending over 15 km from the surface.
- **Dual Gradient Drilling (DGD):**
  - Different drilling fluid densities are used to manage pressure variations in deepwater wells, reducing the risk of kicks and improving well control.
  - **Example:** DGD has been implemented in **Gulf of Mexico deepwater operations**, improving safety and efficiency in ultra-deep drilling projects.
- **Casing While Drilling (CWD):**
  - Integrates casing installation during drilling, reducing the time and costs associated with separate casing operations.
  - **Example:** CWD has been successfully used in **offshore drilling**, where reducing non-productive time is critical for cost efficiency.

By integrating these technological advancements, the industry continues to push the limits of drilling performance, reducing operational risks and enhancing resource recovery from increasingly complex reservoirs.

9.4 PRODUCTION TECHNOLOGIES

Advancements in production technologies are crucial for maximizing hydrocarbon recovery, optimizing well performance, and improving the efficiency of oil and gas extraction. These innovations help extend field life, enhance recovery rates, and enable economic production from challenging reservoirs.

Key Advancements in Production Technologies:

1. **Enhanced Oil Recovery (EOR) Techniques:**
   - **Thermal EOR:** Utilizes heat to reduce oil viscosity and improve flow.
   - **Chemical EOR:** Involves injecting chemicals to alter fluid properties and enhance oil displacement.
   - **Miscible EOR:** Uses gases like **CO_2 or nitrogen** to mix with the oil, reducing its viscosity and improving flow to the wellbore.
   - **Example: CO_2 injection in the Permian Basin** has EOR while simultaneously sequestering greenhouse gases. The injected CO_2 dissolves in crude oil, reducing viscosity and allowing easier extraction.
2. **Subsea Production Systems:**
   - Subsea wells, manifolds, and pipelines allow offshore hydrocarbon production in deepwater and ultra-deepwater environments, reducing the need for surface platforms.

- **Example: North Sea deepwater fields** use subsea production systems to extract hydrocarbons from reservoirs beneath the seafloor, overcoming extreme pressure and temperature challenges.

3. **Artificial Lift Technologies:**
    - **Electric submersible pumps (ESPs):** High-efficiency pumps placed in wells to lift fluids when reservoir pressure declines.
    - **Gas lift systems:** Injects gas into production tubing to lighten the hydrocarbon column and facilitate flow to the surface.
    - **Example: ESPs in mature oil fields** maintain production rates, extending the field's economic life by efficiently lifting fluids to the surface.

Innovations in EOR Techniques:

- **Steam-Assisted Gravity Drainage (SAGD):**
    - Used in heavy oil and bitumen reservoirs, SAGD involves injecting steam to heat the oil, reducing viscosity, and allowing it to drain to a horizontal production well.
    - **Example: Alberta's oil sands projects** employ SAGD, making production viable from previously uneconomical resources.
- **Surfactant-Polymer Flooding:**
    - Surfactants lower the interfacial tension between oil and water, while polymers increase the viscosity of injected water, improving oil displacement efficiency.
    - **Example:** In the Chinese Daqing Field, oil recovery from waterflooded reservoirs is improved by reducing residual oil saturation.
- **Miscible Gas Injection:**
    - Injecting gases like CO_2 or natural gas that mix with crude oil to alter its properties and improve production.
    - **Example: North Sea and Permian Basin fields** have successfully used CO_2 injection, increasing recovery while reducing greenhouse gas emissions through sequestration.

AI-DRIVEN GEOSTEERING

AI-assisted geosteering systems integrate real-time logging data with geological models, allowing drillers to make instant trajectory corrections. This ensures better alignment with target zones and reduces drilling errors in unconventional plays.

By integrating these production technologies, the oil and gas industry continues to enhance recovery efficiency, reduce environmental impact, and optimize field economics.

9.5 RESERVOIR MANAGEMENT TECHNOLOGIES

Reservoir management technologies are essential for optimizing reservoir performance, maximizing hydrocarbon recovery, and extending field life. These innovations help operators make data-driven decisions, enhance production strategies, and minimize inefficiencies.

Key Advancements in Reservoir Management:

1. **Reservoir Simulation:**
   - Advanced modeling techniques and **history matching** help accurately represent reservoir behavior and predict future production.
   - **Example: Prudhoe Bay Field, Alaska,** utilizes reservoir simulation to optimize **waterflooding strategies** by modeling different injection scenarios, leading to EOR and better reservoir pressure maintenance.

SMART WELL TECHNOLOGY—THE NEW FRONTIER

Smart wells enable zonal control using sensors and downhole valves. Operators can selectively open or close reservoir zones without intervention, improving production efficiency and reducing water cut. They've become crucial in complex reservoirs, especially offshore.

2. **Data Analytics and Machine Learning:**
   - These technologies analyze vast amounts of **production and subsurface data** to improve reservoir characterization, detect inefficiencies, and optimize production strategies.
   - **Example:** In the **Permian Basin**, machine learning models are used to **identify production anomalies** and suggest **optimization strategies** based on historical data and well performance trends.
3. **Smart Wells and Intelligent Completion Systems:**
   - **Smart wells** are equipped with **downhole sensors, flow control valves, and fiber optic monitoring systems**, allowing real-time adjustments to optimize production.
   - **Example: North Sea fields** deploy **intelligent completion systems** with **real-time downhole monitoring**, enabling operators to adjust production parameters dynamically to maximize recovery.

> With innovation, even the most mature fields still have potential.
>
> **Amin Nasser**
> *CEO, Saudi Aramco*

Innovations in Reservoir Management:

✓ **Waterflood Optimization:**
   - Designing and optimizing **water injection patterns** to **improve sweep efficiency and enhance oil recovery.**
   - Utilizes **reservoir simulation, real-time monitoring, and machine learning** to adjust injection rates dynamically.
   - **Example: Ghawar Field (Saudi Arabia)** optimizes its **water injection strategy** to sustain pressure and enhance ultimate recovery.[4]

✓ **Reservoir Characterization:**
   - Utilizes **seismic attributes, logging tools, and AI** to **identify sweet spots** for production and **optimize well placement.**

- **Example: Deepwater Gulf of Mexico fields** use **4D seismic surveys** to track reservoir changes over time and optimize drilling locations.

✓ **Enhanced Monitoring Technologies:**

- **Fiber optic sensors, downhole pressure gauges, and remote sensing technologies** enable real-time monitoring of **reservoir conditions, pressure behavior, and fluid movement.**
- **Example: Kashagan Field (Kazakhstan)** deploys an advanced **real-time monitoring system** to detect production inefficiencies and adjust operations accordingly.

By integrating these reservoir management technologies, oil and gas operators can **maximize recovery efficiency, optimize production rates, and minimize operational risks** in conventional and unconventional reservoirs.

TECH INSIGHT—AUTOMATED RIG OPERATIONS

Automated rigs use machine learning to control weight on bit, rotation speed, and fluid flow. The result? Consistent penetration rates and reduced non-productive time (NPT). Major operators report a 15–30% improvement in drilling efficiency.

9.6 DIGITALIZATION AND DATA ANALYTICS

Digitalization is revolutionizing the upstream oil and gas sector by **enhancing efficiency, reducing costs, and improving decision-making** through real-time data processing and advanced analytics. Integrating **big data, cloud computing, AI, and Internet of Things (IoT) technologies** allows companies to optimize exploration, drilling, production, and reservoir management.

> AI in upstream isn't an option anymore — it's a necessity for survival.
>
> *Oil & Gas Technology Review, 2023*

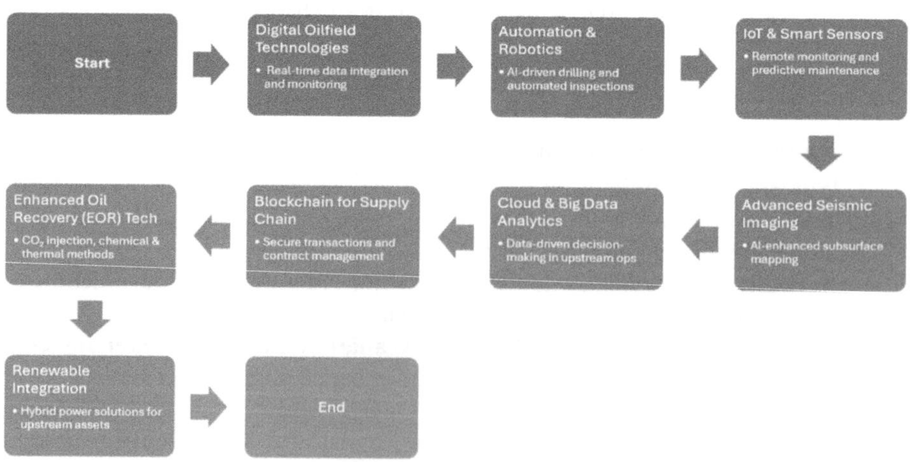

FIGURE 9.2 Flowchart of integrated technologies in upstream operations.

Sequential overview of digital innovations—from real-time monitoring and AI-driven automation to blockchain and renewable integration (Figure 9.2). Highlights how upstream companies deploy layered technologies to enhance efficiency, safety, and sustainability across exploration, drilling, and production.

Did You Know?—Drones in Oil Fields

Drones are now used in flare stack inspections, pipeline monitoring, and methane detection. They minimize risk to human workers and reduce inspection time by over 50%.

The diagram below illustrates how different **digital technologies** interact within upstream operations:

9.6.1 Key Digital Technologies in Upstream Operations

1. **Digital Oilfield Technologies:**
   - **Real-time data integration and monitoring** enhance operational efficiency.
   - **Example:** BP and Shell leverage digital oilfields to monitor **well performance, optimize production, and reduce downtime** using **real-time analytics**.
2. **Automation and Robotics:**
   - AI-driven **automated drilling rigs and robotic inspections** reduce manual labor and **enhance safety**.
   - **Example:** Automated drilling rigs in the **Permian Basin** have **increased drilling speeds by 30%** and **reduced human errors**.
3. **IoT and Smart Sensors:**
   - Remote monitoring of **equipment health, reservoir pressure, and fluid levels** ensures **predictive maintenance**.
   - **Example: Deepwater drilling rigs** use IoT sensors to monitor well pressure in real-time, reducing the risk of **well blowouts**.
4. **Advanced Seismic Imaging:**
   - AI-enhanced **subsurface mapping** improves **reservoir characterization** and **reduces exploration risks**.
   - **Example:** FWI seismic imaging has **increased exploration success rates** in the **North Sea and Gulf of Mexico**.
5. **Cloud and Big Data Analytics:**
   - **Data-driven decision-making** enables **optimized drilling, production forecasting, and asset management**.
   - **Example:** Chevron's cloud-based analytics platform integrates **geophysical, drilling, and production data** to optimize well placement.[5]
6. **Blockchain for Supply Chain and Compliance:**
   - Ensures **secure transactions, contract transparency, and regulatory compliance** in supply chain management.
   - **Example: Smart contracts** using blockchain automate procurement and reduce fraud risks in global oil shipments.
7. **Enhanced Oil Recovery (EOR) Technologies:**
   - **CO_2 injection, chemical, and thermal recovery methods** improve oil displacement and maximize recovery.

- **Example:** CO_2-EOR in the **Permian Basin** has **increased recovery factors by up to 15%** while reducing **carbon emissions**.

8. **Renewable Integration in Upstream Operations:**
    - Hybrid power solutions, including **solar and wind energy**, reduce the carbon footprint of oilfield operations.
    - **Example:** Offshore platforms in the **North Sea** are integrating **renewable energy sources** to reduce reliance on diesel generators.

9.6.2 Benefits of Digitalization in Upstream Operations

✓ **Improved decision-making:** Real-time access to **integrated data** enables better operational strategies and risk assessments.

✓ **Operational efficiency:** AI-driven automation and predictive analytics reduce costs and **optimize drilling, production, and logistics**.

✓ **Predictive maintenance:** AI-powered predictive models identify **potential equipment failures** before they occur, **reducing downtime and repair costs**.

✓ **Enhanced safety and environmental protection:** Digital technologies enable **remote monitoring, automated hazard detection, and emergency response planning**, improving **workplace safety** and **reducing environmental impact**.

9.6.3 Industry Example: Digital Oilfields

Several oil majors have **transformed traditional oilfields into "digital oilfields,"** integrating **big data, IoT, and AI-driven analytics** to improve field performance.

Example:

- **BP's digital oilfield initiatives**[6] have helped optimize drilling operations and reservoir management, **increasing production efficiency by 10–20%**.
- **Chevron uses AI-powered predictive analytics** to detect well-performance anomalies, **reducing non-productive time (NPT) and equipment failures**.

This shift toward **data-driven decision-making and automation** is reshaping the future of upstream operations, ensuring **higher efficiency, safety, and sustainability**.

> The oilfield of the future is intelligent, adaptive, and automated.
>
> **Patrick Pouyanné**
> *CEO, TotalEnergies*

Trend showing inverse correlation between drilling time and adoption of advanced analytical tools (Figure 9.3). As AI, predictive maintenance, and real-time optimization gain traction, average drilling time drops from 100 to under 20 days—highlighting the transformative impact of digitalization on upstream efficiency.

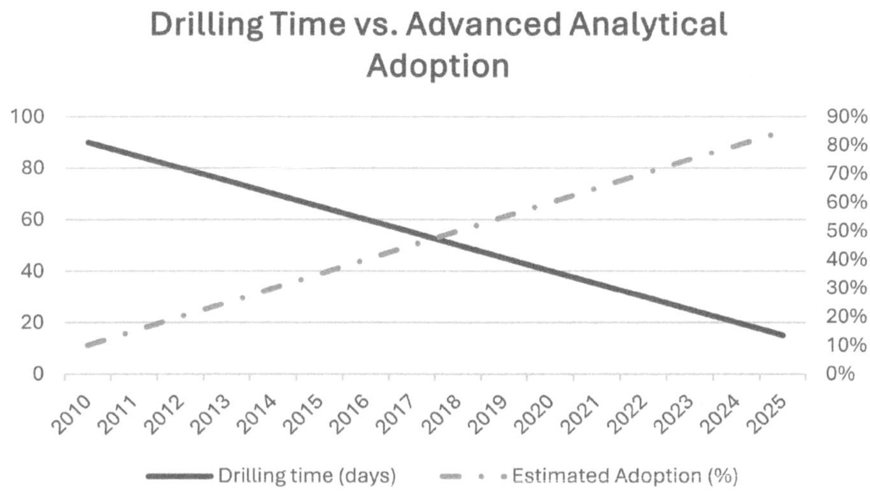

FIGURE 9.3 Impact of advanced analytics on drilling time (2010–2025).

9.7 FUTURE TRENDS IN UPSTREAM TECHNOLOGY

Several emerging technologies hold promise for further transforming the upstream sector:

- **Nanotechnology:** Applications in drilling, production, and reservoir management, such as nano-scale sensors and EOR fluids.
  - **Example:** Nanoparticles have been used to improve the efficiency of chemical EOR processes and enhance oil recovery. Nanotechnology can also be used to create advanced drilling fluids that improve wellbore stability and reduce friction.
- **Biotechnology:** Microbial EOR and biosensors for reservoir characterization.
  - **Example:** Microbial EOR techniques have been used to enhance oil recovery by altering reservoir properties and improving fluid flow. Biosensors can be deployed in wells to monitor reservoir conditions and detect changes in fluid composition.
- **Additive manufacturing (3D printing):** Rapid prototyping and customized equipment manufacturing.
  - **Example:** 3D printing has been used to create customized drilling tools and components, reducing lead times and manufacturing costs. Additive manufacturing enables the production of complex, high-performance parts that are difficult to manufacture using traditional methods.

Additional Emerging Technologies:

- **Virtual and augmented reality (VR/AR):** VR and AR technologies are being used for training, remote operations, and visualization of subsurface data. These technologies provide immersive experiences that enhance learning and decision-making.

- **Hydraulic fracturing innovations:** Advances in hydraulic fracturing techniques, such as waterless fracking and the use of environmentally friendly fracturing fluids, are being developed to reduce environmental impact and improve efficiency.
- **Energy storage solutions:** Innovations in energy storage technologies, such as advanced batteries and flywheel systems, are being explored to improve energy management in upstream operations. Energy storage solutions enable the integration of renewable energy sources and enhance the reliability of the power supply.

Collaborative Efforts and Research:

- **Industry-academia partnerships:** Collaboration between industry and academic institutions to conduct research and develop innovative technologies. These partnerships leverage academic expertise and industry resources to drive technological advancements.
- **Research consortia:** Formation of research consortia that bring together multiple companies and research organizations to tackle usual challenges and accelerate technology development. Research consortia facilitate knowledge sharing and reduce duplication of efforts.
- **Government support:** Government funding and support for research and development initiatives in the upstream sector. Government programs provide financial incentives and resources to encourage innovation and technology adoption.

9.8 CASE STUDIES ON EMERGING TECHNOLOGIES IN UPSTREAM OPERATIONS

To illustrate how technological advancements are transforming upstream operations, we present case studies of companies implementing **cutting-edge innovations** in **nanotechnology, digitalization, EOR, and automation**.

9.8.1 Case Study 1: Shell's Use of Digital Twins for Offshore Asset Optimization

Background:

Shell has been at the forefront of **digital transformation**, integrating **digital twins** to enhance asset management and optimize offshore operations.

Technology Implemented:

- **Digital twin technology**: Virtual replicas of offshore platforms provide real-time **operational insights, predictive maintenance, and scenario modeling**.
- **AI and machine learning**: AI-powered analytics predict **equipment failures** before they occur, reducing downtime.
- **IoT and smart sensors**: Connected sensors **continuously monitor pressure, temperature, and flow rates**, ensuring efficient production.

Results and Impact:

- **20% reduction in unplanned downtime** due to predictive maintenance.
- **Improved safety** by allowing remote monitoring instead of manual inspections in hazardous environments.
- **Enhanced decision-making** with real-time data analytics, leading to a **5% increase in production efficiency**.

9.8.2 Case Study 2: Saudi Aramco's Nanotechnology for Enhanced Oil Recovery (EOR)

Background:

Saudi Aramco has been investing in **nanotechnology** to improve **EOR** and increase production from mature reservoirs.

Technology Implemented:

- **Nanoparticle-based EOR fluids**: Aramco developed **nano-sized surfactants** to improve **oil displacement** and **reduce interfacial tension** between oil and water.
- **Smart nano-sensors**: Deployed in reservoirs to monitor **fluid movement** and **reservoir characteristics** in real time.

Results and Impact:

- **25% increase in oil recovery rates** in pilot projects.
- **Reduced water cut** (water produced along with oil) due to better fluid mobility control.
- **Lower chemical usage**, reducing **environmental footprint** and operational costs.

9.8.3 Case Study 3: ExxonMobil's AI-Driven Predictive Maintenance for Drilling Operations

Background:

ExxonMobil is leveraging **AI and machine learning** to optimize **drilling operations and reduce downtime**.

Technology Implemented:

- **AI-enhanced drilling automation**: AI algorithms analyze **drilling data in real time**, adjusting parameters to **reduce non-productive time**.
- **Predictive maintenance models**: Machine learning predicts **failures in drilling equipment**, allowing preventive action before breakdowns.
- **Cloud computing and big data analytics**: Centralized data processing for **real-time decision-making** across multiple rigs.

Results and Impact:

- **30% faster drilling** in complex wells by optimizing parameters in real-time.
- **Reduced equipment failures by 40%**, cutting costs on unplanned repairs.
- **Improved well integrity** through AI-driven **early detection of pressure anomalies**.

9.8.4 Case Study 4: Chevron's Renewable Energy Integration in Upstream Operations

Background:

Chevron is integrating **renewable energy solutions** into upstream operations to **reduce carbon emissions and improve sustainability**.

Technology Implemented:

- **Hybrid power systems**: **Solar and wind energy** supplement power for remote upstream assets.
- **Battery storage solutions**: Advanced energy storage for a **stable power supply** in fluctuating renewable generation.
- **Electrification of oilfield operations**: Using **grid electricity instead of diesel generators** to power operations, reducing emissions.

Results and Impact:

- **25% reduction in carbon emissions** at test sites.
- **Lower fuel consumption**, resulting in **cost savings of $2 million annually**.
- **Increased energy reliability**, reducing dependence on fluctuating oil prices.

9.9 CHAPTER SUMMARY

This chapter explored how **technological advancements are shaping the future of upstream petroleum operations**, focusing on **exploration, drilling, production, reservoir management, and digitalization**.

Key Takeaways:

1. **Digitalization and AI** are revolutionizing **data management, predictive maintenance, and operational efficiency**.
2. **Nanotechnology and advanced EOR techniques** are **enhancing hydrocarbon recovery** while reducing costs and environmental impact.
3. **Drilling automation and robotics** are improving **safety and efficiency**, reducing human error in complex operations.
4. **Renewable energy integration** is helping the industry **lower emissions and reduce reliance on fossil fuel-based power sources**.
5. **Case studies from Shell, Saudi Aramco, ExxonMobil, and Chevron** demonstrate how leading companies are deploying cutting-edge technologies to enhance productivity and sustainability.

As the Oil and Gas industry evolves, **technological innovation** will remain a driving force in improving efficiency, minimizing environmental impact, and addressing challenges in a dynamic energy landscape.

9.9.1 Next Chapter: The Economics of Technological Innovation in Upstream Operations

The next chapter will examine the **economic implications of adopting new technologies**, focusing on **cost-benefit analysis, return on investment (ROI), and long-term sustainability in the upstream sector**.

NOTES

1. IEA, *World Energy Investment 2023*, on the rising role of AI and seismic innovations in upstream projects.
2. Wood Mackenzie, *Gulf of Mexico Deepwater Outlook 2023.*
3. SPE Technical Paper SPE-202487-MS, "Machine Learning in North Sea Exploration Efficiency."
4. Saudi Aramco, *Ghawar Field Production Optimization Report*, 2022.
5. Chevron, *Investor Day 2022—Data Analytics and Asset Performance.*
6. BP, *Digital Transformation in Upstream Operations*, 2021.

10 Global Market Dynamics

10.1 INTRODUCTION TO GLOBAL ENERGY MARKETS

The global energy market is a **complex and interconnected system** that balances **supply and demand** across different regions and fuel types. The market encompasses various **energy sources** such as **oil, natural gas, coal, nuclear power, and renewables**, all playing crucial roles in fueling **transportation, industrial processes, and electricity generation**.

10.1.1 Key Players in Global Energy Markets

Several major organizations and companies **shape energy policies, production, and pricing dynamics**:

1. **Organization of the Petroleum Exporting Countries (OPEC)**
   OPEC is an **intergovernmental organization** coordinating **oil production policies** among member countries to influence global oil supply and prices.
   - **Impact on the market:** OPEC **sets production quotas** for member nations to stabilize or influence oil prices.
   - **Example:** In **2016, OPEC+ (OPEC plus Russia and other non-OPEC allies) agreed to cut oil production**, leading to **a recovery in oil prices** after a severe decline caused by an oversupply in global markets.
2. **International Energy Agency (IEA)**
   The IEA is an **intergovernmental organization** that provides energy policy advice, **market analysis, and strategic energy security and sustainability recommendations**.
   - **Impact on the market:** The IEA **monitors supply and demand trends**, provides **long-term energy outlooks**, and suggests **policy measures** to improve efficiency and security.
   - **Example:** The IEA's **World Energy Outlook report**[1] is widely used to predict **global energy demand growth, emissions trends, and technology advancements**.
3. **National Oil Companies (NOCs)**
   NOCs are **state-owned oil and gas companies** that control a significant share of **global reserves** and **production**. Many NOCs dominate the energy policies of their respective countries.
   - **Impact on the market:** NOCs **prioritize domestic energy security** and **government revenue generation** over short-term profits.
   - **Example: Saudi Aramco (Saudi Arabia), Gazprom (Russia), and Petrobras (Brazil)** are some of the most influential **NOCs controlling large portions of oil and gas reserves**.

DOI: 10.1201/9781003750345-13

4. **International Oil Companies (IOCs)**
   IOCs are **privately owned, multinational corporations** that operate across the **upstream, midstream, and downstream** oil and gas sectors.
   - **Impact on the market:** IOCs **invest in exploration, production, refining, and distribution**, contributing to **technological advancements** and **capital-intensive projects** worldwide.
   - **Example: ExxonMobil, Shell, BP, Chevron, and TotalEnergies** operate **globally** and influence **energy markets through investments, partnerships, and innovation**.

10.1.2 Energy Market Dynamics

1. **Supply Chain**
   The **energy market supply chain** includes:
   - **Exploration and production:** Identifying and extracting fossil fuel reserves.
   - **Transportation:** Moving crude oil, natural gas, and refined products via **pipelines, LNG tankers, and shipping routes**.
   - **Refining and processing:** Converting raw hydrocarbons into **usable fuels (gasoline, diesel, LNG, etc.).**
   - **Distribution and consumption:** Delivering energy to **power plants, industries, and consumers.**

   Efficient supply chain management ensures energy security and meets **global demand fluctuations**.

FAST FACT—WHO HOLDS THE POWER IN ENERGY PRICING?

OPEC accounts for ~35% of global oil production but controls ~70% of proven reserves. When they move, markets listen.

2. **Energy Prices and Market Volatility**
   Energy prices are influenced by:
   - **Supply and demand:** Imbalances (e.g., oversupply or shortages) cause **price swings**.
   - **Geopolitical events:** Conflicts, trade restrictions, and **sanctions** impact **energy exports/imports**.
   - **Technological advancements:** Innovations in **renewables, fracking, and hydrogen energy** disrupt traditional market patterns.
   - **Regulatory policies:** Government decisions on **taxation, subsidies, emissions caps, and carbon pricing** affect energy costs.
3. **Energy Policies and Regulations**
   Governments influence **energy markets** through the following:
   - **Taxation and subsidies:** Fuel taxes, renewable subsidies, and incentives for cleaner energy sources.
   - **Environmental standards:** Regulations targeting **carbon emissions and pollution control**.

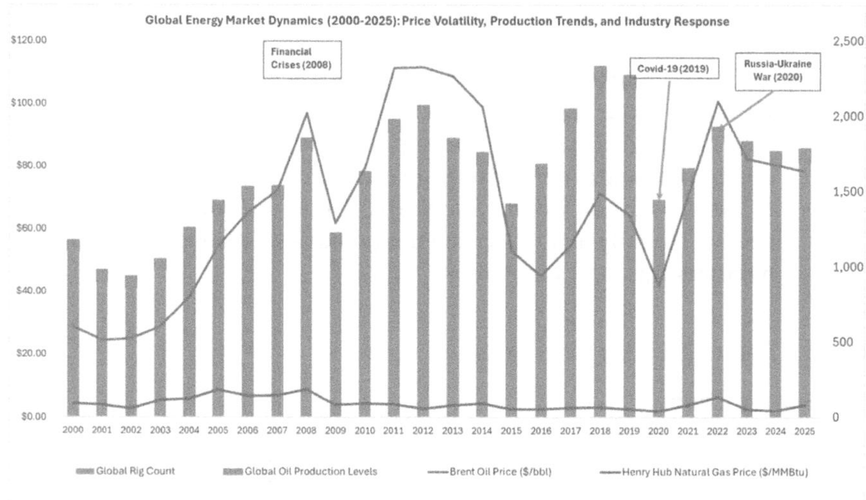

FIGURE 10.1 Global energy market dynamics (2000–2025): Price volatility, production trends, and industry response.

- **Strategic reserves:** Governments maintain **emergency crude oil stockpiles** to **stabilize supply disruptions** (e.g., the **U.S. Strategic Petroleum Reserve**).

Historical trends in Brent oil prices, Henry Hub gas prices, global rig count, and oil production volumes. Highlights how major events—2008 financial crisis, COVID-19,

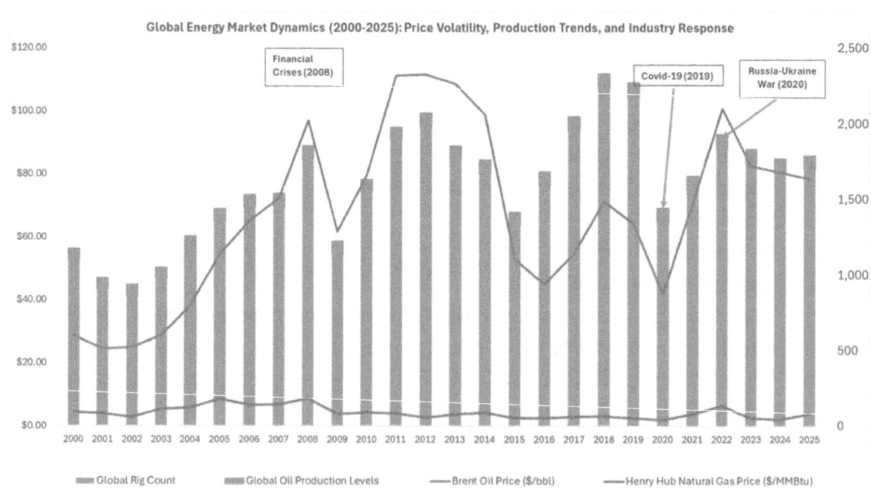

FIGURE 10.2 Global energy market dynamics (2000–2025): Price volatility, production trends, and industry response.

and the Russia–Ukraine war—triggered sharp market shifts, underscoring the upstream sector's sensitivity to geopolitical and macroeconomic shocks (Figures 10.1 and 10.2).

10.2 GEOPOLITICAL FACTORS INFLUENCING THE UPSTREAM SECTOR

Geopolitical factors are crucial in shaping the **upstream petroleum industry**, impacting **exploration, production, investment, and global energy security**. These factors influence **market stability, trade flows, and long-term strategic decisions** in the energy sector.

> The geography of oil is the geography of power.
>
> **Daniel Yergin**
> *Energy Historian*

10.2.1 Key Geopolitical Factors

1. **Political Stability and Security**
   - Political instability, conflicts, and security threats can **disrupt oil and gas production**, leading to **supply shortages** and **price volatility**.
   - **Example: Libya's political unrest and civil war (2011–present)** have resulted in repeated disruptions in **oil production**, affecting **global supply and price stability**.
2. **Geopolitical Conflicts and Sanctions**
   - Wars, **international disputes, and trade restrictions** can cause **supply disruptions**, leading to **sharp price fluctuations** in energy markets.
   - **Example: U.S. sanctions on Iran (2018)** have **restricted Iranian oil exports**, reduced **global supply,** and contributed to **oil price volatility**.
3. **Resource Nationalism**
   - Some governments seek **greater control over their energy resources**, limiting access for **international oil companies (IOCs)** and affecting **foreign investments**.
   - **Example: Venezuela's oil industry nationalization (2000s)** reduced foreign participation, reducing **investment, production decline, and economic instability**.
4. **Energy Security and Trade Relations**
   - Energy security concerns drive **diplomatic relations, trade agreements, and infrastructure projects**, influencing **energy supply chains**.
   - **Example:** The **Nord Stream 2 pipeline (Russia to Germany)** has faced geopolitical tensions as **Europe balances energy dependence on Russia with security concerns**.

> Strait of Hormuz Attacks Send Oil Prices Soaring
>
> *Financial Times, June 2019*

Geopolitical Factor	Impact on the Upstream Sector
Wars & Conflicts	Supply disruptions, production halts, infrastructure damage
Sanctions & Trade Restrictions	Limited access to markets, investment risks, reduced exports
Regulatory & Political Changes	Increased operational costs, uncertain investment climate
Terrorism & Security Threats	Higher security costs, disruptions in exploration and drilling
Resource Nationalism	Expropriation risks, restricted access for foreign firms

FIGURE 10.3 Geopolitical risk factors and their impact on upstream operations.

10.2.2 Impact of Geopolitical Factors on the Upstream Sector

Summary of key geopolitical threats—wars, sanctions, regulatory shifts, terrorism, and resource nationalism—and their operational consequences. Highlights how upstream companies face supply disruptions, investment uncertainty, and elevated security costs in politically volatile regions (Figure 10.3).

10.2.3 Strategies for Managing Geopolitical Risks

To **mitigate geopolitical risks**, oil and gas companies adopt several strategic approaches:

1. **Diversification of Investments**
   - Spreading investments across **multiple regions and asset types** reduces exposure to **regional instability**.
   - **Example: BP and Shell** invest in both **Middle Eastern and North American** assets to balance risks.
2. **Comprehensive Risk Assessments**
   - Conducting **political, economic, and social risk analyses** before entering new markets.
   - **Example:** Companies use **geopolitical forecasting models** to assess investment risks in high-risk countries.
3. **Stakeholder Engagement**
   - Building **relationships with local governments, communities, and international partners** to strengthen operational security.
   - **Example:** African companies engage with **local authorities and communities** to ensure stability and social acceptance.
4. **Political Risk Insurance**
   - **Insurance policies** protect companies against **expropriation, nationalization, contract breaches, and political violence**.
   - **Example: Multilateral Investment Guarantee Agency (MIGA)** provides political risk insurance for energy projects in emerging markets.

10.3 INTERNATIONAL INVESTMENT TRENDS

Foreign direct investment (**FDI**) is key to upstream oil and gas development, funding **exploration, production, and infrastructure projects** worldwide. A region's attractiveness for investment depends on factors such as **political stability, regulatory transparency, fiscal policies, and resource potential**. In recent years, the

Factor	Impact on Investment	Example
Political Risk	Political instability, conflicts, or sudden regulatory changes deter investment. Stable governments with investor-friendly policies attract FDI.	**Libya & Venezuela:** Frequent conflicts and expropriation policies led to reduced foreign investment.
Regulatory Environment	Clear, transparent, and predictable regulations create a stable business climate, reducing investor uncertainty.	**Norway:** A well-defined regulatory framework and strong legal protections have attracted sustained offshore investment.
Fiscal Terms	Competitive tax regimes, royalty rates, and production-sharing agreements (PSAs) impact investment attractiveness.	**U.S. Permian Basin:** Low taxes, flexible lease terms, and infrastructure availability have encouraged investment in shale oil.
Resource Potential	Investors prioritize regions with **large, high-quality oil and gas reserves**, as well as favorable geological conditions.	**Brazil's Offshore Pre-Salt Fields:** Massive hydrocarbon potential has drawn billions in FDI from global oil majors.
Infrastructure & Market Access	The presence of pipelines, export terminals, and refineries reduces logistical costs and enhances investment feasibility.	**Middle East & U.S. Gulf Coast:** Strong energy infrastructure supports large-scale investments.
Technological Capabilities	Countries with advanced **drilling, production, and data analytics technologies** can attract more investment.	**North Sea (UK & Norway):** Adoption of **digital twins, AI-based exploration, and subsea robotics** has made the region more competitive.

FIGURE 10.4 Key factors influencing upstream investment decisions.

industry has also witnessed a shift in investment trends, with growing interest in digitalization, low-carbon technologies, and diversification into renewable energy sources.

10.3.1 Key Factors Influencing Investment Decisions

Investors carefully assess the **risks and rewards** of investing in a given region. The following factors significantly influence upstream investment decisions (Figure 10.4).

Comparison of political, regulatory, fiscal, and geological drivers shaping foreign direct investment (FDI) in upstream oil and gas. Highlights how stable governance, competitive fiscal terms, and resource potential attract capital, while instability and infrastructure gaps deter long-term commitments.

10.3.2 Investment Flows: Conventional vs. Unconventional Resources

Investment in upstream projects is divided between **conventional (offshore/onshore) and unconventional (shale, oil sands, LNG)** resources. While conventional resources still dominate, unconventional projects are **gaining investment due to technology improvements and favorable economics**.

- **Conventional Oil and Gas:**
  - Deepwater projects remain key investment targets, with **technological advancements** making them more viable.
  - **Example: Offshore Guyana's Stabroek Block**– ExxonMobil, Hess, and CNOOC have invested in deepwater discoveries.
- **Unconventional Resources (Shale and LNG):**
  - **Hydraulic fracturing (fracking)** and **horizontal drilling** have unlocked vast reserves of **shale oil and gas**, especially in the U.S.
  - **Example:** The **U.S. Shale Boom** in the **Permian Basin, Eagle Ford, and Bakken** transformed global oil supply dynamics, attracting **trillions in investment**.

10.3.3 Regional Investment Trends

Different regions attract investment based on **resource potential, infrastructure, and geopolitical risks**.

1. **Middle East and North Africa (MENA)**
   Advantages:
   - Vast conventional reserves with low production costs.
   - Well-established infrastructure for crude oil exports.

   Challenges:
   - **Geopolitical instability** (e.g., Yemen, Libya, Iraq).
   - **OPEC+ production agreements** influence investment decisions.

   Example: The UAE's **Abu Dhabi National Oil Company (ADNOC)** is attracting international investment through partnerships in offshore fields.
2. **North America (U.S. and Canada)**
   Advantages:
   - **Leader in shale oil and gas** development, supported by advanced technology.
   - **Favorable fiscal terms** and strong legal frameworks for investors.

   Challenges:
   - **Regulatory shifts** (e.g., methane emission rules, drilling bans on federal land).
   - **Infrastructure constraints** (pipeline bottlenecks).

   Example: The **U.S. Gulf Coast LNG export projects** have attracted global investors looking to tap into growing Asian demand.
3. **South America (Brazil and Argentina)**
   Advantages:
   - **Offshore pre-salt fields in Brazil** are among the world's largest new oil plays.
   - **Argentina's Vaca Muerta shale** has high potential for unconventional production.

   Challenges:
   - **Regulatory uncertainty** and political instability in some nations.
   - **Infrastructure limitations** slowing project development.

   Example: Petrobras and ExxonMobil are leading investments in **Brazil's offshore oil fields**.
4. **Africa (Mozambique, Angola, Nigeria)**
   Advantages:
   - **Emerging offshore plays** and **LNG potential** (Mozambique and Tanzania).
   - Increasing international partnerships to boost exploration.

   Challenges:
   - **Corruption and governance issues** impact investor confidence.
   - **Security risks** (e.g., attacks on oil facilities in Nigeria).

   Example: TotalEnergies' Mozambique LNG project is a major FDI initiative but faced delays due to **security concerns**.

5. **Asia-Pacific (Australia, China, Indonesia)**
   Advantages:
   - **LNG investments are booming** due to rising Asian energy demand.
   - **A stable business environment in Australia that attracts** long-term investment.

   Challenges:
   - **China's increasing energy self-sufficiency** reduces reliance on imports.
   - **High costs of offshore development** in remote areas.

   Example: Australia's Gorgon LNG project, backed by Chevron, is among the **largest LNG developments globally**.

10.3.4 Emerging Investment Trends

1. **Energy Transition and Low-Carbon Investments**
   - Growing **pressure to reduce carbon emissions** is shifting capital toward **renewables, hydrogen, and carbon capture projects**.

   Example: BP, Shell, and TotalEnergies are actively **investing in wind, solar, and carbon capture technologies** as part of their long-term strategies.
2. **Digitalization and Smart Oilfields**
   - Investments in **AI-driven analytics, digital twins, cloud computing, and automation** enhance operational efficiency.

   Example: Digital twin technology allows real-time monitoring of **offshore oil platforms**, reducing operational risks and improving decision-making.
3. **ESG (Environmental, Social, and Governance) Criteria**
   - Investors are prioritizing **projects with strong ESG commitments**, focusing on:
     - **Sustainable energy practices**
     - **Carbon footprint reduction**
     - **Stakeholder engagement with local communities**

   Example: Norway's Equinor invests in **offshore wind projects** to align with **ESG investment trends**.

INVESTMENT SHIFT: FROM BARRELS TO BYTES

Digital twins, AI-driven forecasting, and smart oilfields are now central to cost reduction and investment planning.

10.3.5 Investment Outlook and Future Trends

The global investment landscape is shifting due to **energy transition policies, geopolitical risks, and technological innovations**. Future trends include:

- **Increasing FDI in natural gas and LNG** as a **bridge fuel** to renewable energy.
- **New exploration frontiers** emerging in **Africa and deepwater South America**.

- **Greater emphasis on carbon neutrality and emissions reduction** in upstream projects.
- **More joint ventures and industry-academia partnerships** to drive **technological breakthroughs**.

10.4 MARKET OUTLOOK AND PRICE VOLATILITY

Forecasting oil and gas prices is a complex and challenging task. Various factors influence prices, including supply-demand dynamics, geopolitical events, economic growth trends, technological advancements, and regulatory changes. These factors interact unpredictably, creating **price volatility** and significantly impacting upstream investment and production decisions.

OPEC+ Output Surprise Leaves Traders Scrambling

CNBC, April 2023

10.4.1 Key Factors Influencing Oil and Gas Prices

Several **interconnected drivers** affect the global pricing of oil and natural gas. Understanding these factors helps market participants make informed decisions.

10.4.1.1 Supply and Demand Dynamics

Global supply and demand fundamentals heavily drive oil and gas prices. Major influences include:

- **Economic growth:** Expanding economies require more energy, increasing demand for oil and gas. Conversely, economic slowdowns reduce consumption.
- **Technological advancements:** Improvements in efficiency and new extraction techniques can increase supply and reduce production costs, influencing prices.
- **Geopolitical events:** Wars, sanctions, and trade restrictions can disrupt supply chains and affect availability and price levels.
- **Shale revolution:** The rise of **shale oil and gas production**, particularly in the U.S., has **transformed global supply dynamics**, impacting OPEC's pricing power.[2]

Graphical representation of the intersection between global oil supply and demand curves (Figure 10.5). Illustrates how price per barrel influences quantity traded, with equilibrium occurring near $80/bbl and 300 MMBBL/D. Highlights the balancing forces that shape upstream pricing, investment decisions, and production strategy.

Example: The **COVID-19 pandemic** in 2020 **collapsed oil demand**, causing

THE 2020 NEGATIVE OIL PRICE EXPLAINED

In April 2020, storage capacity maxed out during the pandemic. WTI futures dropped to –$37.63/barrel as traders paid buyers to take crude.

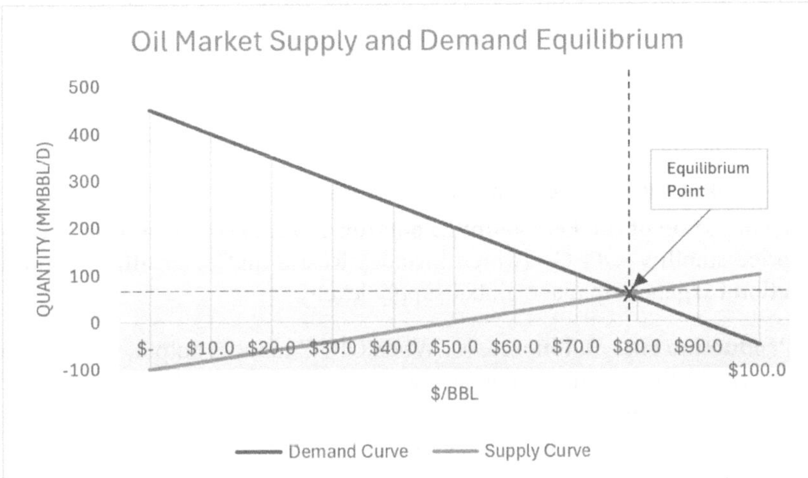

FIGURE 10.5 Oil market supply and demand equilibrium.

Brent crude prices to fall below **$20 per barrel** and even reach negative prices in the WTI futures market.

- As economies recovered, demand surged, pushing oil prices above **$80 per barrel** in late 2021.

10.4.1.2 Geopolitical Events and Market Disruptions

Political instability, sanctions, wars, and policy changes can disrupt **global oil supply**, leading to price swings.

Examples of Geopolitical Price Shocks:

Summary of major geopolitical disruptions—from the 1973 embargo to the Russia–Ukraine conflict—and their influence on global crude prices (Figure 10.6). Highlights

Event	Impact on Prices
1973 Oil Embargo	OPEC's embargo led to a 400% price increase in global crude markets.
1990 Gulf War	Iraq's invasion of Kuwait caused oil prices to rise by 70%.
2011 Arab Spring	Political instability in Libya and Egypt disrupted supply, causing price volatility.
2014-2016 Oil Price Collapse	Increased U.S. shale production and OPEC's decision not to cut output led to prices plummeting from **$100 to $30 per barrel.**
2022 Russia-Ukraine Conflict	Sanctions on Russian oil and gas caused supply disruptions, pushing Brent crude prices above **$130 per barrel.**

FIGURE 10.6 Historical geopolitical events and their impact on oil prices.

how wars, sanctions, and political instability trigger supply shocks, price volatility, and strategic shifts in upstream investment and production.

Russia-Ukraine War Sparks Global Energy Supply Crisis

Wall Street Journal, February 2022

10.4.1.3 OPEC and OPEC+ Policy Decisions

The **Organization of the Petroleum Exporting Countries (OPEC)** plays a key role in oil price stability. OPEC+ (which includes Russia and other allies) coordinates **production targets** to manage global supply levels.

- **Production cuts and increases:** When OPEC reduces output, prices generally rise. When production is increased, prices tend to fall.
- **Quotas and compliance:** Some members struggle to meet output quotas, affecting overall supply expectations.

Example:

- In **April 2020**, OPEC+ **agreed to the largest production cut (9.7 million barrels per day)** to counter COVID-19's demand collapse.
- In contrast, **increased output in 2021–2022** helped stabilize post-pandemic supply concerns.

OPEC+ Faces Credibility Crisis Amid Compliance Failures

10.4.1.4 Technological Advancements in Production

Exploration, drilling, and extraction innovations have impacted global oil supply and pricing.

- **Hydraulic fracturing and horizontal drilling:** The U.S. **shale boom** significantly increased global oil output, reducing reliance on OPEC.
- **Deepwater and arctic drilling:** Advances in **subsea production** have unlocked large offshore reserves.
- **Enhanced oil recovery (EOR):** CO_2 injection, steam flooding, and chemical EOR have improved recovery rates in mature fields.

Example:

- The rise of **U.S. shale oil production** from 2010 to 2014 led to an **oil price crash** in 2014–2016, as excess supply flooded the market.

U.S. Shale Reshapes Global Energy Pricing Models

The Economist, September 2021

10.4.1.5 Regulatory and Environmental Policies

Governments and regulatory agencies impose **taxation, carbon pricing, and emissions standards** that impact production costs.

- **Carbon taxes and emission controls:** Policies to reduce greenhouse gases increase **compliance costs** for oil and gas producers.

- **Fracking bans and pipeline regulations:** Restrictions on hydraulic fracturing and pipeline construction affect **domestic and global supply**.
- **Renewable energy policies:** Investment in solar, wind, and hydrogen projects **diverts capital** from upstream oil and gas projects.

Example:

- **Europe's Green Deal** and **U.S. methane regulations** have pushed energy companies to diversify toward lower-carbon alternatives.

10.4.2 Understanding Price Volatility

Price volatility refers to **sharp and unpredictable fluctuations** in oil and gas prices, creating uncertainty for producers, consumers, and investors.

10.4.2.1 Factors Contributing to Price Volatility

Summary of major price influencers—OPEC output shifts, extreme weather, storage constraints, and speculative trading (Figure 10.7). Highlights how physical disruptions and financial market behavior jointly shape upstream pricing, risk exposure, and hedging strategies.

Example:

- The **2021 Texas Winter Storm** caused power outages, affecting U.S. shale output and leading to short-term price spikes.

Historical price trajectories for Brent crude and Henry Hub gas, annotated with major global disruptions—2008 financial crisis, COVID-19, and the Russia–Ukraine war (Figure 10.8). Highlights the cyclical nature of energy markets and the sensitivity of upstream economics to geopolitical and macroeconomic shocks.

10.4.3 Strategies for Managing Price Volatility

Oil and gas companies adopt risk management strategies, such as hedging, diversification, and operational flexibility, to navigate price uncertainty.

Factor	Impact on Prices
OPEC Output Adjustments	Sudden production cuts or increases can shift the global supply balance.
Extreme Weather Events	Hurricanes, wildfires, and freezing temperatures can shut down production infrastructure.
Storage Capacity Constraints	Limited storage space can force producers to sell at low prices (e.g., **the WTI oil price turned negative in April 2020**).
Speculative Trading	Futures trading by hedge funds and institutional investors can drive price swings.

FIGURE 10.7 Key drivers of oil price volatility in global markets.

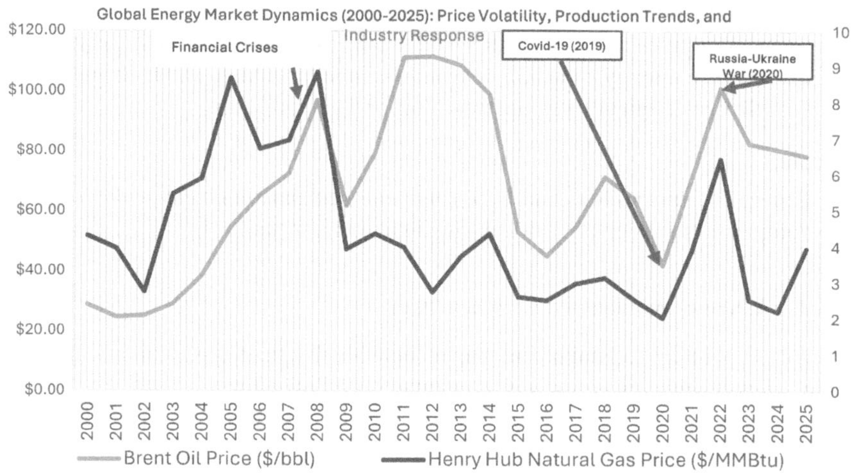

FIGURE 10.8 Brent oil and Henry Hub natural gas price trends (2000–2025).

1. **Hedging Strategies**
    - Companies use **futures contracts, options, and swaps** to **lock in oil and gas prices** and reduce exposure to volatility.
    - **Example:** Airlines hedge jet fuel costs using **oil futures**, while upstream producers hedge **production prices**.
2. **Portfolio Diversification**
    - Investing in **multiple energy sources (oil, gas, renewables)** reduces reliance on volatile commodity prices.
    - **Example:** Major oil companies like **BP, Shell, and TotalEnergies** have expanded into **wind, solar, and hydrogen** to mitigate long-term fossil fuel risks.
3. **Operational Flexibility**
    - Companies **scale production up or down** based on price trends.
    - **Example:** When oil prices **dropped in 2020**, shale producers in the **Permian Basin** shut down drilling rigs to reduce losses.
4. **Cost Optimization and Efficiency**
    - **Automation, digitalization, and AI-driven predictive analytics** help companies **cut costs** and **increase resilience** to price fluctuations.
    - **Example: Saudi Aramco and ExxonMobil** use **AI-powered predictive maintenance** to reduce **unplanned downtime**.

10.4.4 Future Market Outlook

The future of global energy prices depends on **economic recovery, geopolitical stability, and energy transition policies**.

Short-Term (2024–2027)

Upside Potential:

- **China's economic recovery** could drive higher demand.
- **OPEC+ production controls** may support prices above $80/barrel.

Downside Risks:

- **Recession fears** could reduce energy demand.
- **Shale production increases** in the U.S. could create oversupply.

10.4.5 Long-Term (2027–2040)

Key Trends:

- **Gradual shift towards renewables** could impact demand for fossil fuels.
- **Carbon capture and storage (CCS) investments** may keep oil and gas viable under stricter regulations.
- **Natural gas and LNG demand** is expected to rise as a **bridge fuel** in the energy transition.

Example:

- The **IEA forecasts that oil demand will peak by 2030**, but gas demand will remain strong due to **industrial use and power generation**.

IEA Declares Fossil Fuel Peak in Sight by 2030

The Guardian, October 2023

10.4.6 Conclusion

The global oil and gas market remains **highly dynamic**, with prices shaped by **supply-demand fundamentals, geopolitical events, and energy policies**. As price volatility remains a key challenge, **risk management strategies such as hedging, diversification, and operational flexibility** will be **critical for upstream success**.

10.5 THE ENERGY TRANSITION AND ITS IMPACT ON UPSTREAM

The global shift towards renewable energy sources is having a profound impact on the oil and gas industry. The energy transition is driven by concerns about climate change, environmental sustainability, and energy security.

> The energy companies of tomorrow won't just produce oil; they'll produce solutions.
>
> **Patrick Pouyanné**
> *CEO, TotalEnergies*

Key Drivers of the Energy Transition:

- **Climate change:** Growing awareness of the impacts of climate change and the need to reduce greenhouse gas emissions are driving the transition to cleaner energy sources. Governments, businesses, and consumers are increasingly prioritizing sustainability and environmental responsibility.
  - **Example:** The Paris Agreement, an international treaty on climate change, aims to limit global warming to well below 2°C above pre-industrial levels by reducing greenhouse gas emissions.

- **Technological advancements:** Advances in renewable energy technologies, such as solar, wind, and energy storage, have made them more competitive with traditional fossil fuels. Improvements in efficiency, cost reductions, and scalability have accelerated the adoption of renewable energy.
  - **Example:** The cost of solar photovoltaic (PV) systems has decreased significantly over the past decade, making solar energy one of the fastest-growing renewable energy sources.
- **Policy and regulation:** Government policies and regulations are promoting the adoption of renewable energy and incentivizing the reduction of carbon emissions. These policies include subsidies, tax credits, emissions trading schemes, and renewable energy mandates.
  - **Example:** The European Union's Green Deal aims to make Europe the first climate-neutral continent by 2050, with policies to reduce emissions, increase renewable energy deployment, and promote energy efficiency.

Oil Majors Slash CAPEX as Prices Tumble Below $30

Bloomberg, March 2020

Impact on the Oil and Gas Industry:

- **Reduced demand:** The increasing penetration of renewable energy technologies, such as solar and wind power, is gradually reducing the demand for fossil fuels. As the energy mix shifts towards cleaner sources, the role of oil and gas in power generation and transportation is diminishing.
  - **Example:** The growth of electric vehicles (EVs) and the adoption of renewable energy for electricity generation are reducing the demand for gasoline and diesel.
- **Diversification:** Oil and gas companies are diversifying their portfolios to include renewable energy projects and low-carbon technologies. This diversification helps mitigate the risks associated with the energy transition and positions companies for long-term sustainability.
  - **Example:** Major oil companies like Shell, BP, and TotalEnergies are investing in renewable energy projects, such as wind, solar, and hydrogen, as part of their strategies to transition to cleaner energy sources.
- **Natural gas as a transition fuel:** Natural gas is often seen as a transition fuel, playing a role in displacing coal-fired power generation and reducing carbon emissions. Natural gas emits fewer greenhouse gases compared to coal and oil, making it a cleaner option for power generation and heating.
  - **Example:** The use of natural gas for power generation in the United States has increased in recent years, contributing to a decline in coal consumption and a reduction in carbon emissions.

TotalEnergies Shifts $5B from Oil to Renewables

Financial Times, November 2022

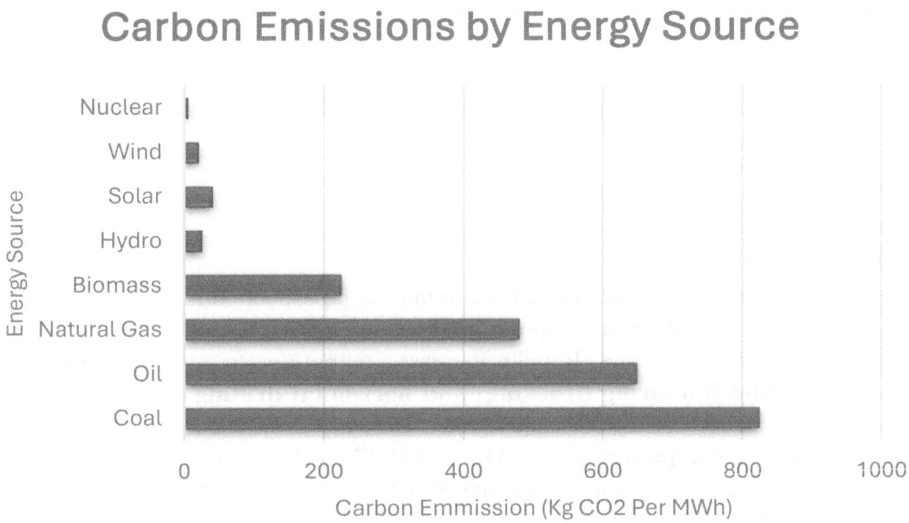

FIGURE 10.9 Carbon emissions by energy source (Kg CO_2 per MWh).

Carbon Capture, Utilization, and Storage (CCUS): CCUS technologies are being explored as a way to reduce the carbon footprint of fossil fuel use. CCUS involves capturing carbon dioxide emissions from industrial processes and power plants, transporting it, and storing it underground or utilizing it in various applications.[3]

- **Example:** The Gorgon LNG project in Australia includes a CCUS component, where CO_2 is captured and injected into a deep reservoir for long-term storage, reducing the project's overall carbon emissions.

Comparison of emissions intensity across eight energy sources (Figure 10.9). Highlights the environmental advantage of nuclear, wind, solar, and hydro over fossil fuels—especially coal and oil—underscoring the strategic imperative for upstream diversification and low-carbon investment

10.6 CASE STUDIES OF MARKET DYNAMICS

10.6.1 Case Study 1: The 2014–2016 Oil Price Crash

The oil price crash from 2014 to 2016 profoundly impacted global energy markets, reshaping investment strategies and forcing companies to adapt to a prolonged low-price environment.

Event Overview:

- In mid-2014, global oil prices dramatically declined, falling from **$115 per barrel (Brent)** in June 2014 to **below $30 per barrel** by early 2016.

- This sharp drop was driven by a **combination of factors**, including:
  - A surge in U.S. **shale oil production** added millions of barrels per day to global supply.
  - Slower **global economic growth**, particularly in China, leading to **weaker demand**.
  - **OPEC's decision** not to cut production in response to falling prices exacerbates the supply glut.

Key Players Involved:

- **OPEC and Saudi Arabia**: Chose to maintain production levels to protect market share rather than support prices.
- **U.S. shale producers**: Rapidly increased output due to advancements in **hydraulic fracturing (fracking) and horizontal drilling**, making them major contributors to the supply glut.
- **Major oil companies (ExxonMobil, BP, Shell, Chevron)**: Faced declining revenues, leading to **massive CAPEX cuts, layoffs, and project cancellations**.

Impact on Oil and Gas Markets:

- **Massive job losses**: The industry saw over **400,000 layoffs worldwide**.
- **Investment slumps**: Global upstream investments dropped by **25% in 2015** and another **24% in 2016**, making it one of the sharpest declines in industry history.
- **Bankruptcies and mergers**: Many U.S. shale producers went bankrupt or were forced into mergers.
- **Cost-cutting measures**: Companies reduced costs by renegotiating contracts, improving efficiency, and delaying capital projects.

Lessons Learned:

- **Vulnerability to supply shocks**: The crisis demonstrated how technological advancements (shale oil) could disrupt long-standing supply structures.
- **The importance of hedging and cost efficiency: Companies that** hedged prices or operated with lower breakeven costs survived better.
- **OPEC's new approach**: After initially resisting cuts, **OPEC+ was formed in 2016**, leading to coordinated production agreements between OPEC and Russia.

10.6.2 Case Study 2: The COVID-19 Pandemic and Oil Market Disruption

The COVID-19 pandemic in 2020 caused an **unprecedented collapse in global oil demand**, creating a market shock unlike any other in modern history.

Event Overview:

- The **pandemic triggered global lockdowns**, halting economic activity and travel. As a result, global oil demand fell by nearly **30 million barrels per day (bpd)** in April 2020.

- On April 20, **WTI crude oil futures turned negative (–$37.63 per barrel)** for the first time in history due to a **lack of storage space**, forcing traders to pay buyers to take oil off their hands.
- OPEC+ responded with the **most significant production cut in history (9.7 million bpd)** to stabilize prices, but the recovery took months.

Key Players Involved:

- **OPEC+ (Saudi Arabia, Russia, etc.)**: Initially engaged in a price war but later agreed to historic output cuts.
- **Governments and central banks**: Introduced stimulus measures and strategic petroleum reserve purchases to stabilize markets.
- **Oil companies and service providers**: Faced severe financial strain, leading to **global CAPEX reductions of over 30%** and cost-cutting measures.

Impact on Oil and Gas Markets:

- **Severe price crash**: Brent crude fell to **$19 per barrel**, while WTI saw negative prices.
- **Industry-wide shutdowns**: Refineries shut down, drilling activity plummeted, and investment in new projects collapsed.
- **Record layoffs**: Over **1 million jobs were lost** across the oil and gas supply chain.
- **Restructuring and bankruptcies**: Many shale producers (e.g., **Chesapeake Energy**) filed for bankruptcy.

Lessons Learned:

- **Market reliance on economic activity**: The pandemic showed the oil market's extreme vulnerability to external shocks.
- **Storage limitations matter**: The crisis highlighted the importance of **storage capacity planning** and oversupply risks.
- **Acceleration of digitalization**: Remote operations, AI-driven maintenance, and automation became critical for companies to **cut costs and maintain efficiency**.

10.6.3 Case Study 3: The Shale Revolution in the United States

The U.S. shale revolution disrupted global energy markets, making the country a leading oil and gas producer and altering traditional energy power dynamics.

Event Overview:

- The widespread adoption of **hydraulic fracturing (fracking) and horizontal drilling** in the early 2000s unlocked vast oil and natural gas reserves from shale formations.
- U.S. crude oil production grew from **5 million bpd in 2008** to over **13 million bpd by 2019**, making the U.S. the **world's top oil producer**.
- Natural gas production also surged, dramatically shifting global LNG markets and reducing coal consumption.

Key Players Involved:

- **Independent shale producers (Pioneer, EOG Resources, Devon Energy)**: Pioneered cost-effective drilling and completion techniques.
- **Major oil companies (ExxonMobil, Chevron, BP, Shell)**: Initially skeptical but invested heavily in U.S. shale.
- **Government and regulators**: Policies supporting private land drilling, minimal regulation, and infrastructure investment contributed to shale growth.

Impact on Oil and Gas Markets:

- **Increased global supply**: The U.S. became a major oil exporter, reducing reliance on foreign crude.
- **Oil price pressure**: The flood of shale oil contributed to **global price volatility**, including the 2014–2016 crash.
- **Natural gas dominance**: The surge in U.S. gas production helped **replace coal in power generation**, lowering carbon emissions.
- **Pipeline and LNG Boom**: New pipeline infrastructure and LNG export terminals transformed the U.S. into a **global gas supplier**.

Lessons Learned:

- **Technology-driven disruptions**: Innovations in drilling techniques can reshape global markets.
- **Cost efficiency is key**: Shale companies had to improve cost structures to survive downturns.
- **Infrastructure bottlenecks matter**: Pipeline and export constraints slowed shale's full potential.

10.6.4 Summary of Case Studies

These case studies demonstrate the **complex interplay between market forces, geopolitics, and technological advancements** in the upstream petroleum sector (Figure 10.10).

Overview of three pivotal events—the 2014–2016 oil price crash, the 2020 COVID-19 collapse, and the U.S. shale revolution. Highlights the causes, consequences, and strategic takeaways from each disruption, emphasizing cost efficiency, digital resilience, and the transformative role of technology in upstream market dynamics.

Case Study	Key Event	Main Impact	Lessons Learned
2014-2016 Oil Price Crash	Oversupply from shale & OPEC strategy	Collapse of investment, bankruptcies, CAPEX cuts	Cost efficiency, market adaptability, OPEC+ cooperation
2020 COVID-19 Oil Collapse	Demand shock from pandemic	Negative oil prices, production shutdowns, layoffs	Importance of storage, digitalization, and hedging strategies
U.S. Shale Revolution	Rise of hydraulic fracturing	U.S. energy independence, price pressure, LNG growth[4]	Tech-driven disruption, need for efficiency, infrastructure importance

FIGURE 10.10 Summary of major upstream disruptions and strategic lessons.

10.6.5 Final Thoughts

The upstream sector is **highly cyclical and influenced by external shocks**, but companies embracing innovation, managing costs efficiently, and diversifying operations can successfully navigate these challenges. Understanding past market dynamics helps **anticipate future trends and risks**, ensuring resilience in a volatile industry.

10.7 CHAPTER SUMMARY

This chapter explored the **global market dynamics** that influence the upstream petroleum sector, highlighting the role of key players, geopolitical risks, investment trends, price volatility, and the **ongoing energy transition**. Through case studies, we examined real-world examples of how **geopolitical events, economic disruptions, and technological advancements** have shaped industry outcomes.

10.7.1 Key Takeaways

1. **Global Energy Markets Are Highly Complex**
   - The interaction between **supply and demand, investment cycles, and geopolitical shifts** makes the upstream industry inherently volatile.
   - Energy markets are influenced by **economic trends, technological innovations, and government policies**, requiring companies to stay adaptable.
2. **Geopolitical Factors Significantly Impact the Upstream Sector**
   - Political stability, **resource nationalism, trade disputes, and sanctions** can affect exploration and production activities.
   - Companies operating in politically sensitive regions must adopt **risk mitigation strategies**, including **portfolio diversification, stakeholder engagement, and insurance mechanisms**.
3. **Foreign Investment is Essential, But Risks Vary by Region**
   - Investment flows depend on **regulatory stability, fiscal terms, and infrastructure readiness**.
   - While **mature markets (e.g., Norway, U.S.)** attract steady investment, emerging markets offer **high potential but have elevated risks**.
4. **Price Volatility is a Constant Challenge**
   - The **2014–2016 price crash, the COVID-19 pandemic, and OPEC policy shifts** illustrate how prices fluctuate due to macroeconomic and geopolitical shocks.
   - Companies use **hedging, diversified asset portfolios, and operational flexibility** to manage price risks and remain competitive.
5. **The Energy Transition is Reshaping Investment and Strategy**
   - The global push for **low-carbon energy** forces traditional oil and gas companies to rethink their strategies.
   - Major players are diversifying into **renewable energy, hydrogen, carbon capture, and natural gas** as part of their adaptation plans.

10.7.2 Looking Ahead

As the **global energy sector evolves**, upstream companies must balance **short-term profitability with long-term sustainability**. Understanding **market trends, geopolitical risks, and emerging energy technologies** will be essential for maintaining a competitive edge.

NOTES

1. International Energy Agency (IEA), *World Energy Outlook 2023.*
2. U.S. Energy Information Administration (EIA), *Annual Energy Outlook 2023.*
3. TotalEnergies, *Sustainability Report 2022.*

11 Upstream Project Finance

11.1 INTRODUCTION TO UPSTREAM PROJECT FINANCE

Upstream petroleum projects require **substantial capital investment, lengthy development timelines, and meticulous financial planning** to ensure profitability. Unlike many industries, the upstream sector must commit substantial funds **before** generating revenue, making financing and investment analysis essential.

11.1.1 Key Financial Considerations in Upstream Projects

1. **Investment appraisal**: Assessing the economic viability of a project, determining whether to proceed based on expected returns and risks.
2. **Funding strategies**: Identifying appropriate capital sources (e.g., internal financing, debt, or joint ventures) to finance project development.
3. **Risk management**: Mitigating exposure to financial, technical, and geopolitical risks that could impact project success.

11.1.2 Unique Financial Characteristics of Upstream Projects

Upstream investments differ from typical capital expenditures in other industries due to the following factors:

- **High Capital Intensity:**
  - Upstream projects require **massive upfront investment** in **exploration, drilling, field development, and infrastructure** before any revenue is generated.
  - Costs often reach **billions of dollars**, especially for complex projects like deepwater exploration or liquefied natural gas (LNG) developments.
  - Example: The **Kashagan Field** in Kazakhstan required over **$50 billion**[1] **in investment** due to its harsh operating environment and high-pressure reservoirs.
- **Long Lead Times and Uncertain Payback Periods:**
  - Development cycles can **span 5–15 years** from discovery to full-scale production.
  - Even after production begins, full cost recovery may take **decades**, especially in **offshore and unconventional projects**.
  - Example: Brazil's **pre-salt offshore fields** required nearly **a decade of investment** before achieving peak production.
- **High Risk and Uncertainty:**
  - Exploration success rates are **low**, and **geological uncertainty** remains a significant challenge.

DOI: 10.1201/9781003750345-14

- Economic factors such as **commodity price volatility, fiscal regime changes, and environmental regulations** can affect profitability.
- Example: The **2014–2016 oil price crash** forced many companies to **write off investments** in high-cost projects.

Examples of High-Capital Upstream Projects:

- **Deepwater Offshore Drilling:**
  - **Requires** specialized drilling rigs, **FPSO (Floating Production Storage and Offloading) units**, and high-cost infrastructure.
  - **Investment scale:** Often exceeds **$5–10 billion per project**.
  - **Example:** The **Johan Sverdrup field (Norway)** required multi-billion-dollar investment in **phased development** to optimize spending.
- **Shale Oil and Gas Development:**
  - **Relies on** horizontal drilling and hydraulic fracturing.
  - **Investment scale: Capital-intensive** but offers **shorter payback periods** compared to offshore fields.
  - **Example:** The **Permian Basin (USA)** has seen **over $100 billion in cumulative investments** from multiple operators.

11.1.3 The Role of Financial Planning in Upstream Projects

Financial planning ensures projects are:

✓ **Economically viable**: Revenue projections and cost control strategies are aligned.
✓ **Adequately funded**: Securing financing through equity, debt, or partnerships.
✓ **Resilient to risks**: Implementing hedging strategies and flexible cost structures to mitigate market downturns.

Effective financial planning is **not just about securing capital—it is about maximizing value** through cost efficiency, strategic partnerships, and **adaptive investment strategies** that align with market conditions.

11.2 CAPITAL BUDGETING TECHNIQUES

Capital budgeting is the **foundation of investment decision-making** in the upstream oil and gas sector. Given the **large capital requirements, long payback periods, and inherent risks** of upstream projects, companies use **rigorous financial evaluation techniques** to ensure that capital is allocated efficiently.

Key Capital Budgeting Methods in Upstream Investment:

1. **Discounted Cash Flow (DCF) Analysis**
   - The most commonly used method to evaluate **long-term investments** in upstream projects.
   - Incorporates the **time value of money (TVM)** to determine the **present value** of expected future cash flows.

◆ **Key DCF Metrics:**

- **Net Present Value (NPV):**

$$NPV = \Sigma \frac{C_t}{(1+r)^t} - C_0$$

  - Ct = Cash inflow at time t
  - r = Discount rate
  - $C0$ = Initial investment
  - If **NPV > 0**, the project is expected to create value. If **NPV < 0**, the project is not economically viable.
  - **Example:** A deepwater drilling project with an initial investment of **$1 billion** and projected future cash flows of **$200 million per year for 10 years** would require NPV calculations to determine feasibility.

> ## DCF: WHAT CAN GO WRONG?
>
> While widely used, DCF models can be misleading if:
>
> - Price forecasts are overly optimistic
> - Discount rates don't reflect risk
> - CAPEX/OPEX assumptions ignore inflation
>
> → Lesson: A good model is only as good as its inputs.

- **Internal Rate of Return (IRR):**
  - The discount rate at which **NPV = 0** (i.e., the project's break-even return).
  - If **IRR > company's hurdle rate (cost of capital)**, the project is attractive.
  - **Example:** The project is considered viable if an oil field development project has an IRR of 15%, and the company's required return is 12% (Figure 11.1).

2. **Payback Period**
   - Measures how quickly an investment recovers its **initial capital outlay**.
   - **Formula:**

$$Payback\ Period = \frac{Initial\ Investment}{Annual\ Cash\ Inflows}$$

   - **Pros:**
     - Provides a **simple measure of liquidity and risk.**
     - Useful for evaluating **short-cycle investments**, such as shale oil projects.

Investment Option	Initial CAPEX (MMUSD	OPEX per Barrel	Break-even Price (USD/Barrel)	IRR (%)	NPV (MMUSD) 10% (Discount rate)	Risk Level
Deepwater Project	5000	12	50	12%	1,200	High
Shale Development	1500	8	40	18%	800	Medium
Onshore Conventional	800	5	30	22%	500	Low
Enhanced Oil Recovery (EOR)	2000	10	45	15%	700	Medium
LNG Terminal	2500	7	35	17%	900	Medium-High

FIGURE 11.1 Comparative investment metrics for upstream project types: Benchmarking CAPEX, OPEX, break-even prices, IRR, NPV, and risk levels across five upstream investment options.

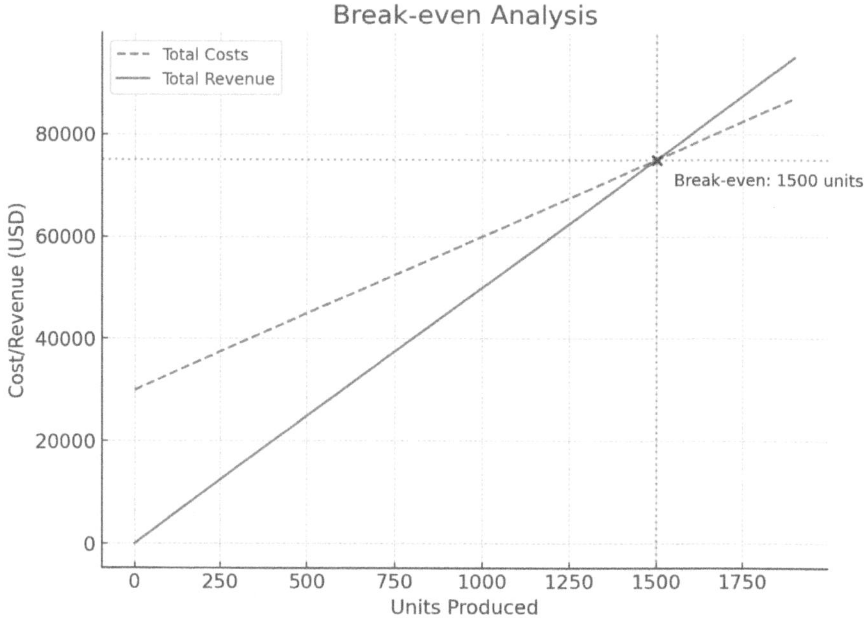

FIGURE 11.2 Break-even analysis graph: Visualizing the production level at which total revenue equals total costs, highlighting the break-even point.

- **Cons:**
  - Does not account for the **time value of money (TVM)**.
  - Ignores cash flows after the payback period.
- **Example:** A **shale oil drilling project** requiring **$50 million** in upfront investment with expected **annual net cash inflows of $10 million** would have a **payback period of 5 years** (Figure 11.2).

3. **Profitability Index (PI)**
   - Measures the **relative profitability** of a project by comparing the present value of future cash flows to the initial investment.
   - **Formula:**

$$PI = \frac{Present\,Value\,of\,Future\,Cash\,Flows}{Initial\,Investment}$$

   - If **PI > 1**, the project is considered **economically viable**.
   - Used for ranking projects **when capital is limited**.
   - **Example:** If an offshore gas field has a **PI of 1.3**, this means that for every **$1 invested, the project generates $1.30 in value**.

Capital Rationing in Upstream Projects:

- **Capital rationing** occurs when companies have **limited investment funds** and must **prioritize projects** based on financial returns and strategic importance.

- Companies allocate capital to **projects with the highest NPV, IRR, or PI**, while considering risk factors such as:
  - **Exploration risk** (geological uncertainty).
  - **Operational risk** (cost overruns, delays).
  - **Market risk** (oil price fluctuations).
  - **Regulatory risk** (government policies, environmental restrictions).
- **Example:** A company with **$2 billion in available capital** may choose to fund **two offshore projects with high NPV** over a single deepwater development with a longer payback period.

11.2.1 Example of Capital Budgeting in Upstream

Consider an **offshore oil development** with:

- **Initial investment: $500 million**
- **Annual cash inflows: $100 million** for 10 years
- **Discount rate: 10%**

Using **capital budgeting techniques**:

- **NPV:** The project's total value, adjusted for the time value of money.
- **IRR:** The project's break-even rate of return.
- **Payback Period: 5 years ($500M ÷ $100M/year)**.
- **PI:** If PI > **1**, the project generates a positive return.

Key Takeaways:

✓ **DCF-based evaluations (NPV and IRR) are essential for long-term decision-making.**

✓ **The payback period is useful for short-cycle projects like shale oil drilling.**

✓ **PI helps prioritize projects when capital is constrained.**

✓ **Capital rationing ensures funds are allocated to the most valuable projects** (Figure 11.3).

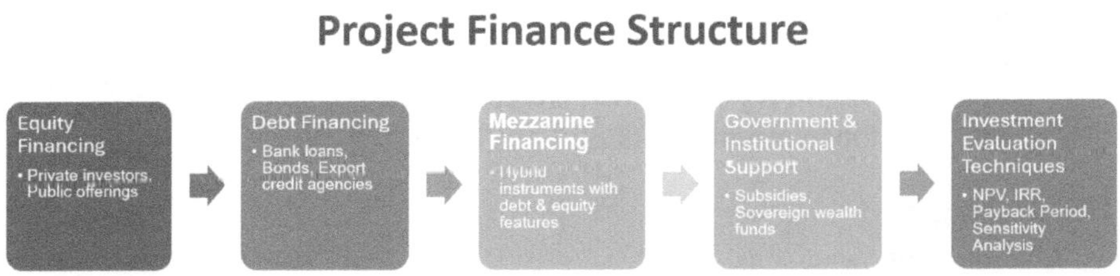

FIGURE 11.3 Equity vs. debt comparison: Key differences in repayment, ownership dilution, cost of capital, and flexibility.

11.3 FUNDING OPTIONS FOR EXPLORATION AND PRODUCTION

Exploration and production (E&P) projects in the upstream petroleum sector require **significant capital investment**, making **strategic funding decisions** critical for project success. Companies use various funding options, each with distinct **advantages, risks, and implications** for ownership and financial flexibility.

⚖ EQUITY VS. DEBT IN NUMBERS

Criteria	Equity	Debt
Repayment	No	Yes (Interest + Principal)
Ownership Dilution	Yes	No
Cost of Capital	High	Low (Typically)
Risk to Lenders	Low	High
Flexibility	Medium	Low

11.3.1 Primary Funding Sources for Upstream Projects

11.3.1.1 Equity Financing

Raising capital by **issuing shares** to investors. This is often used in high-risk projects such as **exploration** and early-stage development.

- **Types of Equity Financing:**
  - **Venture capital**: Used for early-stage exploration startups.
  - **Private equity**: Involves institutional investors in funding later-stage development.
  - **Public equity (stock issuance)**: Listed companies can raise capital through stock markets.

Example:

- **Kosmos Energy**, an independent oil and gas company, raised **venture capital** to fund its offshore exploration activities in West Africa. Investors received **equity stakes** in exchange for financing.

Advantages:

✓ No obligation to repay debt.
✓ Suitable for **high-risk** exploration projects.
✓ Can attract long-term investors.

Disadvantages:

✗ Dilution of ownership.
✗ May result in **loss of control** over company decisions.
✗ Higher cost of capital compared to debt.

11.3.1.2 Debt Financing

Raising capital by **borrowing money** from banks or issuing **bonds**. Used primarily for **development and production** phases, where projects generate predictable cash flows.

- **Types of Debt Financing:**
  - **Bank loans**: Direct lending from financial institutions.
  - **Corporate bonds**: Companies issue debt securities to investors.
  - **Project finance loans**: A loan secured by the future **cash flows of the project**.

Example:

- **Saudi Aramco** issued **$12 billion in bonds** in 2019 to fund expansion projects. Investors lent money in exchange for **fixed-interest payments**.

Advantages:

✓ Retains ownership.
✓ Interest payments are **tax-deductible**.
✓ Lower cost of capital compared to equity.

Disadvantages:

✗ Requires **regular interest payments**, adding financial burden.
✗ Increases the company's **financial leverage and risk**.
✗ Limited borrowing capacity based on creditworthiness.

11.3.1.3 Hybrid Financing (Debt + Equity)

A combination of debt and equity financing to **balance risks and returns**. Common structures include:

- **Convertible bonds**: Bonds that convert into company shares later.
- **Mezzanine financing**: A mix of debt and equity, offering flexible repayment structures.

Example:

- **TotalEnergies** issued **convertible bonds**, allowing investors to switch from debt to equity at a later stage. This provided flexibility for both the company and investors.

Advantages:

✓ Balances **risk and return** between debt and equity.
✓ Can attract **institutional investors** seeking both yield and potential equity upside.

Disadvantages:

✗ More complex structure.
✗ Can be **costlier** than traditional debt.

11.3.1.4 Joint Ventures and Farm-Out Agreements

Partnering with other companies to **share costs, expertise, and risks**.

- **Joint ventures (JVs)**: Two or more companies co-invest in a project, sharing ownership and responsibilities.
- **Farm-outs**: One company **sells part of its interest** in an exploration block to another company in exchange for funding.

Example:

- In **2012, BP and Reliance Industries Limited (RIL)** formed a **joint venture** to explore and develop offshore oil and gas fields in India. BP contributed **technical expertise**, while RIL provided **local knowledge and resources**.

Advantages:

✓ Reduces capital investment for each partner.
✓ Share **technical expertise and operational risks**.
✓ Allows access to new markets or regions.

Disadvantages:

✗ Potential **conflicts of interest** between partners.
✗ Shared decision-making may slow progress.
✗ Complex agreements requiring legal and regulatory approvals (Figure 11.4).

11.3.2 Comparison of Upstream Funding Options

See Figure 11.4.

11.3.3 Selecting the Right Funding Strategy

Choosing the optimal funding strategy depends on:

✓ **Project phase:** Exploration requires equity, while production can be debt-financed.

Funding Type	Advantages	Disadvantages	Best for
Equity Financing	No repayment obligation supports high-risk projects.	Dilutes ownership and reduces control.	Early-stage exploration, high-risk projects.
Debt Financing	Lower cost of capital retains ownership.	Requires interest payments, increases financial risk.	Development and production phases.
Hybrid Financing	Offers a balance between debt and equity.	Complex structure, higher cost than traditional debt.	Large-scale projects with long payback periods.
Joint Ventures & Farm-Outs	Shares costs and expertise reduce financial exposure.	Potential conflicts and complex legal agreements.	Large-scale offshore projects, deepwater drilling.

FIGURE 11.4 Comparison of upstream funding options: Advantages, disadvantages, and best-fit applications for each financing method.

- ✓ **Risk profile:** High-risk projects benefit from equity or joint ventures to spread risk.
- ✓ **Market conditions:** Interest rates, oil prices, and investor appetite affect financing choices.
- ✓ **Company objectives:** Some companies prioritize ownership retention, while others seek external capital.

Key Takeaways:

- ✓ **Upstream projects require significant capital investments**, making financing decisions crucial.
- ✓ **Equity financing** is best for high-risk projects but leads to **ownership dilution**.
- ✓ **Debt financing** retains ownership but **increases financial obligations**.
- ✓ **Hybrid financing** provides flexibility but **adds complexity**.
- ✓ **Joint ventures and farm-outs** share costs and risks but **require legal agreements**.

11.4 FINANCIAL RISK ANALYSIS AND INVESTMENT EVALUATION

Financial risk analysis is **essential for evaluating upstream projects**, ensuring informed decision-making, and **mitigating potential financial losses**. Several techniques help **assess project uncertainties** and determine **the economic viability** of investments.

11.4.1 Key Techniques for Financial Risk Analysis

11.4.1.1 Sensitivity Analysis and Scenario Planning

Sensitivity analysis evaluates how **changes in key assumptions** (such as oil prices, production rates, or costs) impact a project's financial outcomes.

- **Scenario planning**: Assesses different market conditions to estimate **best-case, worst-case, and base-case scenarios**.

Example:

- A **deepwater exploration project** might analyze sensitivity by testing oil price assumptions at **$40, $60, and $80 per barrel** to determine how price volatility affects Net Present Value (NPV).
- A **scenario planning model** could assess the impact of a **geopolitical crisis** on production shutdowns or cost escalations.

Why It's Important:

- ✓ Helps identify **key risk factors** that influence project profitability.
- ✓ Supports **decision-making** under uncertainty by evaluating multiple future conditions.
- ✓ Improves risk **mitigation planning** and investment readiness (Figure 11.5).

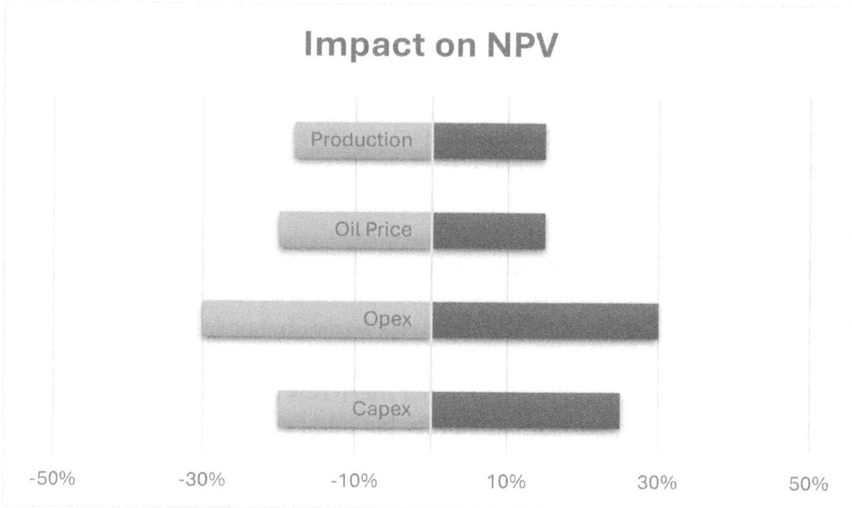

FIGURE 11.5 Investment metrics for upstream project evaluation: Summary of financial indicators used to assess project viability, including NPV, IRR, Payback, and PI.

11.4.1.2 Risk-Adjusted Discount Rates

In discounted cash flow (DCF) analysis, a higher discount rate is applied to riskier projects to reflect potential uncertainties.

- Projects with **high exploration risks** require a higher discount rate.
- Mature, low-risk producing fields **use lower discount rates** due to stable cash flows.

Example:

- A **high-risk exploration project** in an **untested deepwater basin** may use a **15% discount rate**, while a **mature oil field with stable production** may use an **8% discount rate**.

Why It's Important:

✓ Ensures that investors **account for risk when valuing a project**.
✓ Prevents **overestimating the attractiveness** of uncertain investments.
✓ Helps rank projects based on their **risk-return tradeoff**.

11.4.1.3 Monte Carlo Simulation

Monte Carlo simulation uses **probability modeling** to estimate a **range of possible project outcomes**.

- Instead of **one static financial projection**, it **runs thousands of simulations** by randomly changing variables (e.g., oil prices, costs, and production rates).
- The result is a **probability distribution of financial outcomes**, showing the **likelihood of success or failure**.

Example:

A **deepwater drilling project** may use **Monte Carlo simulations** to model different **oil price scenarios**[2] and **reservoir performance variations**, estimating the probability of a **positive NPV**.

Why It's Important:

✓ Quantifies **risk exposure** by generating **multiple potential outcomes.**

✓ Helps investors **understand the probability of financial success or failure**.

✓ Aids in **investment decision-making by providing a full risk spectrum**.

11.4.2 Investment Evaluation Metrics

1. **Net Present Value (NPV)**
   - Measures project **profitability** by calculating the present value of future cash flows.
   - A **positive NPV** indicates a profitable project.
   - **Formula:**

$$NPV = \sum \frac{Ct}{(1+r)^t} - C_0$$

   - where:
     - **Ct** = Cash flow in period **t**
     - ***r*** = Discount rate
     - ***C*0** = Initial investment
     - **t** = Time period

   Example:

   - A project requiring an **investment of $200M**, generating **$50M annual cash flow for 10 years** at an **8% discount rate**, will have its **NPV calculated** to determine investment viability.

   ✓ **If NPV > 0**: Invest

   ✗ **If NPV < 0**: Reject

2. **Internal Rate of Return (IRR)**
   - The **discount rate at which NPV = 0.**
   - Represents the project's **expected return** and helps compare **different investment opportunities**.
   - **Higher IRR** = More attractive investment.

   ✓ **If IRR > Required Return Rate**: Accept Project

   ✗ **If IRR < Required Return Rate**: Reject Project

3. **Payback Period**
   - The time required to **recover the initial investment**.
   - **Shorter payback = Lower risk**.

   ✓ Useful for measuring **liquidity and capital recovery speed**.

   ✗ **Does not account for the time value of money** or cash flows after payback.

Example:

- A **$200M project** generating **$50M per year** has a payback period of **4 years**.

4. **Profitability Index (PI)**
   - **Formula:**

$$PI = \frac{Present\ Value\ of\ Future\ Cash\ Flows}{Initial\ Investment}$$

   - **If PI > 1**: Project is profitable.
   - **If PI < 1**: Project is not viable.

Example:

- **NPV of future cash flows = $280M**
- **Initial investment = $200M**
- **PI = 280/200 = 1.4** (Profitable)

11.4.3 Example: Evaluating an Upstream Project

Project Assumptions:

- **Initial investment**: $200M
- **Annual cash flows**: $50M for 10 years
- **Discount rate**: 8%
- **Oil price scenarios**: $40, $60, $80 per barrel

Financial Risk Analysis:

✓ **Sensitivity Analysis**: Varies **oil prices**, calculating NPV for each price level.
✓ **Monte Carlo Simulation**: Models **multiple price and cost variations** to predict risk probabilities.
✓ **Risk-Adjusted Discount Rate**: Higher discount rates are used for riskier investments (Figure 11.6).

Investment Metrics Results

Decision:

✓ The project is **financially viable**, given its **positive NPV, high IRR, and strong PI**.
✓ Risk mitigation via **hedging oil price exposure** and **cost optimization** is recommended.

Metric	Value
NPV	$120M (Profitable)
IRR	14.5% (Above cost of capital)
Payback Period	4 years (Fast recovery)
Profitability Index (PI)	1.4 (Good investment)

FIGURE 11.6 IRR benchmarks by region: Typical hurdle rates for upstream projects across global regions, reflecting risk-adjusted return expectations.

Key Takeaways:

✓ **Financial risk analysis helps upstream companies assess investment viability.**
✓ **Sensitivity analysis** identifies **key risk factors** affecting financial performance.
✓ **Monte Carlo simulation quantifies uncertainty** and estimates the probability of success.
✓ **Investment evaluation metrics (NPV, IRR, Payback, PI) determine project feasibility.**
✓ **A balanced financial strategy** incorporates **risk-adjusted discount rates** and **scenario planning** to enhance decision-making.

11.5 INVESTMENT DECISION-MAKING IN THE UPSTREAM SECTOR

Investment decisions in the **upstream oil and gas sector** are **strategic, capital-intensive, and risk-sensitive**. Companies must evaluate multiple **technical, economic, and geopolitical factors** before committing capital to projects that can span **decades** and require **billions of dollars** in investment.

11.5.1 Key Factors Influencing Investment Decisions

1. **Strategic Alignment**
   - Projects must align with a company's **long-term vision, core competencies**, and **market positioning**.
   - Companies prioritize projects that support their objectives, such as **expanding reserves, improving efficiency, or entering new markets**.

   Example:
   - A **deepwater exploration-focused company** may **prioritize offshore drilling projects** over shale development because it aligns with its expertise.
   - A company aiming for a **low-carbon transition** may invest in **natural gas over oil** or explore **carbon capture and storage (CCS) projects**.

✓ Ensures **focus on core strengths** and **sustainable business growth**.
✗ Avoids **diversification into unprofitable or high-risk ventures**.

$10 Billion Question: Can Upstream Projects Survive in a 2°C World?

As the energy transition accelerates, project financiers face a new dilemma — how to fund long-life assets in a world that may not need them. ESG ratings and future-proofing take center stage.

2. **Portfolio Management and Diversification**
   - Companies **balance their investment portfolio** by diversifying across:
     - **Geographical regions** (e.g., Middle East vs. North America).
     - **Resource types** (e.g., conventional vs. unconventional).
     - **Project maturities** (e.g., exploration vs. producing assets).

- **Diversification reduces risk** from **any investment's market downturns, political instability, or cost overruns.**

Example:

- **ExxonMobil** maintains a **mix of shale, offshore, and LNG projects** to hedge against fluctuations in oil prices and production costs.
- **BP** has diversified into **renewables** while continuing upstream oil and gas investments, hedging against the energy transition.

✓ **Protects against market volatility** and **economic downturns.**

✓ Balances **high-risk, high-reward exploration** with **low-risk production assets.**

FAST FACT—IRR BENCHMARKS BY REGION

Region	Typical IRR Hurdle Rate
North America (Shale)	10–12%
Deepwater Offshore	14–18%
Frontier/High-Risk	18–22%
NOC-led Middle East	8–10%

3. **Decision Criteria and Hurdle Rates**
   - **Hurdle Rate**: The **minimum acceptable return** (e.g., IRR) required for investment approval.
   - Companies **set hurdle rates** based on:
     - **Cost of capital** (debt vs. equity financing costs[3]).
     - **Project risk level** (e.g., deepwater exploration has a higher risk than onshore production).
     - **Industry benchmarks** (e.g., major oil companies may set a hurdle rate of **12–15% IRR**).

Example:

- **Chevron** may require a **14% IRR** for deepwater drilling projects but accept a **10% IRR** for brownfield (existing field expansion) developments.
- **Shell** might apply **higher discount rates** for investments in **politically unstable countries** due to increased geopolitical risks.

✓ Ensures capital is deployed **only in high-return, strategically aligned projects.**

✗ Prevents investing in **low-profit or excessively risky ventures.**

4. **Economic and Market Conditions**
   - **Oil and gas prices**, **supply-demand forecasts**, **inflation**, and **cost structures** heavily influence investment decisions.
   - Companies adjust investment strategies **based on market cycles**:
     - **High oil prices**: More investment in **new exploration and high-cost projects.**
     - **Low oil prices**: Cost-cutting, efficiency improvements, and focus on **low-cost production assets.**

Example:

- In **2020**, during **COVID-19 and oil price collapse**, **many companies cut capital spending** and delayed investments.
- In **2022**, with oil prices exceeding **$100 per barrel**, companies **accelerated upstream investments** to capitalize on high prices.

✓ **Adapts to changing economic realities** and **optimizes capital allocation.**
✓ Prevents **overinvestment in low-margin projects** during downturns.

11.5.2 Investment Decision Process

1. **Project Identification**
   - Identifying investment opportunities that match strategic **growth objectives** and **market demand.**
   - Evaluation of **resource potential, technical feasibility, and economic attractiveness.**

✓ **Example:** A company evaluating **shale gas opportunities in Argentina's Vaca Muerta basin** may compare its potential against North American shale plays.

2. **Feasibility Studies**
   - Conducting **technical and economic feasibility assessments** to estimate:
     - **Geological risk** (reservoir size, quality).
     - **Technical requirements** (drilling complexity, required infrastructure).
     - **Environmental and regulatory constraints** (permits, emission limits).

✓ **Example:** An operator evaluating an **Arctic drilling project** would assess the **technical feasibility of operations in extreme conditions.**

3. **Economic Evaluation**
   - Performing **capital budgeting techniques**:
     - **Net Present Value (NPV)**: Measures project profitability.
     - **Internal Rate of Return (IRR)**: Determines investment attractiveness.
     - **Sensitivity and Monte Carlo Analysis**: Assesses risks (e.g., oil price fluctuations).

✓ **Example:** A new **deepwater field** requires a **minimum oil price of $60 per barrel** to be profitable—if price forecasts remain above this level, investment is justified.

4. **Risk Assessment**
   - **Identifying risks and mitigation strategies**, such as:
     - **Technical risks** (well failures, cost overruns).
     - **Economic risks** (commodity price volatility).
     - **Political risks** (government expropriation, regulatory changes).

✓ **Example:** Investing in **Venezuela's oil sector** carries **high political risk**, requiring **higher returns** or risk mitigation strategies like **political risk insurance.**

5. **Final Approval and Investment Decision**
   - **Project proposal is reviewed** by senior management/board for **final approval.**
   - **Factors considered**:
     - Alignment with **company strategy.**
     - **Financial viability** (NPV, IRR, payback period).
     - **Risk assessment** and **contingency planning.**

✓ **Example:** A company investing in an **LNG project in Qatar** may seek approval based on:
- **Low break-even costs** ($40 per barrel oil equivalent).
- **Stable regulatory environment**.
- **Projected long-term demand for natural gas**.

11.5.3 Example: Investment Decision in a Deepwater Project

Investment Proposal:

- **Project**: Offshore deepwater oil field in **Brazil**.
- **Investment required**: $2 billion.
- **Projected production**: 200,000 barrels per day.
- **Expected IRR**: 18%.
- **NPV calculation**: $750 million at a 10% discount rate.
- **Risks**: Regulatory delays, cost overruns, and oil price fluctuations.

Decision-Making Process:

Strategic fit: Expands deepwater portfolio in a high-growth region.
Feasibility study: Geological data confirms strong reservoir potential.
Economic viability: High NPV, IRR exceeds hurdle rate (15%).
Risk mitigation: Partnering with local companies reduces regulatory risks.
✓ **Investment approved**: The company commits **capital and resources to project development**.

11.5.4 Key Takeaways

✓ **Investment decisions in upstream oil and gas require balancing risks and returns**.
✓ **Portfolio diversification** reduces exposure to commodity price swings.
✓ **Economic metrics (NPV, IRR, sensitivity analysis) drive decision-making.**
✓ **Market conditions and geopolitical risks impact investment timing.**
✓ **Projects must align with corporate strategy to ensure long-term value creation.**

11.6 CASE STUDIES OF UPSTREAM PROJECT FINANCE

Upstream oil and gas projects require **substantial investment, complex financing structures, and strategic risk management**. The following case studies highlight how financing was secured, the challenges faced, and the key lessons learned from major upstream developments.

11.6.1 Case Study 1: The Kashagan Field (Kazakhstan)

The **Kashagan Field**, located in the Caspian Sea, is one of the **largest offshore oil discoveries** in recent history. Its development required **massive investment** and **cutting-edge technology** to overcome extreme operational challenges.

Financing Structure Used:

- **Equity contributions:** The project was backed by a **consortium of major oil companies**, including:
  - **Eni (Operator)**
  - **ExxonMobil**
  - **Shell**
  - **TotalEnergies**
  - **CNPC** (China National Petroleum Corporation)
  - **Inpex & KazMunayGas** (Kazakhstan's state-owned company)
- **Debt financing:** Additional funding was secured through **export credit agencies (ECAs)** and **bank loans** to help finance infrastructure development.

$55 Billion Later, Kazakhstan Strikes Oil – Kashagan Finally Produces

After years of delay, extreme drilling conditions, and cost overruns, Kazakhstan's Kashagan Field delivers oil at last — but at what financial cost? Investors debate whether it's a success story or a cautionary tale.

Challenges Faced:

- **Technical risks**: The field is located at a **depth of 4,200m** and has **extremely high reservoir pressure (~800 bar)** and **hydrogen sulfide (H_2S) contamination**, requiring specialized production equipment.
- **Logistical and environmental challenges**: Operations in the **harsh Caspian climate** were further complicated by **icy winters**, **shallow waters**, and **strict environmental regulations**.
- **Delays and cost overruns**: Originally expected to start production in **2005**, the project faced multiple delays and eventually started in **2016**, with costs exceeding **$55 billion**—more than **three times the initial estimate**.

Outcomes:

- **Production began in 2016**, making **Kazakhstan a major oil exporter**.
- Kashagan's expected **plateau production** is over **400,000 barrels per day**, with further expansion planned.
- Despite **delays and high costs**, the project has **long-term economic significance** for Kazakhstan.

Lessons Learned:

- **Strong consortium partnerships** can help share **megaprojects' risks, expertise, and costs**.
- **Advanced engineering solutions** are essential for overcoming **high-pressure reservoirs and extreme operating conditions**.
- **Regulatory engagement and environmental management** must be factored into early planning.

11.6.2 Case Study 2: The Tupi (Lula) Field (Brazil)

The **Tupi Field (formerly Lula Field)** is one of Brazil's largest pre-salt oil discoveries. Located in deepwater off the Brazilian coast, it revolutionized **offshore production technology**.

Financing Structure Used[4]:

- **Equity contributions:** The project was financed by a **joint venture** consisting of:
  - **Petrobras (Operator, 65%)**
  - **Shell (25%)**
  - **Galp Energia (10%)**
- **Debt Financing:**
  - Petrobras secured financing from **international banks** and **development institutions**.
  - The company also issued **bonds** to raise capital.

Challenges Faced:

- **Deepwater and pre-salt challenges**: The oil is trapped **7,000m below the seabed**, beneath thick layers of **salt, rock, and water**, making drilling extremely **complex and expensive**.
- **Regulatory uncertainty**: Brazil introduced **local content laws**, requiring companies to source **a percentage of equipment and services domestically**. This led to **cost escalations** and **supply chain delays**.
- **High development costs**: The need for **specialized deepwater drilling rigs** and **CO_2 separation technology** increased investment requirements.

Brazil's Pre-Salt Gamble Pays Off: Deepwater Becomes Deep Value

Once considered a high-risk frontier, Brazil's offshore pre-salt fields now power Petrobras' resurgence. Innovative financing and strategic partnerships turned a geologic prize into an economic win

Outcomes:

- **First production in 2010**, making Brazil one of the world's **top offshore oil producers**.
- The field contributes **over 1 million barrels per day** to **Brazil's oil production**.
- Petrobras became a **global leader in deepwater technology**.

Lessons Learned:

- **Early investments in R&D** (e.g., **high-tech drilling and production systems**) enabled successful pre-salt development.
- **Diversified financing sources** (equity, bonds, and bank loans) provided **financial flexibility**.
- **Collaboration with international partners** helped **mitigate technical and financial risks**.

11.6.3 Case Study 3: The Johan Sverdrup Field (Norway)

The **Johan Sverdrup Field**, located in the North Sea, is one of the **largest oil discoveries in Norway in decades**. It is a **highly efficient** and **low-carbon** oil field that is setting new benchmarks for **sustainability in offshore production.**[5]

Financing Structure Used:

- **Equity contributions:** The project was financed through a consortium of:
  - **Equinor (operator, 42.6%)**
  - **Lundin Energy (20%)**
  - **Petoro (17.4%)**
  - **Aker BP (11.6%)**
  - **TotalEnergies (8.4%)**
- **Debt Financing:**
  - Equinor used its **strong cash reserves** and **credit lines** to finance its share of the project.
  - Other partners secured **loans** from **Norwegian banks** and **international lenders**.

Challenges Faced:

- **Complex reservoir geology** → Advanced geological modeling and horizontal drilling techniques are required to maximize oil recovery.
- **Environmental regulations** → Norway has **stringent environmental laws**, requiring Equinor to integrate **low-emission power solutions**.
- **High development costs** → The project required **state-of-the-art offshore platforms, subsea systems, and long-distance pipelines**.

Outcomes:

- **First oil production in 2019**, with an expected field life of **over 50 years**.
- **One of the most carbon-efficient oil fields globally**, using **hydropower from Norway's grid**, reducing **CO_2 emissions by 80%** compared to traditional offshore platforms.
- Significant contributor to **Norway's oil exports** and **government revenues**.

Norway's Green Oil? Johan Sverdrup Breaks Even at $20/bbl

The North Sea's mega-project proves that low-carbon and low-cost can coexist. Electrified platforms, advanced forecasting, and fiscal discipline redefine the economics of offshore oil.

Lessons Learned:

- **Leveraging advanced digitalization** (e.g., AI and automation) can **reduce operational costs and improve efficiency**.
- **Sustainability-focused financing** (low-carbon initiatives) can enhance **regulatory approval and investor confidence**.
- **Collaboration with local government and communities** ensures **long-term project success**.

11.6.4 Summary of Case Studies

These case studies illustrate **key financing strategies, risk management approaches, and success factors** in upstream projects (Figure 11.7).

Project	Financing Approach	Challenges	Key Success Factors
Kashagan (Kazakhstan)	Equity, Debt, Export Credit	Harsh climate, high-pressure reservoir, environmental hurdles	Strong consortium partnerships, risk-sharing, adaptation to extreme conditions
Tupi (Brazil)	Equity, Debt	Deepwater drilling, high CO_2 content, regulatory uncertainty	Advanced technology, financing diversification, strategic partnerships
Johan Sverdrup (Norway)	Equity, Debt	Complex geology, strict environmental laws, high costs	Digitalization, sustainability focus, government collaboration

FIGURE 11.7 Summary of upstream project finance case studies: Comparison of financing approaches, challenges, and success factors across Kashagan, Tupi, and Johan Sverdrup.

11.6.5 Key Takeaways from Case Studies

✓ **Consortium financing structures** help **mitigate risks and share costs** in high-risk projects.

✓ **Debt and equity financing combinations** provide **financial flexibility** and ensure **long-term project viability**.

✓ **Technological innovation** plays a crucial role in overcoming **operational and environmental challenges**.

✓ **Government and regulatory engagement** is critical for securing **licenses, approvals, and long-term stability**.

✓ **Sustainable finance and low-carbon initiatives** are increasingly important for securing **investor support and regulatory compliance**.

11.7 CHAPTER SUMMARY

This chapter provides an overview of financing and investment analysis in the upstream petroleum sector. We discuss capital budgeting techniques, funding options, financial risk analysis, and investment decision-making, and emphasize the importance of sound financial planning and risk management for the success of upstream projects.

Key Takeaways:

- **Capital budgeting:** Techniques such as NPV, IRR, payback period, and profitability index are essential for evaluating the economic viability of upstream projects.

- **Funding options:** Upstream companies can access various funding sources, including equity financing, debt financing, hybrid financing, and joint ventures. Each option has its advantages and disadvantages.
- **Risk analysis:** Financial risk analysis techniques such as sensitivity analysis, scenario planning, and Monte Carlo simulation are crucial for evaluating project risks and making informed investment decisions.
- **Investment decision-making:** Investment decisions are influenced by strategic alignment, portfolio management, economic conditions, and risk tolerance. A structured investment decision process ensures that projects align with the company's objectives and deliver value.

By examining real-world case studies, we demonstrated how effective financial planning, diverse financing structures, and robust risk management contribute to the successful development of upstream projects. The next chapter will discuss contracts and agreements, focusing on the legal and commercial arrangements that underpin upstream operations.

NOTES

1. IEA. World Energy Investment Report, 2023.
2. BP Statistical Review of World Energy, 2023.
3. Moody's Oil and Gas Project Finance Report, 2023.
4. Petrobras Annual Report, 2022.
5. Equinor Investor Presentation, 2022.

12 Contracts and Agreements

12.1 INTRODUCTION TO UPSTREAM CONTRACTS

Contracts are the backbone of the upstream petroleum industry, defining the rights and obligations of the various parties involved in exploration, development, and production activities. They are essential for managing risks, allocating responsibilities, and ensuring the smooth execution of complex projects. Upstream contracts encompass a range of activities, including exploration agreements with host governments and service contracts with specialized companies.

> In the oil industry, contracts are the blueprints of trust. Without them, there is only chaos.
>
> *Anonymous Industry Executive*

Key Types of Upstream Contracts Include:

- **Exploration agreements:** These agreements govern the exploration phase, granting oil and gas companies the right to explore for hydrocarbons in a specific area. They outline the terms and conditions for exploration activities, including the duration of the exploration period, work commitments, and financial obligations.
- **Development contracts:** These contracts cover the development phase, outlining the terms for building infrastructure and bringing a field into production. They include provisions for constructing facilities, drilling production wells, and installing equipment.
- **Production sharing agreements (PSAs):** These agreements define how the produced hydrocarbons are shared between the host government and the oil and gas company. PSAs specify the cost recovery mechanism, profit-sharing arrangement, and fiscal terms.
- **Service contracts:** These contracts cover specific services, such as seismic surveys, drilling, and reservoir engineering. They detail the scope of work, payment terms, and performance standards for the services to be provided.

Key Contractual Provisions Typically Include:

- **Scope of work:** A detailed description of the services or activities to be performed. This provision ensures that all parties understand their responsibilities and the expected deliverables clearly.
- **Payment terms:** How and when payments will be made. Payment terms outline the schedule and conditions for payment, including milestones, progress payments, and penalties for late or overdue payments.

DOI: 10.1201/9781003750345-15

- **Risk allocation:** How risks will be shared between the parties. This provision outlines the allocation of risks associated with technical, economic, political, and environmental factors and specifies measures for mitigating these risks.
- **Liability:** Defining the responsibilities for damages or losses. Liability clauses set out the extent to which each party is responsible for any damages, losses, or injuries arising from the contract.
- **Dispute resolution:** Mechanisms for resolving disagreements. Dispute resolution clauses provide a framework for addressing conflicts, including negotiation, mediation, arbitration, and litigation (Figure 12.1).

Example of key contract provisions: In a production sharing agreement, the scope of work may include the exploration, development, and production of hydrocarbons in a designated area. Payment terms may involve cost recovery and profit sharing, where the oil and gas company recovers its costs from a portion of the produced hydrocarbons, and the remaining production is shared with the host government. Risk allocation provisions may specify that the company bears all exploration risks, while liability clauses outline the responsibilities for any environmental damages. Dispute resolution mechanisms may include arbitration under international rules to resolve any conflicts.

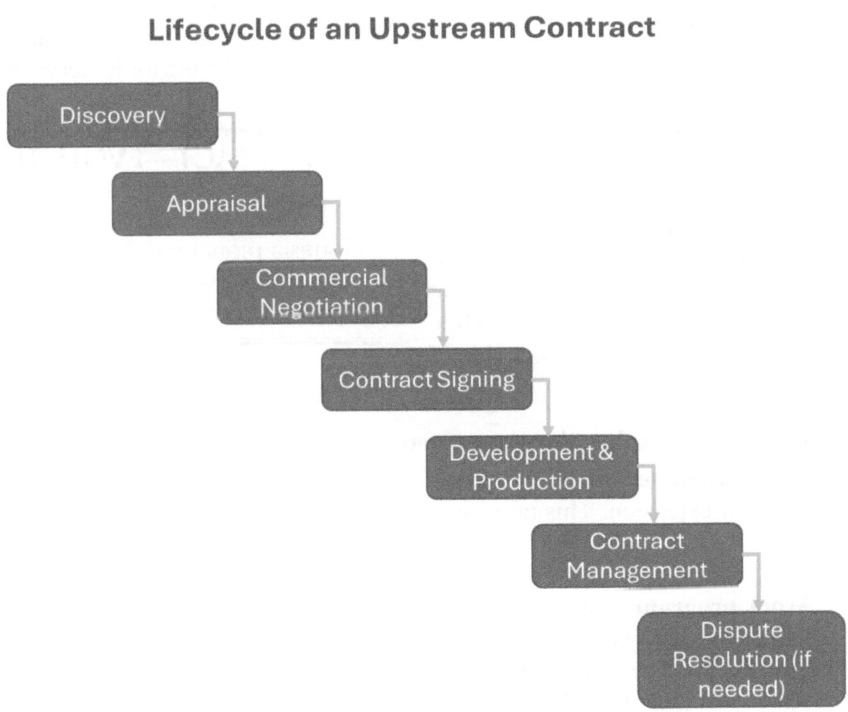

FIGURE 12.1 Lifecycle of an upstream contract: Sequential stages from discovery to dispute resolution, illustrating the contractual journey in upstream petroleum projects.

12.2 EXPLORATION AGREEMENTS

Exploration agreements grant oil and gas companies the right to explore for hydrocarbons in a designated area. Common types of exploration agreements include:

- **Concession agreements:** The host government grants exclusive exploration rights to an oil and gas company, typically for a specified period. The company bears all exploration risk and is responsible for all costs associated with exploration activities. The company may develop and produce the hydrocarbons if a commercial discovery is made.
  - **Example:** In many Middle Eastern countries, concession agreements grant international oil companies (IOCs) exploration rights, allowing them to explore and develop large oil fields.
- **Production sharing contracts (PSCs):** The oil and gas company explores and develops the field, and the production is shared between the company and the host government according to a pre-agreed formula. The company typically bears the exploration risk and is entitled to recover its costs from some of the hydrocarbons that are produced. The remaining production is shared based on the terms of the contract.
  - **Example:** In Indonesia, PSCs[1] are commonly used to regulate exploration and production activities, with the oil and gas company sharing production with the government after cost recovery.
- **Risk service contracts:** The oil and gas company conducts exploration activities on behalf of the host government and is paid a fee for its services, regardless of whether a discovery is made. The government bears the exploration risk and retains ownership of the resources.
  - **Example:** In Mexico, risk service contracts have been used to attract foreign investment in exploration activities while maintaining state ownership of hydrocarbon resources.

> **FAST FACT—EVOLUTION OF PSCS**
>
> Indonesia pioneered the PSC model in the 1960s, becoming the global standard for upstream agreements.

Key Provisions in Exploration Agreements:

- **Exploration period:** The duration of the exploration phase and any conditions for extension. This provision ensures that the company has sufficient time to conduct exploration activities and assess the potential of the designated area.
- **Work program:** The minimum exploration work commitments, including geological and geophysical studies, seismic surveys, and drilling of exploration wells. The work program outlines the specific activities that the company must undertake during the exploration period.
- **Fiscal terms:** The company's financial obligations, including signature bonuses, royalties, taxes, and cost recovery mechanisms. Fiscal terms

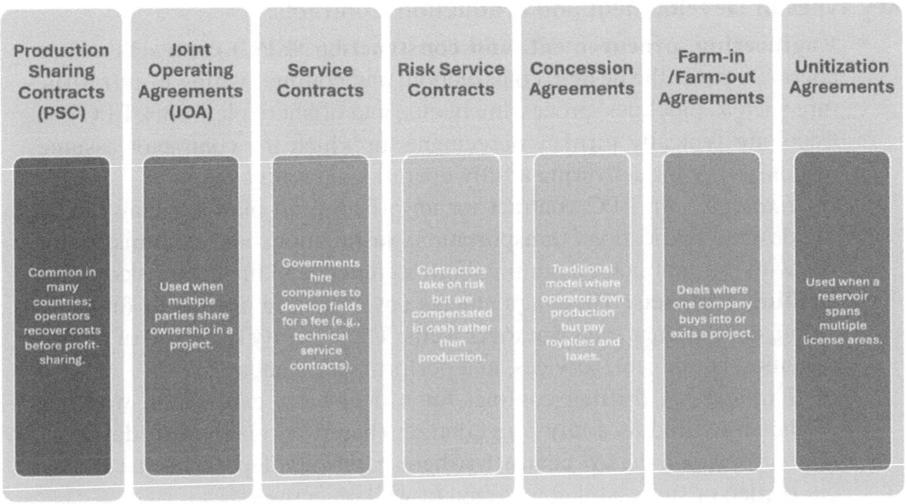

FIGURE 12.2 Comparison of upstream contract types: Overview of key upstream contract structures, highlighting differences in ownership, risk allocation, compensation, and operational control.

determine the economic benefits to the host government and the financial incentives for the company.

- **Relinquishment:** If no commercial discovery is made within a specified period, the company must relinquish a portion of the exploration area. Relinquishment provisions ensure that unproductive areas are returned to the government for potential reallocation.

Example of exploration agreement: A PSC in an African country grants an international oil company the right to explore for hydrocarbons in an offshore block for a period of five years. The company commits to conducting seismic surveys and drilling three exploration wells. If a commercial discovery is made, the company can develop the field and recover its costs from a portion of the production. The remaining production is shared with the host government based on a sliding scale formula, with the government receiving a higher share as production increases. The agreement includes provisions for extending the exploration period, paying royalties and taxes, and relinquishing unproductive areas (Figure 12.2).

12.3 DEVELOPMENT AND PRODUCTION CONTRACTS

Development and production contracts govern the construction of infrastructure and the operation of production facilities. These agreements define the responsibilities of contractors, operators, and service providers, ensuring that projects are completed on time, within budget, and according to technical and environmental standards.

Key Types of Development and Production Contracts:

- **Engineering, procurement, and construction (EPC) contracts**[2]: These contracts cover the design, construction, and commissioning of infrastructure such as pipelines, processing plants, and offshore platforms. EPC contracts are typically turnkey agreements in which the contractor assumes responsibility for delivering a fully operational facility.
  - *Example:* An EPC contract for an offshore oil platform may include design, fabrication, transportation, installation, and commissioning, with the contractor assuming both technical and financial risks.
- **Drilling contracts:** These agreements cover drilling operations for exploration, development, and workover wells. They specify the scope of drilling activities, equipment, services, and performance standards.
  - *Example:* A drilling contract for a deepwater exploration well may be structured as a *day-rate* contract (based on operational days) or a *performance-based* contract (where payments depend on the footage drilled or successful completion of well objectives).
- **Operations and maintenance (O&M) contracts** govern the long-term operation and upkeep of production facilities. They ensure facilities run efficiently, with regular maintenance to prevent equipment failures and downtime.
  - *Example:* An O&M contract for a gas processing plant may require the contractor to provide ongoing inspections, preventive maintenance, and spare parts management to ensure uninterrupted production.

Key Provisions in Development and Production Contracts:

- **Scope of work:** A detailed description of activities, deliverables, and technical requirements.
- **Payment terms:** Defines the schedule and payment conditions, including milestone-based, fixed-cost, or performance-based compensation.
- **Performance standards:** Specifies quality benchmarks, operational efficiency targets, and regulatory compliance requirements.
- **Change orders:** Procedures for managing modifications to the scope of work, including cost adjustments and schedule impacts.
- **Warranty and liability:** Outlines contractor warranties for work quality and liability for defects or failures.
- **Risk allocation:** Specifies which party bears financial, operational, and technical risks, including force majeure provisions and insurance requirements.

Contract Variations and Risk Allocation:

- **EPIC (Engineering, Procurement, Installation, and Commissioning) contracts:** These are a variation of EPC contracts commonly used for offshore projects, in which the contractor is also responsible for installation.
- **EPCI (Engineering, Procurement, Construction, and Installation) contracts:** These are similar to EPIC but with a stronger focus on offshore structure installation.

EPC vs. EPIC vs. EPCI Comparison Table

Criteria	EPC	EPIC	EPCI
Full Form	Engineering, Procurement, Construction	Engineering, Procurement, Installation, Construction	Engineering, Procurement, Construction, Installation
Focus	Surface facilities, infrastructure	Offshore platforms, subsea systems	Integrated surface and subsea execution
Contractor Responsibility	Delivery of constructed asset	Installation included	Installation and construction bundled
Common Usage	Onshore projects	Offshore fields	Offshore mega-projects
Example Project	Desert pipeline network	Shallow offshore platform	Deepwater subsea system

FIGURE 12.3 Comparison of EPC, EPIC, and EPCI contract models: Differentiating upstream project delivery structures by scope, contractor responsibility, and typical usage across onshore and offshore developments.

- **Lump-sum vs. cost-plus contracts:**
  - *Lump-sum (fixed-price)* contracts transfer cost risk to the contractor, providing budget certainty.
  - *Cost-plus* contracts allow for reimbursement of actual costs plus a fee, benefiting complex projects with uncertain costs.

Sustainability Trends in Development and Production Contracts:

- **Environmental commitments:** Many modern contracts include sustainability clauses, such as limits on greenhouse gas emissions, flaring restrictions, and decommissioning obligations.
- **Carbon footprint reduction:** Newer EPC contracts may require using energy-efficient technologies or renewable power sources to reduce environmental impact.
- **Decommissioning obligations:** Operators must plan for asset retirement, ensuring that infrastructure is safely dismantled and environmental restoration is conducted (Figure 12.3).

12.3.1 Example of a Development and Production Contract

An EPC contract for a **gas processing plant in the Middle East** includes provisions for design, procurement of equipment, construction, and commissioning. The contractor must deliver a fully operational facility that meets strict **emission reduction targets** and complies with local **environmental regulations**. Payments are milestone-based, with a **performance bonus** for early completion. The contract includes **change order provisions** to accommodate modifications and a **warranty clause** to ensure operational reliability for the first three years.

12.4 SERVICE CONTRACTS

Service contracts govern specialized technical and operational services required throughout the upstream operations' exploration, development, and production phases. These contracts define **service scope,**

DID YOU KNOW?—THE ROLE OF SERVICE CONTRACTS

In mature fields, over 70% of operational activities are outsourced through specialized service contracts.

quality standards, payment structures, and risk-sharing mechanisms between oil companies and service providers.

Key Types of Service Contracts:

- **Seismic survey contracts:** Cover the acquisition, processing, and interpretation of seismic data to image subsurface geology and identify hydrocarbon reservoirs.
  - *Example:* A seismic survey contract for an offshore block in Southeast Asia includes **3D seismic data acquisition**, processing, and interpretation. The contractor mobilizes survey vessels, deploys seismic equipment, and delivers high-resolution imaging reports to support drilling decisions.
- **Well logging and testing contracts:** Govern data collection from drilled wells, helping determine **reservoir properties, fluid composition, and formation pressure.**
  - *Example:* A well logging contract for a deepwater exploration well requires **real-time data acquisition**, including resistivity, porosity, and permeability measurements. The contract ensures **data accuracy standards** and specifies performance-based payments for successful completion.
- **Reservoir engineering and consulting contracts:** Provide expertise in **reservoir modeling, production optimization, and enhanced oil recovery (EOR) strategies.**
  - *Example:* A reservoir engineering contract for a mature oil field in the Middle East involves conducting **reservoir simulations**, evaluating **EOR techniques**, and advising on **long-term production strategies.**
- **Drilling and well services contracts:** Cover directional drilling, cementing, and hydraulic fracturing activities.
  - *Example:* A well stimulation contract for a shale gas project includes **fracturing fluid preparation, pump operations, and post-treatment analysis.** Performance metrics ensure well productivity improvements.
- **Maintenance and inspection contracts:** Ensure the reliability of **offshore platforms, pipelines, and processing facilities.**
  - *Example:* A predictive maintenance contract for an offshore FPSO (Floating Production Storage and Offloading) vessel requires **periodic inspections using drones, AI-powered analytics, and corrosion monitoring technologies.**

Key Provisions in Service Contracts:

- **Scope of work**: Defines **specific tasks, deliverables, and timeframes**.
- **Payment terms:** Outlines **fixed-price, milestone-based, or performance-based** compensation models.
- **Performance standards:** Establishes **minimum service levels, response times, and data accuracy requirements.**
- **Confidentiality:** Protects **proprietary geological, operational, and financial data.**

- **Termination and liabilities:** Defines **early termination penalties, compensation for completed work, and dispute resolution mechanisms.**

Risk Allocation in Service Contracts:

- **Fixed-price vs. cost-plus contracts:**
  - *Fixed-price contracts* transfer cost risk to the service provider, ensuring budget predictability.
  - *Cost-plus contracts* reimburse actual expenses, suitable for complex, high-risk services.
- **Liability and indemnity clauses:** Protect both parties against **operational failures, equipment damage, or environmental liabilities.**
- **Force majeure:** Outlines protections against **unforeseen events such as hurricanes, geopolitical conflicts, or supply chain disruptions.**

Example of a Service Contract:

A **seismic survey contract** for an **offshore deepwater block** specifies **multi-client data acquisition**, processing deliverables, and interpretation requirements. Payment is milestone-based, linked to **survey completion, data processing, and final report submission.** The contract includes strict **confidentiality terms** to protect proprietary seismic data and **performance guarantees** to ensure **high-resolution imaging standards.** A **liability clause** holds the contractor accountable for equipment failures, and a **force majeure provision** protects both parties in case of extreme weather conditions.

12.5 LEGAL AND REGULATORY CONSIDERATIONS

Upstream contracts are subject to **domestic laws, international regulations, and industry-specific legal frameworks** that influence contract enforcement, dispute resolution,[3] and compliance obligations. Understanding the legal and regulatory landscape is essential for mitigating legal risks, ensuring contractual stability, and maintaining regulatory compliance.[4]

> Contracts don't just protect profits — they protect reputations.
>
> *Leading Arbitration Lawyer*

12.5.1 Key Legal and Regulatory Considerations

1. **Jurisdiction and Governing Law**
   - Defines **which country's legal system applies** to the contract and how disputes will be resolved.
   - Jurisdiction clauses determine **where legal proceedings can occur, ensuring dispute clarity**.
   - Contracts involving international parties may **apply international law, host country law, or a hybrid approach** to address cross-border legal complexities.

Example:

- An **exploration agreement** between a European oil company and an African host government may specify that **local laws govern the contract**, with disputes handled in **domestic courts or international arbitration.**
- Given the UK's well-established oil and gas legal framework, a joint venture agreement in offshore drilling may be governed by UK law.

QUICK TIP—ARBITRATION VS. LITIGATION

Arbitration offers confidentiality and speed compared to lengthy public court cases, making it the preferred choice in upstream disputes.

2. **Dispute Resolution Mechanisms**
   - Upstream contracts commonly include dispute resolution provisions, with options such as:
     - **Arbitration**: Preferred in **international contracts** as it provides a **neutral, confidential, and enforceable** resolution.
     - **Mediation**: A voluntary, **non-binding** dispute resolution method that encourages negotiations before litigation.
     - **Litigation**: Court proceedings are **time-consuming and costly,** but may be required if disputes are unresolved.

Example:

- A **production sharing agreement (PSA)** may require that **disputes be resolved through arbitration under the International Chamber of Commerce (ICC) rules** in **a neutral venue such as London or Singapore.**
- A **drilling contract** may include a **mediation clause** requiring parties to **attempt settlement before arbitration.**

3. **Regulatory Compliance in Upstream Contracts**

Environmental Regulations:

- Upstream contracts must comply with **environmental laws** related to:
  - **Environmental impact assessments (EIA)**: Mandatory before project approval.
  - **Emissions controls**: Compliance with **carbon regulations and pollution control laws.**
  - **Waste management**: Proper disposal and treatment of hazardous materials.

Example:

- An **offshore exploration contract** in Norway may require compliance with **Norwegian Petroleum Directorate (NPD) environmental standards** and **OSPAR marine pollution regulations.**
- An **EPC contract** for an LNG terminal may include **stringent methane emission reduction clauses** to comply with EU sustainability policies.

Health and Safety Regulations:

- Contracts must ensure compliance with **occupational health and safety (HSE) standards** to **protect workers and communities**.
- Companies must establish:
  - **HSE management systems**: Safety protocols, audits, and risk assessments.
  - **Emergency response plans**: Procedures for handling oil spills, gas leaks, and industrial accidents.

Example:

- A **drilling contract** in the U.S. requires adherence to **OSHA (Occupational Safety and Health Administration) regulations** and **API (American Petroleum Institute) safety standards**.

Anti-Corruption and Compliance Laws

- Contracts must ensure **compliance with global anti-bribery and corruption laws**, such as:
  - **U.S. Foreign Corrupt Practices Act (FCPA)**: Prohibits bribery of foreign officials.
  - **UK Bribery Act**: Covers bribery in both public and private sectors.
  - **Extractive Industries Transparency Initiative (EITI)**: Promotes financial transparency in upstream contracts.

Example:

- A **service contract** may require the contractor to **certify compliance with FCPA** and **conduct due diligence on subcontractors** to prevent illegal payments.

4. **Contractual Provisions for Legal Compliance**
   - **Liability and indemnity:** Defines the extent of financial responsibility in case of contractual breaches or regulatory violations.
   - **Force majeure:** Covers unforeseen events (e.g., natural disasters, war, political instability) that **prevent contract performance.**
   - **Termination clauses:** Outlines conditions under which a contract can be terminated **due to non-compliance with legal or regulatory obligations.**

Example:

- An **EPC contract** for an oil refinery in the Middle East includes **force majeure protection** for project delays caused by **geopolitical unrest or regulatory changes.**
- A **joint venture agreement** between an IOC and an NOC includes an **indemnity clause**, ensuring that environmental liabilities are shared **based on operational control.**

12.5.2 Example of Legal and Regulatory Compliance in Upstream Contracts

A **multi-billion-dollar LNG development project in East Africa** requires a combination of **legal, regulatory, and compliance measures** in its contracts:

- ✓ **Governing law:** The contract is governed by **English law**, providing neutrality for international investors.
- ✓ **Dispute resolution:** Arbitration is conducted **under ICC rules in Singapore.**
- ✓ **Environmental Compliance:** The contract mandates adherence to **International Maritime Organization (IMO) regulations** on marine pollution.
- ✓ **Anti-corruption compliance:** Contractors must **certify compliance with FCPA and UK Bribery Act** and submit to **regular audits.**
- ✓ **HSE compliance:** Drilling operations must follow **ISO 45001 safety management standards.**

12.6 NEGOTIATION AND CONTRACT MANAGEMENT

Effective negotiation and contract management are essential in the upstream petroleum industry to ensure agreements are **mutually beneficial, legally sound, and properly executed.** Successful contract management enhances project efficiency, minimizes risks, and prevents costly disputes.

12.6.1 Key Aspects of Negotiation and Contract Management

1. **Negotiation Strategies**
   Successful negotiations require a **deep understanding of all parties' interests, risks, and objectives**. Key negotiation strategies include:
   - **Win-win approach:** Striving for solutions that benefit both parties rather than adopting an adversarial stance.
   - **Risk-reward balancing:** Ensuring both parties assume appropriate levels of risk and reward.
   - **Flexibility:** Recognizing when compromises can strengthen long-term partnerships.

   Example:

   - In negotiating a **Production Sharing Agreement (PSA),** an oil company may seek **higher cost recovery** to mitigate financial risks, while the host government aims to **maximize its revenue share.** Negotiators must balance these objectives to achieve a mutually acceptable agreement.
2. **Contract Drafting and Review**
   - Contracts must be **precise, comprehensive, and legally enforceable** to avoid misunderstandings or disputes.

- Legal teams, financial analysts, and technical experts should **collaborate to draft explicit provisions** addressing all critical aspects, including:
  - **Scope of work**: Defining what is to be delivered.
  - **Financial terms**: Payment structure, taxes, and cost recovery mechanisms.
  - **Liability and risk allocation**: Establishing responsibility for potential issues.

Example:

- A **drilling contract** should include explicit provisions on **equipment requirements, safety protocols, and operational milestones.** Before finalizing the agreement, legal and technical teams must review the document to **ensure compliance with industry regulations and company objectives.**

3. **Contract Administration**
   - Once a contract is signed, **effective administration ensures adherence to agreed terms** throughout its lifecycle.
   - Key administrative tasks include:
     - **Tracking project milestones**: Ensuring timely completion of work.
     - **Managing change orders**: Addressing scope, budget, or schedule modifications.
     - **Ensuring regulatory compliance**: Adhering to environmental, safety, and financial reporting requirements.

Example:

- An **Engineering, Procurement, and Construction (EPC) contract** for an offshore platform must be actively managed to **track project progress, approve design modifications, and monitor costs**. Contract administrators work closely with contractors to **ensure timely and quality project delivery**.

4. **Performance Monitoring and Compliance**
   - Ongoing performance evaluation **identifies inefficiencies, ensures quality, and mitigates risks**.
   - Performance is monitored through:
     - **Key performance indicators (KPIs):** Measuring project success in productivity, cost control, and quality.
     - **Audits and inspections:** Ensuring contractual and regulatory compliance.
     - **Reporting and communication:** Maintaining transparency between parties.

Example:

- In an **Operations and maintenance (O&M) contract** for a production facility, the contractor must **meet safety standards, conduct routine maintenance, and achieve production targets.** Regular performance audits help **identify potential bottlenecks and ensure compliance with safety protocols.**

12.6.2 Example of Negotiation and Contract Management

A **Joint Venture Agreement (JVA) for offshore exploration** between an International Oil Company (IOC) and a National Oil Company (NOC) involves:

- ✓ **Negotiation:** Determining ownership structure, capital contributions, and decision-making authority.
- ✓ **Contract drafting:** Outlining legal responsibilities, financial commitments, and operational frameworks.
- ✓ **Contract administration:** Managing **drilling operations, regulatory approvals, and financial reporting.**
- ✓ **Performance monitoring:** Conducting periodic **reviews to assess compliance with safety, environmental, and production targets.**

12.7 CASE STUDIES OF UPSTREAM CONTRACT DISPUTES

Upstream petroleum contracts are often subject to disputes due to **financial disagreements, cost overruns, regulatory changes, or geopolitical factors.** These case studies highlight real-world disputes, their legal complexities, and the lessons learned for future contract management.

12.7.1 Case Study 1: The Karish and Tanin Gas Fields (Israel)

Background:
The Karish and Tanin gas fields, located offshore Israel, became the focus of a dispute between the **Israeli government and the Greek energy company Energean.** The conflict revolved around gas export agreements and pricing adjustments.

Nature of the Dispute:

- The Israeli government **sought to renegotiate the gas export contracts** to align with market fluctuations and regulatory changes.
- Energean argued that the original contracts were legally binding and should be upheld.

Legal Arguments Involved:

- **Energean's position:** The agreed contract terms should be honored per **contractual obligations and investment stability.**
- **Israeli government's position:** Contracts should be **adjusted to reflect changing market realities and updated regulatory requirements.**

Outcome:

- The dispute was resolved through **negotiation and mediation,** leading to a revised contract that reflected market conditions and regulatory concerns.

Lessons Learned:

- ✓ **Flexibility in contract terms**: Contracts should allow adjustments based on **economic and regulatory changes.**
- ✓ **Negotiation and mediation**: Alternative dispute resolution methods can lead to **mutually beneficial solutions** without escalating to litigation.

12.7.2 CASE STUDY 2: THE KASHAGAN FIELD (KAZAKHSTAN)

Background:
The **Kashagan Field** in Kazakhstan is one of the largest offshore oil discoveries in the Caspian Sea. Its development faced multiple contract disputes between the **Kazakh government and a consortium of IOCs.**

Nature of the Dispute:

- The dispute centered on **cost overruns, project delays, and revisions to the fiscal terms** in the **Production Sharing Agreement (PSA).**
- The **Kazakh government sought to modify fiscal terms** to gain a greater share of profits, citing increased project risks and costs.
- The **IOCs argued that cost overruns** were due to **unforeseen technical challenges** and should be covered under the PSA.

> **CASE INSIGHT—WHY KASHAGAN MATTERS**
>
> The Kashagan field dispute showed the importance of flexible PSAs when projects face massive cost overruns.

Legal Arguments Involved:

- **IOCs' position:** The original **contractual terms should be upheld**, and additional costs should be accounted for within the agreement.
- **Kazakh government's position:** The **contract must be renegotiated** to **reflect the increased project costs** and **protect national economic interests.**

Outcome:

- The dispute was resolved through **renegotiation of the PSA**, where the consortium **agreed to absorb some additional costs** while Kazakhstan provided **certain fiscal concessions** to advance project completion.

Lessons Learned:

✓ **Risk management in upstream contracts**: Large-scale projects require **contingency planning** for **unforeseen costs and delays.**
✓ **Clear communication and flexibility**: Open discussions **between host governments and investors** can **prevent prolonged disputes.**

12.7.3 Case Study 3: The DRC Contract Dispute (Democratic Republic of Congo)

Background:
A major contract dispute arose between the **Democratic Republic of Congo (DRC) and an international oil company** over **fiscal terms and revenue sharing** in a **Production Sharing Agreement (PSA).**

Nature of the Dispute:

- The **DRC government sought to revise contract terms** to increase its share of revenue from oil production.
- The **oil company opposed any changes**, arguing that the original contract **was legally binding and necessary for investment stability.**

Legal Arguments Involved:

- **DRC government's position:** Contracts should be revised to reflect **new market conditions and national economic interests.**
- **Oil company's position:** A legally signed PSA should be **respected as per contract law,** and any changes **could undermine investor confidence.**

Outcome:

- The dispute was settled through **international arbitration.** The arbitrators ruled **in favor of the oil company,** upholding the original PSA terms.
- The **DRC government was required to honor the contract,** reaffirming **the legal enforceability of international agreements.**

Lessons Learned:

✓ **Legal certainty is crucial**: Investors require **stable legal frameworks** for long-term commitments.

✓ **Role of arbitration in upstream disputes**[5]: Arbitration can serve as an **effective mechanism for dispute resolution, ensuring neutrality.**

12.7.4 Summary of Case Studies

These case studies illustrate how **contract disputes in upstream petroleum projects** arise from financial, regulatory, and geopolitical factors.

Key Takeaways:

✓ **Contracts should account for market changes:** Upstream contracts should include **adaptive mechanisms** to accommodate **price fluctuations and regulatory updates.**

✓ **Risk sharing is essential**: Governments and oil companies **must agree on cost-sharing frameworks** to handle **unexpected cost overruns.**

✓ **Legal stability is vital**: **International arbitration ensures** contracts **remain enforceable, boosting investor confidence.**

By integrating **clear risk management strategies, strong legal frameworks, and effective negotiation practices,** companies and governments can **avoid prolonged disputes** and **facilitate project success.**

12.8 CHAPTER SUMMARY

This chapter provides an overview of contracts and agreements in the upstream petroleum industry. We discuss the importance of contracts, the diverse types of contracts used, key contractual provisions, legal and regulatory considerations, negotiation strategies, and contract management. We emphasize the importance of well-drafted contracts and effective contract management for the success of upstream projects.

Key Takeaways:

- **Types of contracts:** Upstream contracts include exploration agreements, development contracts, production sharing agreements, and service contracts. Each type of contract addresses specific aspects of exploration, development, and production activities.
- **Contractual provisions:** Key provisions in upstream contracts include the scope of work, payment terms, risk allocation, liability, and dispute resolution. These provisions ensure that the parties' rights and obligations are clearly defined and risks effectively managed.
- **Legal and regulatory considerations:** Upstream contracts are governed by contract law principles and must comply with relevant laws and regulations. Dispute resolution mechanisms like arbitration, mediation, and litigation provide conflict resolution frameworks.
- **Negotiation and contract management:** Effective negotiation, contract drafting, and contract administration are essential for reaching mutually beneficial agreements and ensuring successful project execution. Performance monitoring and initiative-taking risk management are key to maintaining contract compliance and addressing issues promptly.

By examining real-world case studies, we demonstrated how clear contract terms, effective negotiation, and appropriate dispute resolution mechanisms contribute to successfully managing upstream projects. The next chapter will discuss the future of upstream petroleum economics, focusing on emerging trends, challenges, and opportunities in the industry.

NOTES

1. Production Sharing Contracts (PSCs) were first introduced by Indonesia in the 1960s under the leadership of Pertamina, aiming to retain state control over resources while incentivizing international oil companies. Source: Johnston, D. "International Exploration Economics, Risk, and Contract Analysis," PennWell Books, 1994.
2. According to Wood Mackenzie's 2023 Global Upstream Survey, Engineering, Procurement, and Construction (EPC) contracts now constitute over 60% of major offshore development costs, reflecting a shift toward contractor-driven project execution models.
3. The 2022 Global Arbitration Review reports that nearly 90% of international oil and gas disputes are resolved through arbitration rather than traditional litigation, emphasizing its importance for confidentiality and enforceability.
4. The OSPAR Convention (1992) established stringent environmental protection standards for offshore activities in the Northeast Atlantic, influencing modern regulatory frameworks for marine and oilfield development compliance.
5. The International Chamber of Commerce (ICC) Rules of Arbitration are widely adopted for resolving complex upstream oil and gas disputes, offering neutrality, enforceability, and flexibility to parties in different jurisdictions.

13 Sustainability in Upstream

ESG, Carbon Neutrality, and the Future of Hydrocarbons

13.1 INTRODUCTION: WHY SUSTAINABILITY MATTERS IN UPSTREAM

The upstream petroleum sector has traditionally focused on maximizing hydrocarbon production while balancing costs and operational efficiency. However, growing environmental, social, and governance (ESG) pressures, regulatory shifts, and market demands drive a fundamental transformation. Sustainability is no longer an optional corporate initiative; it has become a critical business imperative.

The industry faces increasing scrutiny regarding its environmental footprint, particularly in greenhouse gas (GHG) emissions, water usage, land impact, and community engagement. Investors, policymakers, and the public now expect oil and gas companies to demonstrate a clear commitment to reducing carbon emissions and integrating sustainable practices into their operations.

This chapter explores key sustainability challenges, strategies for reducing the environmental impact of upstream operations, the role of ESG in investment decisions, and the emerging pathways toward a more sustainable hydrocarbon industry (Figure 13.1).

> Sustainability is no longer about doing less harm. It's about doing more good.
>
> **Jochen Zeitz**
> *CEO, Harley-Davidson*

13.2 KEY SUSTAINABILITY CHALLENGES IN UPSTREAM

Despite significant technological advances and operational efficiency improvements, upstream oil and gas production continues to face a range of complex sustainability challenges. These challenges encompass environmental concerns such as GHG emissions, water usage, and potential impacts on biodiversity. The extraction processes often involve substantial land disruption, leading to habitat loss and soil degradation. Additionally, the industry grapples with managing produced water and waste, necessitating effective disposal methods to prevent contamination of local water sources. Furthermore, the reliance on fossil fuels raises questions about long-term viability in a world shifting toward renewable energy sources, highlighting the

DOI: 10.1201/9781003750345-16

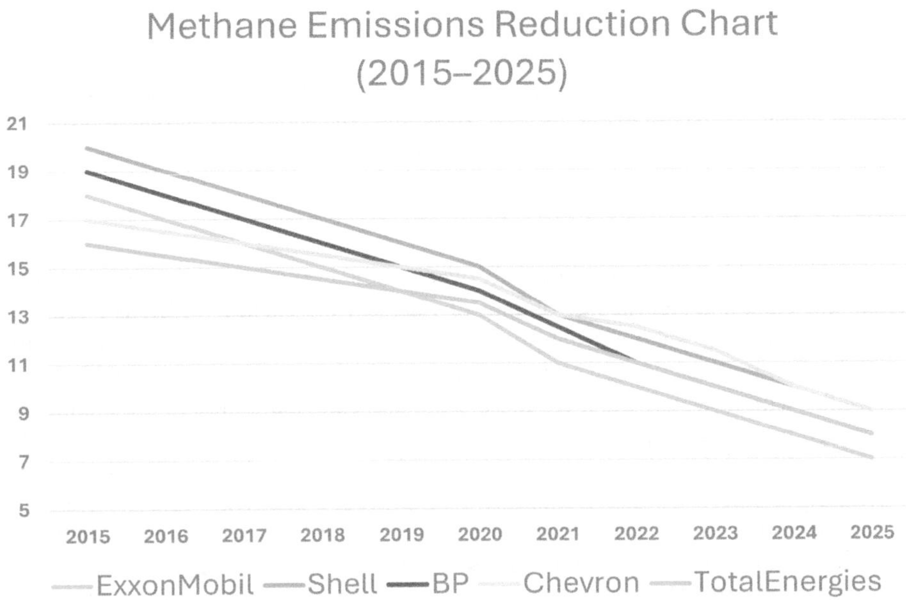

FIGURE 13.1 Methane emissions reduction by major energy companies (2015–2025). Trend analysis of methane emissions across five leading oil and gas firms—ExxonMobil, Shell, BP, Chevron, and TotalEnergies. Highlights industry-wide progress in emissions mitigation, with TotalEnergies showing the steepest decline. Underscores the growing role of environmental stewardship, regulatory compliance, and digital monitoring in upstream sustainability strategy.

need for innovative solutions and practices that promote economic growth and environmental stewardship.

13.2.1 Greenhouse Gas (GHG) Emissions

The extraction, processing, and transportation of hydrocarbons result in substantial emissions of carbon dioxide (CO_2) and methane (CH_4), two of the most significant GHGs contributing to climate change.[1] Specifically, the combustion of fossil fuels during extraction and processing releases vast amounts of CO_2 into the atmosphere, while methane, which is emitted during the extraction of natural gas and oil, has a global warming potential that is over 25 times greater than CO_2 over a 100-year period. Effective management and reduction of these emissions are critical for mitigating the oil and gas industry's impact on global warming. This involves implementing advanced technologies such as carbon capture

THE TRUE COST OF METHANE

Methane traps **84–87 times** more heat than CO_2 over 20 years.

Cutting leaks could avoid **up to 0.3°C** of global warming by 2050.

Quick Impact:

Detecting and fixing upstream methane leaks is the **fastest and cheapest climate win** for oil and gas.

and storage (CCS), improving energy efficiency in operations, and transitioning to renewable energy sources. Additionally, strict regulatory measures and industry best practices aimed at minimizing leaks and flaring can significantly enhance emission reduction efforts. By focusing on these strategies, the industry can play a proactive role in combating climate change and promoting a more sustainable future.

13.2.2 Water Usage and Management

Hydraulic fracturing, enhanced oil recovery (EOR), and various other extraction techniques demand significant volumes of water, often reaching millions of gallons per well. This high water consumption raises serious concerns about water scarcity, especially in regions already grappling with limited freshwater resources. In many cases, the withdrawal of water for these processes can lead to decreased availability for agricultural, municipal, and ecological needs, affecting not only human populations but also local ecosystems.

Moreover, these extraction methods can introduce additional risks of contamination to both surface and groundwater sources. The chemicals used in hydraulic fracturing fluids, as well as the produced water from EOR, can pose hazards if not managed correctly. This can result in harmful pollutants entering the water supply, impacting drinking water quality and aquatic habitats.

Implementing effective water management practices is crucial to addressing these issues. Strategies may include reusing and recycling water in extraction processes, developing more water-efficient technologies, and rigorously monitoring water quality. Communities, regulators, and industry stakeholders must work collaboratively to establish sustainable practices that mitigate the adverse effects of water use in extraction activities, ensuring that water resources remain available for future generations.

13.2.3 Land and Biodiversity Impact

Oil and gas exploration frequently leads to significant disruptions within ecosystems, adversely affecting biodiversity and often contributing to deforestation, particularly in sensitive regions such as rainforests and wetlands. Activities associated with exploration and extraction, including land clearing, road construction, and drilling, can fragment habitats, isolate wildlife populations, and diminish the availability of resources necessary for survival.

The resultant ecological imbalance can trigger a cascade of negative effects, including the loss of plant and animal species, which may take decades or even centuries to recover. Moreover, the chemicals used in extraction processes can contaminate soil and water sources, further threatening both terrestrial and aquatic life forms.

Therefore, protecting and restoring natural habitats is not only essential for wildlife conservation but also for safeguarding the overall health of our planet. Implementing sustainable practices, such as responsible land use planning and habitat restoration initiatives, is crucial for preserving biodiversity and mitigating the long-term environmental damage caused by oil and gas operations. Effective conservation strategies can enhance ecological resilience, promote the recovery of endangered species, and help maintain the essential services provided by healthy ecosystems.

13.2.4 Social and Community Impact

The operations of energy companies significantly influence local communities, especially in developing countries, where the stakes are often higher. Issues such as land rights, economic displacement, and the protection of indigenous peoples' rights have emerged as critical challenges that these companies must address. For instance, the acquisition of land for energy projects can lead to the displacement of entire communities, disrupting their livelihoods and cultural heritage. Therefore, proactive engagement with local populations is essential. This involves not only transparent communication but also meaningful consultation processes that respect the voices and concerns of the community members. By prioritizing the well-being of these communities and integrating their input into project planning and execution, energy companies can cultivate trust and ensure a social license to operate. This approach not only benefits the companies by minimizing conflicts and enhancing reputation but also contributes to the sustainable development of the regions involved.

13.2.5 Regulatory and Compliance Challenges

In recent years, governments across the globe have been implementing increasingly stringent environmental regulations in response to the urgent need for sustainable development and climate action. These regulations often include more rigorous emission limits for GHGs and other pollutants, which require industries to invest in cleaner technologies and modify their operational practices. Furthermore, there is a growing emphasis on transparency, leading to enhanced reporting requirements for sustainability metrics, such as carbon footprints, water usage, and waste generation.

Firms are now expected to provide detailed disclosures on their environmental impact, often following specific frameworks like the Global Reporting Initiative (GRI) or Sustainability Accounting Standards Board (SASB). Compliance with these evolving regulations is not just a legal obligation; it serves as a crucial factor in maintaining both legal and social legitimacy in the eyes of stakeholders, including investors, customers, and regulators. Failure to adhere to these standards can result in significant legal penalties, reputational damage, and loss of market access, underscoring the importance of staying ahead of regulatory trends in an increasingly eco-conscious marketplace.

13.3 STRATEGIES FOR SUSTAINABLE UPSTREAM OPERATIONS

The oil and gas industry is increasingly adopting various strategies to mitigate its environmental impact and align with global sustainability goals. These strategies encompass innovative technologies and practices designed to promote ecological balance and efficiency:

- **Carbon Capture, Utilization, and Storage (CCUS):** This technology captures carbon dioxide (CO_2) emissions produced during oil and gas extraction and processing. The captured CO_2 can be permanently stored underground in geological formations or utilized in EOR processes, where it is injected

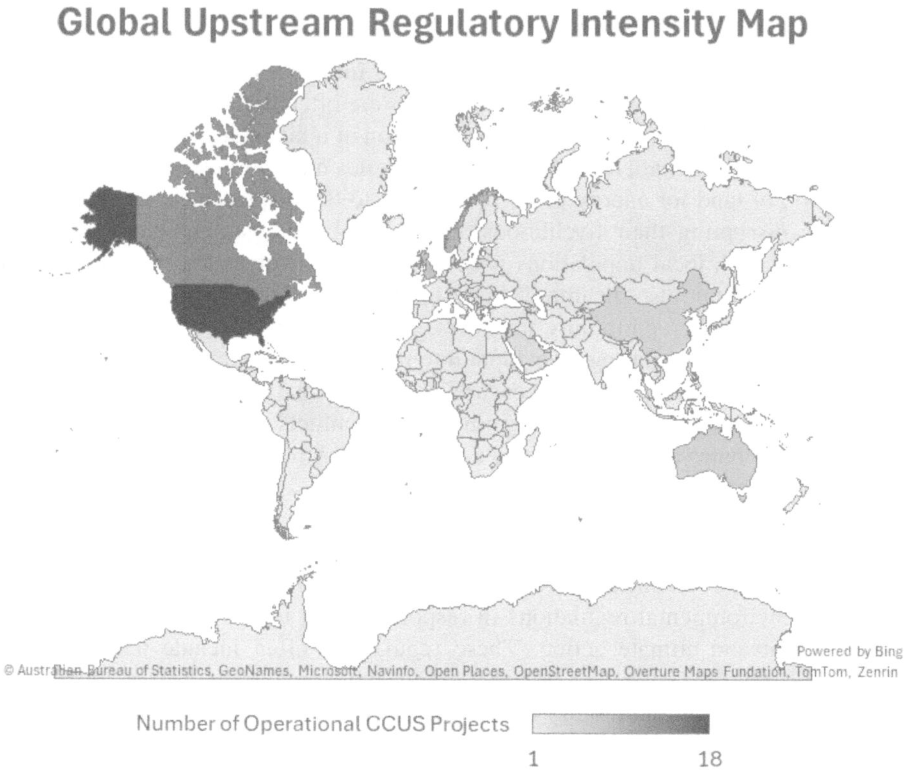

FIGURE 13.2 Regulatory impact on methane emissions reduction (2015–2025). Illustrates the correlation between tightening environmental regulations and declining methane emissions across major energy firms. Highlights how policy frameworks—such as methane intensity targets, carbon pricing, and reporting mandates—drive operational changes and accelerate upstream decarbonization.

back into oil reservoirs to help extract additional crude oil. This approach reduces GHG emissions and enhances oil recovery efficiency (Figure 13.2).

- **Reducing Methane Emissions:** Methane is a potent GHG; its reduction is critical. Companies are implementing comprehensive leak detection and repair (LDAR) programs, including regular inspections using advanced technologies like optical gas imaging cameras. These tools allow for the timely identification and sealing of leaks in pipelines, storage tanks, and other infrastructure. By proactively addressing methane emissions, companies can significantly lower their environmental footprint.
- **Water Recycling and Responsible Usage:** Upstream operations often require significant amounts of freshwater, particularly for hydraulic fracturing. To minimize freshwater consumption, companies are investing in technologies to recycle produced water, which can contain various contaminants. By treating and reusing this water for hydraulic fracturing and other operational needs, the industry can conserve

freshwater resources and reduce the overall environmental impact of water extraction.

- **Renewable Energy Integration in Upstream:** To curtail reliance on fossil fuel-based energy, companies are increasingly integrating renewable energy sources such as solar, wind, and geothermal power into their upstream operations. This shift not only reduces carbon emissions but also enhances energy resilience. On-site renewable energy installations can power drilling rigs, processing facilities, and support infrastructure, thus promoting a more sustainable energy model across operations.
- **Sustainable Decommissioning and Site Restoration:** Responsible decommissioning becomes paramount as wells reach the end of their productive lives. Companies are developing frameworks for safely sealing abandoned wells to prevent leaks and environmental hazards. Furthermore, they are implementing biodiversity restoration programs that focus on rehabilitating the surrounding ecosystem, ensuring that land used for oil and gas activities is returned to a healthier state that supports local flora and fauna.

By embracing these strategies, the oil and gas industry can move toward more sustainable upstream operations, ultimately contributing to a lower environmental impact and a more sustainable future.

13.4 ESG INVESTING AND CORPORATE RESPONSIBILITY

As ESG metrics become increasingly central to investment strategies,[2] oil and gas companies must fundamentally reshape their business models to attract and retain capital from conscientious investors. Sustainable investment trends reflect this shift and include several key areas:

- **Green Financing and ESG-linked Investments:** Companies increasingly issue sustainability-linked bonds and green bonds to finance renewable energy projects and initiatives to reduce carbon emissions. These financial instruments often feature incentives that tie interest rates to the company's performance on ESG criteria, thereby encouraging continuous improvement in sustainable practices.
- **Transparent ESG Reporting:** Investors increasingly demand robust and detailed disclosures regarding companies' carbon footprints, environmental performance, and corporate social responsibility (CSR) initiatives. Companies are expected to adopt standardized reporting frameworks, such as the Global Reporting Initiative (GRI) or the Sustainability Accounting Standards Board (SASB) standards, to provide clarity and accountability about their sustainability efforts and achievements.
- **Industry Initiatives and Partnerships**: Major oil and gas companies are increasingly joining forces through coalitions to advance carbon neutrality goals and address climate change challenges. One notable example is the Oil and Gas Climate Initiative (OGCI), which focuses on reducing methane

FIGURE 13.3 Growth of ESG investment value (2010–2025). Bar chart showing the rise in global ESG investment from $1 trillion in 2010 to nearly $10 trillion by 2025. Highlights the accelerating financial commitment to sustainability, signaling investor preference for low-carbon portfolios, transparent governance, and socially responsible upstream strategies.

emissions and enhancing energy efficiency across the sector. Collaborative efforts among industry players demonstrate a commitment to ESG principles and facilitate knowledge sharing and best practices for sustainable innovation (Figure 13.3).

By actively embracing sustainable investment trends, oil and gas companies can position themselves as responsible investors and align their operations with global sustainability goals. This alignment is crucial as stakeholders, including investors, regulators, and consumers, increasingly demand greater transparency and accountability regarding environmental impacts and corporate governance. By adopting practices such as investing in renewable energy projects, implementing advanced technologies for carbon capture, and enhancing energy efficiency, these companies can demonstrate their commitment to mitigating climate change. Furthermore, establishing robust reporting practices on sustainability metrics will help build trust and strengthen relationships with stakeholders, ultimately enhancing the company's reputation and long-term viability in an evolving market landscape.

13.5 CASE STUDIES IN SUSTAINABLE UPSTREAM OPERATIONS

Implementing sustainability initiatives in the upstream oil and gas sector presents a diverse landscape that varies significantly across regions, influenced by differing regulatory frameworks, environmental conditions, and corporate strategies. For instance, in regions with stringent environmental regulations, companies may invest heavily in advanced technologies such as CCS to minimize GHG emissions. In contrast, areas with less regulatory oversight might focus on voluntary sustainability programs that enhance community relations.

Case studies from various companies illustrate these approaches in action. For example, a leading multinational may adopt a comprehensive water management strategy in arid regions to reduce freshwater usage and minimize the impact on local ecosystems. Other companies, operating in landscapes rich in biodiversity, might implement habitat restoration projects alongside their extraction activities, demonstrating their commitment to environmental stewardship.

Moreover, effective community engagement is critical for the success of these initiatives. Companies that actively involve local communities in the decision-making process often see improved relations and support for their operations. Programs that support local education, healthcare, and infrastructure development foster goodwill and enhance the company's reputation and operational efficiency.

These examples underscore the importance of tailored sustainability efforts that align with specific regional needs and corporate goals. These efforts ultimately contribute to a reduced environmental footprint and stronger community ties.

13.5.1 Case Study 1: Reducing Methane Emissions—Equinor's Norwegian Shelf Project

13.5.1.1 Background

Methane emissions are a major contributor to climate change, with a global warming potential more than 25 times that of CO_2 over a 100-year period. In offshore production, methane leaks can occur due to aging infrastructure, equipment malfunctions, and venting practices. Recognizing the importance of minimizing methane emissions, **Equinor**, one of the world's largest offshore operators, initiated an advanced methane leak detection and repair (LDAR) program in the **Norwegian Continental Shelf**.

13.5.1.2 Implementation

Equinor deployed a combination of **satellite monitoring, airborne sensors, and infrared cameras** to detect and measure methane leaks in real-time. Advanced AI-driven analytics were integrated into its monitoring systems, enabling rapid leak detection and automation of maintenance schedules.

Outcomes:

- Equinor reduced its offshore methane emissions by **50% within five years**, surpassing regulatory requirements set by the Norwegian government.
- The company saved millions of dollars annually by capturing previously wasted methane and using it for power generation.
- The Norwegian government has since adopted Equinor's methodologies as a benchmark for industry-wide methane mitigation strategies.

Key Takeaways:

- Advanced **leak detection and AI-driven analytics** can significantly reduce methane emissions.
- Proactive **maintenance and real-time monitoring** improve operational efficiency and reduce environmental risks.
- Government-industry collaboration can enhance regulatory compliance and innovation in methane management.

13.5.2 Case Study 2: CCUS in Action—The Gorgon Project (Australia)

Background

The **Gorgon LNG Project**, operated by Chevron, is one of the world's largest liquefied natural gas (LNG) operations, located on **Barrow Island, Australia**. Given the carbon-intensive nature of LNG production, Chevron is committed to implementing one of the world's most ambitious carbon capture, utilization, and storage (CCUS) projects.

Implementation

Chevron developed an **integrated CCUS system** that captures CO_2 from natural gas production before it enters the atmosphere. The captured CO_2 is **injected 2 km underground into a deep saline aquifer**, preventing it from contributing to climate change. The Australian government supported the project and aligned it with its national carbon reduction targets.

Outcomes:

- Since operations began, **over 6 million tons of CO_2 have been successfully stored underground**, making it one of the most significant CCS projects globally.
- The project has set a **benchmark for integrating CCS with LNG production**, proving that large-scale emissions reduction is possible.
- The success of the Gorgon CCS initiative has influenced policy discussions about mandatory CCS implementation for future energy projects in Australia.

Key Takeaways:

- Large-scale **CCUS projects can effectively reduce CO_2 emissions** from fossil fuel production.
- Government support and **public-private collaboration** can drive innovation in carbon capture technologies.
- CCS must be **integrated into long-term energy planning** to align with global decarbonization targets.

CARBON-NEUTRAL OIL—MYTH OR MILESTONE?

Occidental aims to deliver the world's **first carbon-neutral oil** using **direct air capture** and **geological storage**.

Reality Check:

"Carbon-neutral" claims often rely on **offset accounting** — real impact depends on **verifiable storage**.

13.5.3 Case Study 3: Community Engagement and Social Responsibility—Shell's Nigeria Initiative

Background

Oil and gas operations in **Nigeria's Niger Delta** have been historically associated with environmental degradation, oil spills, and community conflicts. This has led to **widespread opposition from local communities**, resulting in pipeline sabotage, legal battles, and operational disruptions for companies. Recognizing the need for improved relations, **Shell** launched a **Community Development and Environmental Protection Initiative** to address these issues.

Implementation

Shell implemented a **three-pronged approach** to community engagement:

1. **Environmental Restoration:** Investing in remediation efforts for oil spill-affected areas and implementing **zero-routine flaring policies**.
2. **Education and Economic Development:** Funding **scholarship programs** for local students and creating vocational training centers to develop workforce skills.
3. **Infrastructure Investment:** Partnering with local governments to improve **roads, schools, healthcare facilities, and clean water access** in oil-producing regions.

Outcomes:

- The number of oil spills caused by community sabotage **dropped by 40%** due to improved relations and stakeholder engagement.
- More than **$200 million has been invested in community development projects** since the initiative's inception.
- Shell's commitment to ESG principles has enhanced its reputation and strengthened its long-term **social license to operate** in Nigeria.

Key Takeaways:

- **Stakeholder engagement and social responsibility** are critical for maintaining operational stability.
- Companies must **invest in community welfare** to ensure long-term sustainability and prevent local conflicts.
- Environmental protection initiatives **help rebuild trust and enhance corporate reputation** in oil-producing regions.

13.5.4 Case Study 4: Carbon-Neutral Oil Production—Occidental's Permian Basin

Background

The idea of producing "carbon-neutral oil" was once considered unrealistic, but **Occidental Petroleum (Oxy)** has taken bold steps toward making this concept a reality. The company operates extensively in the **Permian Basin (Texas, USA)** and has leveraged **Carbon Capture, Utilization, and Storage (CCUS)** technology to offset emissions from its operations.

Implementation

Oxy developed the **Permian CCUS Hub**, integrating carbon capture directly into its oil production process. The captured CO_2 is then injected into depleted reservoirs to enhance oil recovery (EOR), effectively **trapping the CO_2 underground** while increasing hydrocarbon extraction efficiency.

Additionally, Oxy **partnered with Airbus and other corporations** to sell "net-zero" crude oil, which is marketed as having **zero net carbon emissions** after accounting for carbon storage.

Outcomes:

- Oxy has **offset millions of tons of CO_2 emissions**, positioning itself as an industry leader in low-carbon oil production.
- The company has secured **long-term contracts** with buyers seeking carbon-neutral crude, opening **new market opportunities**.
- The project demonstrates that **carbon-neutral hydrocarbons are commercially viable**, paving the way for other operators to adopt similar strategies.

Key Takeaways:

- **Carbon-neutral oil production is feasible** with the combination of CCUS and EOR.
- **Market incentives for low-carbon hydrocarbons** are growing, making sustainability a competitive advantage.
- Integrating **decarbonization strategies into core operations** attracts ESG-focused investors and buyers.

13.5.5 Summary of Case Studies

These case studies illustrate **different approaches to sustainability in upstream petroleum operations**. While each company has tailored its strategy to its specific needs and challenges, common themes emerge:

✓ **Advanced technologies** (e.g., CCUS, AI-driven methane monitoring) play a key role in sustainability.
✓ **Strong regulatory support** accelerates the adoption of cleaner practices.
✓ **Community engagement and social responsibility** are essential for maintaining operational stability.
✓ **Carbon-neutral strategies** are gaining traction as viable alternatives for the oil and gas industry's

13.6 THE FUTURE OF SUSTAINABLE UPSTREAM OPERATIONS

13.6.1 Decarbonization Roadmaps

In response to the growing urgency of climate change, companies across various sectors are establishing ambitious targets to achieve net-zero emissions by

2050. These roadmaps outline comprehensive strategies that include substantial investments in renewable energy sources, such as solar, wind, and hydroelectric power, essential for reducing reliance on fossil fuels. Additionally, organizations are increasingly focusing on Carbon Capture, Utilization, and Storage (CCUS) technologies, which help mitigate emissions from industrial processes and contribute to innovative ways of repurposing carbon dioxide. Moreover, enhancing energy efficiency is a critical component of these plans, with measures such as upgrading infrastructure, optimizing production processes, and implementing more intelligent energy management systems. Together, these initiatives create a holistic approach to sustainability and play a fundamental role in the global transition to a low-carbon economy.[3]

13.6.2 Carbon-Neutral Oil and Gas Production

Emerging projects in the oil and gas sector are increasingly focused on achieving carbon neutrality by implementing innovative strategies to offset GHG emissions. These strategies often involve nature-based solutions, such as reforestation, afforestation, and the restoration of wetlands, which serve to absorb carbon dioxide from the atmosphere. Additionally, Carbon Capture, Utilization, and Storage (CCUS) technologies are being integrated into production processes, allowing facilities to capture emissions before they enter the atmosphere and either utilize them in various industrial applications or store them underground permanently.

These initiatives are designed to enable the extraction and production of hydrocarbons with a significantly reduced environmental footprint. By combining traditional extraction methods with advanced sustainable practices, these projects aim to meet global energy demands while addressing climate change and minimizing detrimental impacts on ecosystems. As a result, the industry is moving toward a more sustainable model that prioritizes environmental stewardship alongside economic viability.

13.6.3 The Role of Hydrogen and Alternative Fuels

Blue hydrogen, which is produced through the process of natural gas reforming combined with carbon capture, utilization, and storage (CCUS), is emerging as a significant player in the global energy transition. This process allows for the extraction of hydrogen while minimizing carbon dioxide emissions, thus providing a more sustainable pathway for industries that rely heavily on fossil fuels.

On the other hand, green hydrogen is derived from renewable energy sources, such as wind, solar, or hydroelectric power, through the process of electrolysis, where water is split into hydrogen and oxygen. This method is entirely emissions-free, making green hydrogen a critical component for achieving net-zero goals.

Both blue and green hydrogen are gaining traction in various sectors, including transportation, heating, and industrial processes, as they offer cleaner alternatives to traditional fossil fuels. Their growing adoption is crucial for reducing GHG emissions and enhancing energy security, paving the way for a more sustainable and resilient energy future.

13.6.4 Collaboration with Renewable Energy Companies

Upstream firms are increasingly investing in renewable energy sources such as wind, solar, and biofuels to diversify their portfolios and reduce reliance on fossil fuels. This shift is driven by the growing demand for sustainable energy solutions and the need to meet regulatory requirements for carbon reduction. These companies are forming strategic partnerships with established renewable energy companies, leveraging their technological expertise and industry experience. For instance, collaborations may involve joint ventures to develop new wind farms or solar power plants, as well as investments in research and development of advanced biofuel technologies. Through these initiatives, upstream firms aim to enhance their competitive edge while contributing to a greener energy landscape.

13.7 CONCLUSION: SUSTAINABILITY AS A COMPETITIVE ADVANTAGE

The long-term viability of the oil and gas industry hinges on its ability to align with global sustainability goals, particularly in the face of mounting environmental challenges and shifting consumer expectations. Companies that proactively integrate ESG principles into their operational frameworks mitigate risks and enhance their reputations in an increasingly conscientious marketplace.

To remain competitive, these companies must prioritize initiatives to reduce their carbon footprint. This includes investing in advanced technologies such as CCS, utilizing renewable energy sources in their operations, and transitioning toward biofuels and hydrogen solutions. By adopting innovative low-carbon technologies, firms can meet regulatory requirements and anticipate future legislation to curb GHG emissions.

Furthermore, embracing sustainable practices is pivotal for the upstream sector, which is vital in meeting global energy demand. As the energy landscape evolves, companies that lead in sustainability will find themselves better positioned to gain market share and consumer trust. This involves minimizing the environmental impact of their operations through improved efficiency and waste management, and actively engaging with local communities to enhance their social license to operate.

> Energy companies that fail to adapt will not only lose relevance — they will cease to exist.
>
> *IEA Executive Director, 2022*

In conclusion, the oil and gas industry has the opportunity to redefine its role in the global energy transition. By committing to sustainability, companies can navigate the complexities of today's energy challenges while ensuring their long-term success and relevance in a rapidly changing world.

NOTES

1. IPCC Sixth Assessment Report, 2023
2. Bloomberg Green, 2023 ESG Investment Survey
3. IEA, Net Zero by 2050 Roadmap, 2021.

14 Economic Modeling in Upstream Petroleum

14.1 INTRODUCTION: THE ROLE OF ECONOMIC MODELING IN UPSTREAM

Economic modeling is a fundamental pillar of upstream petroleum economics. It provides a systematic approach for industry professionals to make informed investment decisions, optimize production strategies, and evaluate risks associated with exploration and production (E&P) activities. Given the capital-intensive and inherently uncertain nature of oil and gas projects, economic models represent structured frameworks that enable stakeholders to estimate future revenues, costs, and profitability across various scenarios, including fluctuating market conditions and regulatory changes.

This chapter delves into various economic modeling techniques, such as discounted cash flow (DCF) analysis, scenario analysis, and Monte Carlo simulations. Each technique offers unique insights and applications, catering to specific decision-making needs within the upstream sector. For instance, DCF analysis allows for assessing the time value of money and the intrinsic value of potential investments. In contrast, scenario analysis helps understand the range of possible outcomes based on varying assumptions.

To further highlight the significance of these models, real world case studies will be presented, demonstrating how robust economic analysis underpins successful petroleum projects. These case studies will illustrate the critical role of economic modeling in navigating the complexities of investment choices, adapting to technological advancements, and managing environmental and geopolitical risks. Ultimately, this chapter will underscore the importance of comprehensive economic modeling in driving strategic decisions that optimize resource allocation and enhance profitability within the upstream petroleum sector.

FAST FACT—MONTE CARLO AT WORK

One oil major ran 10,000 simulations before green-lighting a \$6 billion offshore project — a single variable adjustment shifted NPV by \$500 million.

14.2 KEY ECONOMIC MODELING TECHNIQUES IN UPSTREAM

> In petroleum economics, the only certainty is uncertainty.
>
> *Anonymous industry saying*

In the upstream sector, various economic modeling approaches are employed to evaluate investment projects and mitigate associated risks. These techniques are

DOI: 10.1201/9781003750345-17

crucial for making informed decisions in the E&P of oil and gas. Among the predominant methods are.

14.2.1 Discounted Cash Flow (DCF) Analysis

DCF analysis is a cornerstone technique utilized in the financial evaluation of upstream projects. It determines the present value of anticipated future cash flows by discounting them at a specific rate, which reflects the project's risk profile and the cost of capital. The primary metrics derived from DCF include:

- **Net present value (NPV)**: This metric quantifies a project's profitability by calculating the difference between the present value of cash inflows and outflows. A positive NPV signifies that the projected earnings exceed the investment costs, suggesting that the project is economically viable and likely to generate value for stakeholders.
- **Internal rate of return (IRR)**: The IRR is the discount rate that equates the project's NPV to zero. This statistic is crucial in evaluating a project's expected profitability against the company's required rate of return, also known as the hurdle rate. A higher IRR compared to the benchmark indicates a more attractive investment opportunity.
- **Payback period:** This measure estimates the duration required to recover the initial investment in the project. While it does not account for the time value of money, the payback period is helpful in understanding the liquidity and risk level of the investment. Projects with shorter payback periods are generally regarded as less risky (Figure 14.1).

Example: Consider a hypothetical deepwater drilling project located off the coast of Brazil, which necessitates an upfront capital investment of $500 million. Analysts using DCF methodologies project future cash flows and determine an NPV of $200 million based on a discount rate of 10%. This analysis suggests a robust financial outlook for the project, indicating it is likely to yield significant returns over its lifespan, thus making it an attractive investment opportunity in the competitive upstream sector.

Oil's Economic Rollercoaster: $500 Billion in Investment Decisions Hinges on Forecasting Models

Financial Times, March 12, 2024

Region	Average Break-Even Price ($/bbl)
Middle East	25
U.S. Shale	40
Deepwater Brazil	50
Arctic Projects	70
North Sea	55

FIGURE 14.1 Regional break-even oil prices: Comparison of average break-even costs across global upstream regions, highlighting cost competitiveness and investment thresholds.

14.2.2 Probabilistic vs. Deterministic Models

Economic evaluations can be conducted using either deterministic or probabilistic approaches, each with its strengths and weaknesses:

- **Deterministic models:** These models operate on fixed values for crucial inputs such as oil prices, production rates, and capital expenditures (CAPEX). Their straightforward nature makes them easy to understand and implement; however, they cannot inherently account for uncertainty and variability in these critical assumptions. For example, a deterministic model might assume a constant oil price of $60 per barrel over the entire project lifespan, regardless of market dynamics.
- **Probabilistic models (Monte Carlo simulation[1]):** In contrast, probabilistic models incorporate probability distributions for key variables, allowing for a comprehensive analysis of potential outcomes and risks. By utilizing Monte Carlo simulations, these models can simulate thousands of scenarios based on various assumptions, such as fluctuating oil prices, diverse production rates, and changing drilling costs. This approach generates a probability distribution of NPVs, enabling decision-makers to evaluate risk-adjusted returns. For instance, simulating a shale gas project might reveal a wide range of NPVs from negative values to substantial profits, depending on the input variables' scenarios (Figure 14.2).

14.2.3 Real Options Valuation[2] (ROV)

Traditional DCF analysis typically relies on a static project plan, which does not accommodate the fluid nature of real-world market conditions. Real Options Analysis (ROA) enhances project valuation by allowing flexibility to incorporate future decision points. This approach will enable organizations to dynamically model project decisions based on evolving market conditions and projections.

For example, a company operating an offshore oil project might assess the feasibility of deferring, expanding, or abandoning the project based on anticipated fluctuations in oil prices. With ROA, these strategic decisions can be quantified, providing management with insights into the value of maintaining operational flexibility in the face of uncertainty and enabling them to optimize investment strategies as conditions change (Figure 14.3).

Model Type	NPV Range ($M)
Deterministic	$450–$500
Probabilistic (P10-P90)	-$200 to $700

FIGURE 14.2 NPV range comparison: Deterministic vs. probabilistic models. Illustrates how probabilistic modeling captures a broader range of financial outcomes, including potential losses, compared to deterministic projections.

Model Type	Project Value ($Billion)
Traditional DCF	9.8
Real Options Approach	11.7

FIGURE 14.3 Project valuation: Traditional DCF vs. real options approach. Comparison of project value estimates using static DCF modeling versus flexible real options analysis, highlighting the impact of strategic adaptability.

14.2.4 Break-Even Analysis

Break-even analysis is crucial for determining the minimum oil price at which a project remains profitable. This analysis enables companies to assess their economic resilience during periods of price volatility.

For instance, consider a North Sea oil project with a break-even price of $45 per barrel. This analysis suggests that the project is financially viable if market prices stay above this threshold. If oil prices dip below $45, the project would incur losses. Thus, understanding the break-even point helps companies gauge risk, make informed investment decisions, and develop strategies to enhance competitiveness in fluctuating markets.

14.3 APPLICATIONS OF ECONOMIC MODELING IN UPSTREAM

Economic modeling plays a crucial role in several key decision-making[3] areas within the upstream oil and gas sectors:

- **Exploration economics**: Economic modeling facilitates the comprehensive evaluation of the viability of drilling new prospects. This includes quantifying geological success probabilities, analyzing historical drilling data, and estimating expected reserves based on geophysical and geological surveys. Rigorous analysis helps in determining the potential return on investment (ROI) and overall economic feasibility of exploration activities.
- **Field development planning:** In this phase, economic models assess various development scenarios to maximize oil and gas recovery. This involves optimizing capital expenditure (CAPEX) and operational expenditure (OPEX) over the life cycle of the field. By using advanced simulation tools, companies can evaluate different configurations of infrastructure, drilling techniques, and recovery methods to identify the strategy that yields the highest NPV.
- **Production forecasting:** Reliable production forecasting is critical for strategic planning and investment decisions. This is achieved through decline curve analysis[4] and sophisticated reservoir simulation models that consider various factors, including reservoir characteristics, historical production data, and expected decline rates. Such models help anticipate future production rates and guide decisions regarding investments in enhancements or increased drilling activity.

- **Mergers and acquisitions (M&A) analysis:** During M&A activities, economic modeling is employed to conduct thorough asset valuations and assess the potential synergies that could be realized through the acquisition. This involves comprehensive financial modeling, including DCF analysis, to ascertain fair market values and identify risks associated with the integration of assets.
- **Risk assessment and mitigation:** Economic models are essential for incorporating sensitivity analysis and scenario planning, enabling companies to evaluate potential downside risks associated with market fluctuations, regulatory changes, and operational challenges. By simulating different scenarios, organizations can identify critical risk factors and develop robust strategies to mitigate these risks.

Example: Consider a company contemplating the acquisition of an oil field. The organization runs multiple economic scenarios, carefully adjusting for variables such as oil price volatility, production uncertainties,[5] and operational costs. By analyzing these diverse scenarios, the company can make a well-informed decision on whether to proceed with the acquisition, ensuring that all financial implications are thoroughly evaluated before finalizing the deal.

When Oil Models Miss the Mark: Lessons From the 2020 Crash

Reuters, June 9, 2020

14.4 CASE STUDIES IN ECONOMIC MODELING

Economic modeling plays a crucial role in evaluating upstream projects. It is a foundational tool for companies seeking to assess financial viability, optimize investment decisions, and effectively mitigate risks associated with E&P activities. These models integrate various data inputs, including market trends, cost structures, and production forecasts, to comprehensively analyze potential outcomes.

The following case studies highlight several distinct modeling techniques applied in real-world upstream projects, showcasing their effectiveness in guiding strategic decision-making. These examples illustrate how quantitative analyses have been instrumental in evaluating factors such as commodity price fluctuations, regulatory impacts, and technological advancements, ultimately aiding companies to navigate complex economic landscapes and maximize their ROI.

Flexibility is value — it's not just about what you drill, but when and how.

SPE Upstream Economics Panel

From Spreadsheets to Simulations: How Big Oil Bets Are Made

Bloomberg Energy, August 28, 2023

14.4.1 Case Study 1: Deepwater Oil Project Evaluation (West Africa)

A multinational oil and gas company was evaluating an offshore deepwater project in West Africa. Given the significant capital expenditure (CAPEX) required, economic modeling was crucial in determining the project's financial viability.

- **Project overview:** The deepwater development required a total CAPEX of **$3 billion**, with an expected production life of 25 years.
- **Economic evaluation:** To assess profitability, a DCF analysis was conducted using a 12% discount rate. The NPV was estimated at **$1.2 billion**, indicating strong potential for financial success.
- **Risk assessment:** A Monte Carlo simulation was performed to account for uncertainty in oil prices and reservoir performance. This simulation ran **10,000 iterations** with varying inputs for oil price fluctuations, operating costs, and production efficiency.
- **Outcome:** The probability of achieving a **positive NPV** was **85%**, giving the company confidence to invest. The model also highlighted key risks, such as geopolitical instability and regulatory changes, which were incorporated into risk management planning.

This case study demonstrates how **DCF and Monte Carlo simulation** can be used together to assess project viability and quantify risks in deepwater developments.

14.4.2 Case Study 2: Shale Gas Break-Even Analysis (United States)

A shale gas producer in the United States needed to assess the economic feasibility of new drilling operations in an unconventional shale play. Given the volatility of oil and gas prices, determining the break-even price was essential for risk management.

- **Project overview:** The company planned to drill **100 horizontal wells**, each costing **$7 million** to develop.
- **Break-even analysis:** Economic modeling determined that the **average break-even price was $35 per barrel** of oil equivalent.
- **Sensitivity analysis: A sensitivity analysis** was conducted to evaluate how different price scenarios would impact the project's profitability:
  - At **$70 per barrel**, the project showed **significant profitability**, with an IRR of **28%**.
  - If prices fell to **$40 per barrel**, profitability would be **marginal**, forcing the company to reconsider its capital allocation.
  - At **$30 per barrel**, the project would generate negative returns, leading to potential well shut-ins.
- **Outcome:** Based on these insights, the company developed a **hedging strategy** to lock in favorable prices for future production, ensuring financial stability even if market prices declined.

This case study highlights the importance of **break-even analysis and sensitivity modeling** in the highly price-sensitive shale sector.

14.4.3 Case Study 3: LNG Export Project Real Options Analysis (Qatar)

A liquefied natural gas (LNG) developer in **Qatar** was considering investing in an **LNG export terminal** but faced **uncertainty in global demand** due to competition from new LNG suppliers and shifting energy policies in key markets.

- **Project overview:** The proposed LNG terminal required a **$12 billion investment** and was expected to export **15 million tons of LNG per annum (MTPA)**.
- **Investment uncertainty:** Due to fluctuating LNG demand and policy changes in major importing countries, committing the entire investment upfront posed a significant risk.
- **Real options analysis (ROA):** Instead of an immediate full-scale commitment, a **phased investment approach** was modeled using **ROA**, allowing flexibility based on market conditions:
  1. **Phase 1:** Initial capacity of **5 MTPA**, requiring **$4 billion** investment.
  2. **Phase 2:** Expansion to **10 MTPA** if demand conditions are favorable.
  3. **Phase 3:** Full-scale **15 MTPA** production only if long-term LNG contracts justify further investment.
- **Outcome:** This approach provided the **option to defer or expand investment**, reducing exposure to demand risks while capturing upside potential. The model showed that a **staged expansion strategy increased overall project value by 20%** compared to an immediate full-scale investment.

This case study illustrates how **Real Options Analysis (ROA)** allows companies to **adapt investment strategies dynamically**, reducing risk in large-scale infrastructure projects.

INDUSTRY INSIGHT—REAL OPTIONS SAVE BILLIONS

Deferring a final investment decision (FID) using real options saved an LNG project in Australia over $2 billion during a global demand collapse.

14.5 CHALLENGES IN ECONOMIC MODELING

While essential, economic modeling within upstream petroleum projects is fraught with challenges that can significantly impact decision-making and strategic planning. These challenges primarily arise from market volatility, regulatory complexities, and uncertainties in geological reserve estimations.

> Crisis Forecasting: Why the Best Models Still Get It Wrong
>
> *The Economist, November 5, 2022*

14.5.1 Oil Price Volatility

One of the most critical challenges in upstream economic modeling is the unpredictability of future oil and gas prices. Such volatility can profoundly influence project feasibility, complicating the creation of precise long-term financial projections.

Example: The unprecedented decline in oil prices from \$60 to below \$20 per barrel in early 2020, driven mainly by a collapse in demand due to the COVID-19 pandemic and an oversupply of oil, rendered many previously reliable economic models inaccurate. This forced numerous companies to reevaluate their investment priorities and postpone major development projects, leading to substantial financial setbacks.

Solution: Implementing scenario analysis and stochastic modeling techniques, such as Monte Carlo simulations, allows companies to prepare for potential price variability by assessing project resilience in future price environments. This involves simulating thousands of scenarios, incorporating different price paths, and providing a more comprehensive risk assessment.

14.5.1.1 Uncertainty in Geological Estimates

Economic models heavily depend on accurate forecasts of reserves and production rates, which are inherently uncertain. Overestimating reserves can inflate project valuations, whereas underestimating might result in missed investment opportunities.

Example: A deepwater oil project off the coast of Brazil was initially estimated to have recoverable reserves of 2 billion barrels. However, only 1.3 billion barrels were confirmed after three years of exploratory drilling, resulting in substantial financial losses and a complete restructuring of the project's production plans.

Solution: Utilizing probabilistic reserve estimates—represented as P10 (low estimate), P50 (best estimate), and P90 (high estimate)—will help account for geological uncertainty. This approach offers a more realistic range of possible outcomes, allowing for informed decision-making.

14.5.2 Regulatory and Taxation Complexity

The complexity of fiscal regimes across different oil-producing nations can significantly influence the economic viability of projects. Variations in taxation policies, royalties, and profit-sharing agreements are vital components that can dramatically affect cash flow models.

Example: In Norway, oil companies must comply with a steep 78% marginal tax rate, which can hamper profitability, especially for smaller operators. Conversely, more favorable tax structures and an emphasis on reducing regulatory burdens have made investment opportunities considerably more appealing in the United States.

Solution: Economic models must incorporate detailed country-specific fiscal terms, including various forms of profit-sharing, royalties, and corporate taxes. This will ensure accurate financial assessments that account for regional economic landscapes.

14.5.3 Technology and Cost Assumptions

Assumptions regarding future cost reductions derived from technological advancements can often be overly optimistic and speculative. While emerging technologies may promise reduced drilling costs, enhanced recovery rates, and operational efficiencies through automation, their impact can carry considerable uncertainty.

Example: Many firms have previously overestimated the potential cost savings from automated drilling technologies in unconventional plays. This leads to overly optimistic projections of the break-even costs that ultimately required significant adjustments as actual performance data came in.

Solution: Conducting a sensitivity analysis on CAPEX, OPEX, and efficiency gains will yield a more robust economic model. This analysis should explore a range of cost scenarios, thereby ensuring that potential variances are adequately considered.

14.5.3.1 Conclusion: Enhancing Economic Modeling for Upstream Success

Economic modeling is indispensable for guiding upstream investment decisions in a highly uncertain environment. However, these models must constantly evolve to effectively tackle the challenges presented by market fluctuations and integrate advanced, AI-driven forecasting methodologies. Future trends in economic modeling could include:

- **Integration of AI and machine learning:** For real-time data analysis and predictive modeling. These technologies can analyze vast datasets to identify patterns and trends, providing more accurate forecasts.
- **Carbon pricing and ESG factors:** To incorporate sustainability considerations into investment decisions, reflecting the increasing importance of environmental, social, and governance factors in the energy sector.
- **Advanced scenario planning:** To strengthen risk mitigation strategies by systematically evaluating potential market, regulatory, and technological shifts.
- **Digital twin simulations:** Creating real-time economic models based on actual field performance, allowing companies to visualize and optimize production processes and economic outcomes interactively.

By leveraging these advancements in technology and modeling approaches, upstream companies can enhance their profitability, reduce risk, and make more informed, data-driven decisions in an increasingly volatile energy landscape.

14.6 CONCLUSION: THE FUTURE OF ECONOMIC MODELING IN UPSTREAM

The field of economic modeling in the upstream oil and gas sector is undergoing significant transformation as the industry confronts an array of new challenges and emerging opportunities. Future trends that are expected to shape this evolution include:

- **Integration of AI and machine learning:** Incorporating advanced AI algorithms and machine learning techniques will revolutionize forecasting accuracy and risk assessment. These technologies can analyze vast datasets from geological surveys, production histories, and market dynamics, leading to more precise predictions of oil and gas prices, production rates, and potential disruptions.
- **Advanced scenario planning:** Companies increasingly utilize sophisticated methods such as Monte Carlo simulations and real options analysis to

navigate the uncertainties inherent in the energy market. These tools allow firms to explore various scenarios, accounting for geopolitical instability, regulatory changes, and technological advancements, thus enabling better preparedness for fluctuating market conditions.

- **Carbon pricing and ESG factors:** As environmental, social, and governance (ESG) considerations become more prominent in investment decisions, future economic models must integrate carbon pricing and sustainability metrics. This means quantifying the potential financial impact of carbon emissions on operations and investments, aligning profitability with environmentally responsible practices, and regulatory requirements.
- **Dynamic decision-making models:** AI-driven decision trees and adaptive algorithms will facilitate real-time project plans and strategy adjustments. These models enable companies to pivot rapidly in response to changing conditions by continuously processing new data and outcomes, ensuring more resilient and agile project management.

By leveraging these cutting-edge modeling techniques, upstream oil and gas companies can significantly enhance their profitability, effectively manage risks, and make more informed investment decisions amidst an increasingly complex and volatile energy landscape. Embracing these innovations supports operational efficiency and aligns with the broader transition toward sustainable energy practices, ultimately positioning firms for long-term success.

14.6.1 Summary of Key Takeaways

- Economic modeling is essential for evaluating upstream projects, managing risk, and optimizing decision-making.
- DCF analysis, Monte Carlo simulations, and Real Options Analysis are key methodologies used in upstream petroleum economics.
- Economic models are applied in exploration planning, field development, M&A decisions, and price risk mitigation.
- Case studies demonstrate the real-world impact of robust economic modeling on investment outcomes.
- Future advancements will integrate AI, ESG metrics, and dynamic modeling for more resilient decision-making.

NOTES

1. McKinsey & Company, Oil & Gas Practice Report 2023, Application of Probabilistic Modeling in Project Risk Analysis.
2. Society of Petroleum Engineers (SPE), *Real Options in Upstream Oil and Gas*, 2022—for real options valuation methods.
3. McKinsey & Company, Oil & Gas Practice Report 2023—for modeling in investment decision-making.
4. Schlumberger, Petroleum Economics Fundamentals, 2022—for decline curve modeling practices.
5. BP, Statistical Review of World Energy 2023, Data Sheet on Oil Production Volatility (2010–2022).

15 Discounted Cash Flow (DCF) Alternatives in Upstream Oil and Gas

15.1 INTRODUCTION: THE NEED FOR ALTERNATIVES TO DCF

Discounted Cash Flow (DCF) analysis has long been regarded as the gold standard for evaluating upstream petroleum projects, primarily due to its systematic methodology in assessing the value of future cash flows. By discounting these expected cash flows at a predetermined rate—often based on the weighted average cost of capital—investors can gauge the current value of future earnings generated by oil and gas ventures. This approach allows for a clearer understanding of the potential profitability of a project.

However, DCF analysis is not without its limitations. One major drawback is its reliance on various assumptions, including projected oil prices, production rates, and operational costs, which can introduce significant uncertainty. Furthermore, DCF may lack the flexibility to account for changing market conditions and unexpected geopolitical events that often affect the energy sector. The inherent time sensitivity in investment decisions also poses challenges, as the timing of cash flows can dramatically alter the perceived value of a project.

In light of these limitations, many industry professionals advocate alternative valuation methods that complement or enhance the insights provided by DCF. Techniques such as real options analysis and scenario planning enable a more dynamic evaluation by considering various potential future scenarios, thereby offering a nuanced perspective on investment decision-making in an increasingly volatile oil and gas landscape. These methods can provide a more robust framework for assessing risk and opportunity, ultimately leading to more informed and strategic investment choices.

This chapter explores the **limitations of DCF analysis**, introduces **alternative valuation techniques**, and discusses how these methods can enhance decision-making in upstream oil and gas projects (Figure 15.1).

15.2 LIMITATIONS OF DCF ANALYSIS IN UPSTREAM INVESTMENTS

While the DCF method remains a widely utilized valuation approach in the upstream sector of the oil and gas industry, its effectiveness is frequently questioned. This skepticism arises primarily from several critical factors: first, the inherent price volatility of crude oil and natural gas, which can dramatically impact cash flow

DOI: 10.1201/9781003750345-18

Evaluation Criteria	Discounted Cash Flow (DCF)	Real Options Valuation (ROV)	Monte Carlo Simulation	Decision Tree Analysis
Flexibility in Decision	*No*	**Yes**	**Yes**	**Yes**
Risk Quantification	*No*	**Yes**	**Yes**	**Yes**
Sensitivity to	Low	High	High	Medium
Strategic Value	*No*	**Yes**	**Yes**	**Yes**
Ease of	High	Low	Low	High
Best Use Case	Stable, mature assets	Frontier, high-risk projects	Price-sensitive investments	Phased infrastructure builds

FIGURE 15.1 Comparative evaluation of DCF and alternative valuation methods: Comparison of key valuation techniques across critical decision-making dimensions in upstream oil and gas.

projections; second, the uncertainty surrounding reserve estimates, which are often subject to technical and geological risks; and third, the constantly evolving regulatory landscape that can impose unforeseen costs or operational restrictions. These factors create challenges in accurately forecasting long-term financial performance, making DCF less reliable as a standalone valuation tool in this dynamic environment. Key limitations include:

1. **Sensitivity to Discount Rate Selection**
   The **choice of discount rate** significantly affects project valuation. Slight variations in the discount rate can drastically change the **Net Present Value (NPV)**, leading to **overly optimistic or conservative projections**.
   - **Example:** A deepwater oil project in Brazil with an NPV of $500 million at a **10% discount rate** may show an NPV of only **$150 million at 15%**, causing **investment hesitation**.
   - **Alternative:** Sensitivity analysis and scenario planning help identify how discount rate variations impact project viability.
2. **Inability to Capture Investment Flexibility**
   DCF assumes a **fixed investment path**, failing to account for real-world **flexibility** in delaying, expanding, or abandoning projects.
   - **Example:** A shale gas operator may **delay drilling** until gas prices recover, but DCF does not capture the financial benefits of waiting.
   - **Alternative: Real options analysis (ROA)** incorporates flexibility into project valuation.
3. **Poor Handling of Uncertainty and Risk**
   Upstream projects involve **geological uncertainties, regulatory changes, and price fluctuations**. DCF relies on single-point estimates for cash flows, **oversimplifying risk exposure**.
   - **Example:** A North Sea oil field is expected to produce **200 million barrels**, but **subsurface uncertainty** could mean anywhere from **150 million to 250 million**.
   - **Alternative: Monte Carlo simulation** models a range of possible outcomes to **quantify uncertainty**.

4. **Short-Term Bias and Neglect of Strategic Value**
   DCF **aggressively discounts future cash flows, making long-term benefits seem insignificant. Thus, many** high-potential projects appear **unattractive** when they provide **strategic value.**
   - **Example:** LNG export terminals require **massive upfront investment** but provide **stable long-term revenues**. A pure DCF analysis may undervalue such projects.
   - **Alternative: Game theory and competitive analysis** account for **strategic positioning** in long-term investments.

15.3 ALTERNATIVE VALUATION METHODS

Several alternative methods offer more comprehensive frameworks for economic evaluation to address the limitations of DCF analysis. These approaches include **Real Options Valuation (ROV)**, which allows for the assessment of potential future opportunities and the flexibility to adapt to changing market conditions, and **Economic Value Added (EVA)**, which focuses on the true economic profit of a project by considering both the cost of capital and the returns generated.

Additionally, **Multi-Criteria Decision Analysis (MCDA)** incorporates qualitative factors and stakeholder preferences, providing a more holistic view of project viability beyond financial metrics. Each method enhances the evaluation process by integrating a broader range of variables, ultimately leading to more informed decision-making. These include:

15.3.1 Real Options Valuation (ROV)

ROV[1] is a sophisticated analytical framework that treats upstream investments, such as in oil and gas exploration or technology development, as financial options. This approach enables companies to assess the economic value of strategic decisions, including the timing and scale of project investments. By incorporating market volatility, price fluctuations, and potential for future opportunities, ROV allows businesses to make informed choices about delaying, expanding, contracting, or abandoning projects. This flexibility is crucial in dynamic markets, where conditions can change rapidly, influencing investment initiatives' risk and return profile. By quantifying these choices, ROV provides a more nuanced understanding of investment potential and helps organizations optimize their capital allocation strategies.

> In uncertain environments, value lies not in prediction but in flexibility.
>
> **Stewart Myers**
> *MIT Sloan*

- **Key Features:**
  - Models' **decision flexibility**
  - Accounts for **market volatility**
  - Improves **capital allocation strategies**

- **Example:**
  - A company evaluating **Arctic drilling** can assign a **real option value** to **deferring the project** until oil prices rise above $90 per barrel.
  - Traditional DCF would classify this as a **negative NPV investment**, while **ROA highlights the strategic benefit of waiting**.

15.3.2 Monte Carlo Simulation

Monte Carlo simulation[2] is a powerful statistical technique that incorporates **probability distributions** into **cash flow projections**, enabling companies to model and analyze a vast array of potential financial outcomes. By simulating thousands of scenarios with varying assumptions about key variables, such as market conditions, interest rates, and operational costs, businesses can not only assess the likelihood of different cash flow scenarios but also quantify the associated risks. This approach allows organizations to make more informed decisions, understand the impact of uncertainty on their financial forecasts, and strategically plan for potential adverse events or opportunities, ultimately leading to more resilient financial management (Figure 15.2).

North Sea Operator Writes Down $2 Billion After Price Miss

Bloomberg, November 2015

The best way to cope with uncertainty is not to eliminate it, but to understand it.

Nassim Nicholas Taleb

- **Key Features:**
  - Generates **thousands of possible outcomes**
  - Helps determine the **probability of financial success or failure**
  - Provides a **risk-adjusted view** of project value

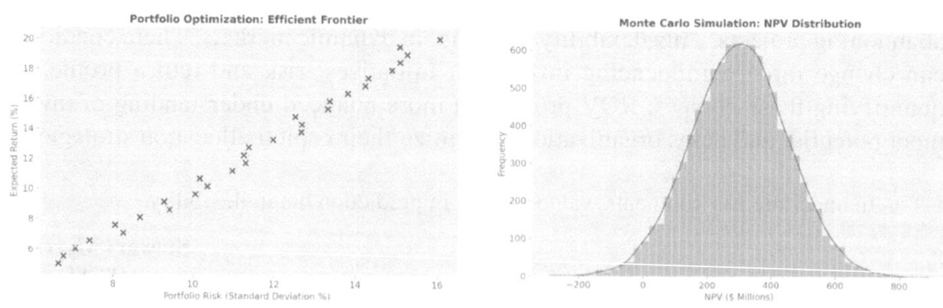

FIGURE 15.2 Portfolio optimization and Monte Carlo simulation in upstream valuation. (Left) Efficient frontier showing optimal risk-return combinations across upstream assets. (Right) NPV distribution from Monte Carlo simulation, illustrating probabilistic outcomes and investment uncertainty.

- **Example:**
  - An offshore drilling project runs **10,000 simulations** for oil prices, **ranging from $40 to $120 per barrel**.
  - The model calculates a **75% probability** of positive NPV, providing **better risk insight than static DCF**.

15.3.3 Decision Tree Analysis

Decision trees are a powerful tool for mapping out **investment pathways** by visually representing potential scenarios and branching decisions that investors may encounter. Each tree branch illustrates a different alternative or outcome based on specific criteria, allowing for a structured analysis of choices such as risk levels, market conditions, and expected returns. This approach not only helps visualize the potential consequences of each decision but also aids in evaluating the probabilities and impacts associated with different investment strategies, ultimately guiding investors toward more informed and strategic choices.

- **Key Features:**
  - Breaks down complex projects into **manageable stages**
  - Quantifies risk and reward at **each decision point**
  - Helps structure **staged investments**
- **Example:**
  - A liquefied natural gas (LNG) terminal can be developed in **phases**:
    - **Phase 1:** Small-scale export facility ($2B investment)
    - **Phase 2:** Expand to full scale operations ($8B investment)
  - Decision trees help evaluate **optimal expansion paths** under various market conditions.

15.3.4 Break-Even and Sensitivity Analysis

Break-even analysis is a financial assessment that identifies the **minimum oil price necessary** for a project to cover its total costs, ensuring profitability. This analysis takes into account both fixed costs, such as infrastructure and development expenses, and variable costs, including operational expenditures and production costs. By calculating the **break-even point**, project managers can evaluate the sensitivity of profitability to fluctuations in oil prices, enabling more informed decision-making regarding investment and operational strategies in the volatile energy market. Understanding this threshold is crucial for assessing the viability of an oil project and making strategic adjustments to enhance economic performance (Figure 15.3).

THE BREAK-EVEN MIRAGE

Break-even price ≠: Safe investment

Many shale plays boast $35–40 breakevens, but suffer from steep decline curves, high reinvestment rates, and hidden costs.

- Always pair break-even with sensitivity or MCS to reveal hidden fragilities.

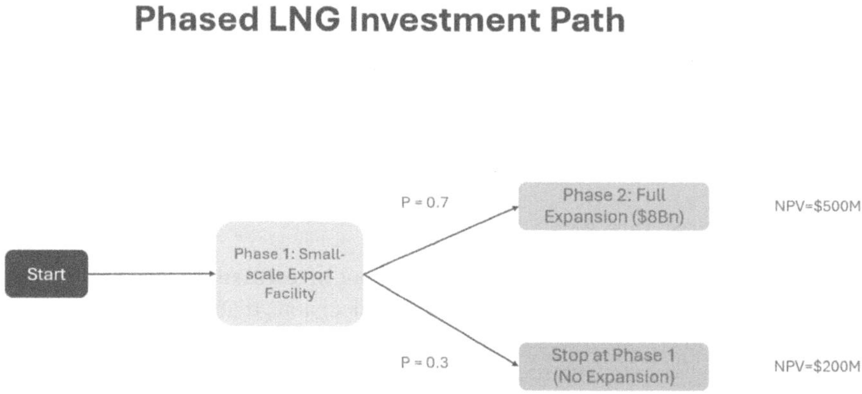

FIGURE 15.3 Phased LNG investment path using real options analysis. Decision tree illustrating staged investment in LNG infrastructure, with probability-weighted outcomes and NPV impact. Highlights strategic flexibility in deferring or expanding capital commitments based on market signals.

- **Key Features:**
  - Identifies **vulnerability to price fluctuations**
  - Helps guide **risk management strategies**
  - Supports **hedging and cost control efforts**
- **Example:**
  - A shale gas well has a **break-even price of $40 per barrel**.
  - If oil prices fall below $40, production **may be halted**.
  - Sensitivity analysis helps **anticipate price swings and optimize production schedules**.

WHAT RAROC TELLS YOU THAT IRR DOESN'T

RAROC vs. IRR

While IRR shows a project's return relative to investment, RAROC tells you how much return you're getting per unit of economic risk.

- A high-IRR project can be rejected under RAROC if its risk exposure is too high.
- Useful in constrained capital environments like NOCs or corporate portfolio reviews.

15.3.5 RAROC (Risk-Adjusted Return on Capital)

RAROC provides a structured framework for evaluating upstream oil and gas investments by comparing expected returns against the economic capital at risk. Unlike traditional financial metrics, which often focus on absolute returns, RAROC adjusts profitability based on the project's riskiness, making it ideal for comparing assets with differing uncertainty levels or capital intensities. This makes it especially valuable for non-operating partners and corporate investment committees looking to prioritize opportunities within a limited capital envelope. RAROC aligns investment selection with broader portfolio risk appetite and ensures that capital is deployed to projects that generate actual economic value per unit of risk.

- **Key Features:**
  - Quantifies return per unit of risk capital
  - Useful for comparing projects with unequal risk or size
  - Supports capital allocation and portfolio balancing decisions
- **Example:**
  Two offshore developments are under review:
  - **Project A**: $1B investment, 20% expected return, low reservoir risk
  - **Project B**: $500M investment, 25% expected return, but high geological and fiscal uncertainty
- Although Project B has a higher nominal return, its RAROC is lower due to elevated risk.
- Project A is preferred under a risk-adjusted capital strategy.

15.3.6 EVA (Economic Value Added)

EVA[3] is a performance metric that captures the true economic profitability of an upstream project by subtracting the cost of capital from the net operating profit after tax (**NOPAT**). Unlike accounting-based measures such as net income, EVA highlights whether a project genuinely creates value or merely covers costs. This approach encourages capital discipline and aligns investment decisions with shareholder value creation, especially in capital-intensive sectors like oil and gas. By quantifying value creation in dollar terms, EVA is a practical tool for evaluating asset performance, benchmarking operators, and screening marginal developments.

HOW EVA DIFFERS FROM ROCE

EVA vs. ROCE (Return on Capital Employed) ROCE = EBIT/Capital Employed EVA subtracts the full cost of capital from profit.

- ROCE can look great even when shareholders are not compensated for risk.
- EVA ensures capital truly generates value.

- Key Features:
  - Measures value creation beyond the cost of capital
  - Encourages efficient use of capital and long-term thinking
  - Useful for benchmarking across operators or business units
- Example:
  - A mature offshore oil field generates a yearly NOPAT of $150 million.
  - The capital employed is $900 million, and the WACC is 10%.
  - EVA = $150M – ($900M × 10%) = $60M
  - This shows the project adds $60M of value per year above its cost of capital.

15.3.7 Scenario-Based Valuation

Scenario-based valuation evaluates upstream projects across a set of well-defined, discrete future states, such as high-price, base-case, and low-price scenarios, rather

Scenario	Price ($/bl	NPV ($M)	?robability	Weighted	NPV
1	High Case	100.00	600.00	0.20	120.00
2	Base Case	75.00	300.00	0.60	180.00
3	Low Case	50.00	(50.00)	0.20	(10.00)

FIGURE 15.4 Scenario-Based valuation of upstream project. Weighted NPV analysis across high, base, and low oil price scenarios. Highlights how probability-adjusted outcomes support strategic investment decisions under market uncertainty.

than relying on a single deterministic outcome. This method enables decision-makers to visualize the impact of key uncertainties (like oil prices, development costs, or delays) under structured assumptions. While not as granular as a Monte Carlo Simulation, it provides clarity and simplicity when communicating with executives or boards. It also supports stress testing and can be paired with probability weighting to approximate risk-adjusted outcomes without complex simulations (Figure 15.4).

Oil Plunges to $30 – Dozens of Projects Shelved Overnight

Financial Times, March 2020

- Key Features:
  - Evaluates discrete "what-if" scenarios with explicit assumptions
  - Useful for stress testing and board-level strategy reviews
  - Helps highlight key value drivers under varying conditions
- Example:
  - An exploration asset is evaluated under three oil price scenarios:
    - High Case: $100/bbl → NPV = $600M
    - Base Case: $75/bbl → NPV = $300M
    - Low Case: $50/bbl → NPV = –$50M
  - The company assigns probabilities: 20%/60%/20%
  - Expected NPV = (0.2 × 600) + (0.6 × 300) + (0.2 × –50) = $305M
  - This provides a more balanced perspective than a single-case analysis.

15.3.8 Multi-Criteria Decision Analysis (MCDA)

MCDA is a structured approach for evaluating upstream projects when decisions must consider quantitative and qualitative factors. It is particularly useful when trade-offs are involved, such as choosing between a high-NPV project with environmental concerns and a lower-return project with stronger social alignment. By assigning weights to multiple criteria (e.g., financial return, emissions, geopolitical risk, strategic fit), MCDA helps investors and policymakers rank competing options transparently and consistently. It also supports consensus-building in multi-stakeholder environments.

- **Key Features:**
  - Combines financial and non-financial criteria into one evaluation
  - Supports transparent, structured decision-making
  - Useful in ESG-sensitive, regulatory, or complex political settings

- **Example:**
  - A national oil company evaluates two gas field development options:
    - **Field A**: High IRR but located in a conflict-prone region
    - **Field B**: Moderate IRR but strong infrastructure and environmental performance
  - Criteria: IRR (40%), Political Risk (30%), ESG Alignment (30%)
  - MCDA scoring leads to **Field B** ranking higher overall, despite lower financial returns.

15.3.9 Portfolio Optimization

Portfolio optimization in upstream oil and gas focuses on balancing risk and return across a company's asset base rather than evaluating projects in isolation. Using techniques such as mean-variance analysis, efficient frontier modeling, or risk envelope frameworks, it allows companies to identify which combination of projects offers the best risk-adjusted performance. This approach is especially critical for capital-constrained organizations or those seeking to align their portfolio with evolving strategic goals (e.g., decarbonization, regional focus). It also highlights underperforming or non-core assets for possible divestment (Figure 15.5).

Majors Exit Risky Blocks in Africa, Focus on Near-Infrastructure Plays

Reuters, July 2023

- **Key Features:**
  - Evaluates overall portfolio performance, not just individual projects
  - Helps identify diversification benefits and risk concentration
  - Supports capital allocation, asset rotation, and long-term planning

Efficient Frontier Table - Upstream Portfolio Assets

Asset	Expected Return (%)	Risk (Std Dev%)	Region
Asset A	6.2	6.5	MENA
Asset B	7.8	7.4	Asia
Asset C	9.1	8.3	Asia
Asset D	10.3	9.2	North Sea
Asset E	11.5	10	GOM
Asset F	12.7	10.8	Brazil
Asset G	13.9	11.7	West Africa
Asset H	15	12.5	US Onshore
Asset I	16.2	13.4	Asutralia
Asset J	17.4	14.2	Frontier

FIGURE 15.5 Efficient frontier table: Upstream portfolio assets. Expected return and risk metrics for ten upstream assets across global regions. Supports portfolio optimization by identifying optimal combinations of return and volatility.

- **Example:**
  - An IOC has 10 global upstream assets varying in size, region, and risk profile.
  - Using optimization tools, it finds that selling two high-risk frontier assets and reinvesting in lower-risk, near-infrastructure projects improves expected returns by 15% while reducing downside risk by 25%.
  - This repositioning strengthens the portfolio's resilience to oil price shocks.

15.4 CASE STUDIES: APPLYING DCF ALTERNATIVES IN UPSTREAM PROJECTS

15.4.1 Case Study 1: Real Options in Arctic Drilling

A major oil company planned a **$5 billion Arctic exploration project** but was hesitant due to oil price uncertainty. Using **Real Options Analysis (ROA):**

- The company calculated that **waiting 5 years** would increase project value by **$1.2 billion** if oil prices exceeded $90 per barrel.
- ROA justified a **delayed investment decision**, avoiding financial losses during low-price periods.

15.4.2 Case Study 2: Monte Carlo Simulation in Deepwater Oil

A **West African deepwater project** required a **$4 billion investment**. Traditional DCF suggested a **50% probability of success**. Monte Carlo simulation revealed:

- A **30% probability of negative NPV** due to oil price risk.
- A **70% probability of NPV > $800 million** under favorable conditions.
- This insight helped structure **contingency plans and risk-mitigation strategies**.

15.4.3 Case Study 3: Break-Even Analysis in U.S. Shale

A shale gas producer analyzed whether to invest in **100 new wells**. Using **break-even modeling**:

- At **$45 per barrel**, the project had a **positive NPV**.
- At **$35 per barrel**, only **30% of wells remained viable**.
- This led to **targeted drilling**, focusing on **high-margin locations**.

15.5 CONCLUSION: ENHANCING UPSTREAM DECISION-MAKING WITH ADVANCED ECONOMIC MODELS

While DCF analysis remains a foundational tool in upstream project valuation, the energy sector's evolving complexity and volatility demand more sophisticated approaches. To capture value, manage uncertainty, and support capital efficiency, upstream companies must integrate a broader set of modern economic evaluation tools.

> Risk comes from not knowing what you're doing.
>
> **Warren Buffett**

- **Real options valuation (ROV):** This method incorporates flexibility into investment planning, allowing companies to delay, scale, or abandon projects based on evolving market signals. It helps capture upside potential and mitigate downside risk in uncertain environments.
- **Monte Carlo simulation (MCS):** Quantifies uncertainty by simulating various outcomes for key input variables. It produces probability-weighted financial metrics and supports risk-informed decisions across project phases.
- **Decision tree analysis:** This technique structures complex investment decisions into branches and nodes, mapping out various outcomes, probabilities, and decision points. It is ideal for phased developments and regulatory-sensitive assets.
- **Break-even and sensitivity analysis** help identify financial thresholds and key value drivers. They support cost control, stress testing, and strategic hedging decisions under varying price or cost environments.
- **RAROC (risk-adjusted return on capital):** Measures project return per unit of risk capital. It aligns investment decisions with corporate risk appetite and enables consistent comparisons across assets of different sizes and uncertainty profiles.
- **Economic value added (EVA):** This measure highlights whether a project truly creates shareholder value after covering its full cost of capital. It promotes capital efficiency and long-term value generation beyond accounting profits.
- **Scenario-based valuation:** This approach offers clear, structured analysis across distinct future cases (e.g., high/low oil prices, delayed development). It enhances transparency and supports decision-making in environments with discrete strategic choices.
- **Multi-criteria decision analysis (MCDA):** This method integrates non-financial factors such as ESG, political risk, and social impact. It is particularly useful when decisions must balance commercial, environmental, and strategic trade-offs.
- **Portfolio optimization:** Evaluates the risk-return profile of the asset portfolio as a whole. It helps prioritize capital allocation, reduce concentration risk, and align project selection with long-term strategic goals.

Integrating these advanced tools enables upstream companies to move beyond static financial metrics and toward a more dynamic, risk-aware, and value-driven approach to investment evaluation. By doing so, firms enhance decision quality and ensure better alignment with shareholder expectations, energy transition goals, and global market uncertainties.

NOTES

1. Real Options Valuation is based on financial option theory first formalized by Black-Scholes (1973), adapted for project evaluation by Trigeorgis and others in the 1990s.
2. Named after the Monte Carlo Casino in Monaco, this technique was developed by scientists working on the Manhattan Project in the 1940s to model nuclear physics problems.
3. EVA was popularized by Stern Stewart & Co. as a performance metric for corporate finance in the 1990s.

16 Energy Policy and Regulation

16.1 INTRODUCTION: THE ROLE OF ENERGY POLICY IN UPSTREAM OIL AND GAS

Energy policies and regulations play a crucial role in shaping the global upstream petroleum sector by directly influencing investment decisions, operational practices, and the overall sustainability of the industry. Governments, regulatory bodies, and international organizations establish comprehensive frameworks designed to balance the often conflicting priorities of resource development, environmental protection, and energy security.

These frameworks encompass a range of measures, including licensing requirements, tax incentives, environmental assessments, and emissions regulations. For instance, countries may implement stringent environmental regulations to minimize ecological damage from exploration and extraction activities, while simultaneously offering tax breaks to attract foreign investments in oil and gas projects.

Moreover, international agreements—such as the Paris Agreement—encourage nations to commit to reducing carbon emissions and transitioning to renewable energy sources, thereby influencing domestic policies governing fossil fuel production. This interplay between regulation and policy not only impacts the operational practices of oil companies but also drives innovation in cleaner technologies and more sustainable practices within the sector. Ultimately, effective energy policies aim to ensure a reliable energy supply while safeguarding natural resources for future generations.

This chapter explores the key components of upstream energy policy, the impact of regulations on the oil and gas industry, and how companies navigate these evolving frameworks. It also examines case studies of regulatory successes and challenges in different regions.

16.2 KEY ELEMENTS OF UPSTREAM ENERGY POLICY

Energy policies play a critical role in shaping the legal and economic framework that governs oil and gas exploration, production, and resource management. These policies establish regulations that dictate how resources are discovered, extracted, and utilized, ensuring environmental protection and sustainable practices. They also define the licensing processes for exploration and production, setting standards for safety and efficiency. Additionally, energy policies often include provisions for taxation, investment incentives, and the allocation of royalties, which significantly influence the economic viability of energy projects. By balancing the interests of stakeholders, including governments, corporations, and local communities, these

DOI: 10.1201/9781003750345-19

policies aim to foster responsible resource management while addressing energy security and climate change challenges. The key elements include:

1. **Licensing and Exploration Rights**
   Governments exert control over access to hydrocarbon resources through a structured framework that includes licensing rounds, production-sharing agreements (PSAs), and concession contracts. Licensing rounds are competitive processes where oil companies bid for the right to explore and extract hydrocarbons in specific areas, ensuring that the government can select the most qualified operators. PSAs outline the terms under which oil companies can operate, specifying how production will be shared between the government and the company, allowing for a transparent division of both costs and revenues based on pre-agreed ratios. Concession contracts grant companies the right to extract resources for a certain period, often involving royalties and taxes that ensure the government receives a fair share of the income generated. Collectively, these agreements are designed to promote fair competition among companies while also safeguarding national interests, ensuring that resource extraction translates into significant revenue generation for the government and the local economy.
   - **Example:** The UK's licensing rounds under the Oil and Gas Authority (OGA) allow companies to bid for offshore exploration blocks, balancing industry growth with environmental considerations.
2. **Fiscal Regimes and Taxation**
   Oil and gas projects operate under a range of fiscal regimes designed to balance the interests of governments and investors. These regimes typically include royalties, which are payments made to the government based on a percentage of the production value; corporate taxes, which are levied on the companies' profits[1]; and profit-sharing agreements, where the profits are divided between the government and the operating companies after recovering the investment costs (Figure 16.1).

   Comparison of effective tax rates across Norway, Nigeria, the U.S., and Qatar. Highlights key attributes of fiscal regimes—predictability, competitiveness, and transparency—and their role in balancing government take and investor incentives. Norway's model exemplifies high taxation paired with policy stability and robust development activity.

   > **WHAT MAKES A GOOD FISCAL REGIME?**
   >
   > - Predictability
   > - Competitive rates
   > - Transparent cost recovery
   > - Aligned incentives (e.g., early cost recovery)

   Governments often structure these tax frameworks to create an attractive investment climate, encouraging private sector capital while also ensuring that a fair portion of the revenues generated from these natural resources is allocated to public services and infrastructure. This intricate balancing act aims to promote sustainable economic growth in resource-rich regions, ensuring that local communities benefit from the wealth generated by their natural resources. Additionally,

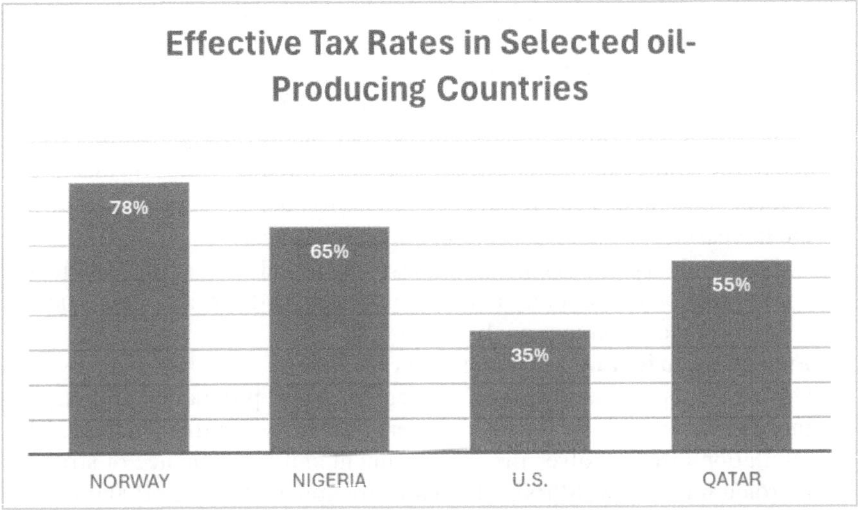

FIGURE 16.1 Effective tax rates and fiscal regime characteristics in oil-producing countries.

the specific terms of these fiscal regimes can vary significantly from one jurisdiction to another, reflecting local policy priorities, economic conditions, and the level of resource development activity.

- **Example:** Norway's petroleum tax regime imposes a high tax rate (78%) but provides cost recovery incentives, making it an attractive yet heavily regulated environment.

3. **Environmental and Safety Regulations**
Regulatory frameworks play a crucial role in enforcing stringent environmental standards, safety protocols, and emissions limits designed to mitigate the adverse effects of upstream operations in sectors such as oil, gas, and mining. These regulations ensure that companies adopt best practices to minimize environmental degradation and protect public health. Compliance with these legal requirements is not only essential for obtaining necessary permits to initiate and continue operations but also vital for maintaining a social license[2] to operate—a concept that emphasizes the importance of gaining and retaining community trust and support. By adhering to these standards, companies demonstrate their commitment to sustainable development, corporate responsibility, and environmental stewardship, which are increasingly vital for long-term operational success.
    - **Example:** The U.S. Bureau of Safety and Environmental Enforcement (BSEE) enforces strict offshore drilling regulations to prevent disasters like the Deepwater Horizon spill.
4. **Local Content and Social Responsibility**
Regulatory frameworks play a crucial role in enforcing stringent environmental standards, safety protocols, and emissions limits designed to mitigate the adverse effects of upstream operations in sectors such as oil, gas, and mining. These regulations ensure that companies adopt best practices to

minimize environmental degradation and protect public health. Compliance with these legal requirements is not only essential for obtaining necessary permits to initiate and continue operations but also vital for maintaining a social license to operate—a concept that emphasizes the importance of gaining and retaining community trust and support. By adhering to these standards, companies demonstrate their commitment to sustainable development, corporate responsibility, and environmental stewardship, which are increasingly vital for long-term operational success.

- **Example:** Brazil's Petrobras follows strict local content rules, requiring international operators to source materials and services domestically, boosting economic growth.

5. **Energy Security and Strategic Reserves**
Governments play a crucial role in regulating oil production and managing reserves to ensure long-term energy security for their nations. This regulation often involves the establishment and maintenance of strategic petroleum reserves (SPRs), which are stockpiles of crude oil and refined products that can be tapped during supply disruptions or crises. These reserves are typically sufficient to meet a country's emergency needs for several months. Additionally, policies may include stringent regulations on oil exports to control domestic supply and prices, preserve resources for future generations, and maintain a stable energy market. Governments may also implement measures to promote energy efficiency and transition to alternative energy sources, further enhancing their strategies for sustainable energy management.
   - **Example:** The U.S. maintains the Strategic Petroleum Reserve (SPR), a massive stockpile of crude oil used during supply disruptions.

16.3 THE IMPACT OF ENERGY REGULATIONS ON UPSTREAM INVESTMENT

Regulatory stability and predictability are essential factors for attracting upstream oil and gas investment. A well-defined regulatory framework enables investors to accurately assess the long-term viability and potential returns of their projects. Excessive regulation, characterized by convoluted processes, frequent changes, and high compliance costs, can deter potential investors, as it creates uncertainty and increases the risk associated with capital expenditures. Conversely, insufficient oversight may lead to significant environmental issues, such as oil spills or habitat destruction, as well as financial repercussions, including costly legal battles and damage to the company's reputation. Therefore, striking an appropriate balance between regulation and flexibility is crucial to fostering a healthy investment climate while ensuring environmental protection and financial integrity.

> Regulations are not roadblocks. They are the guardrails for sustainable growth.
>
> **Fatih Birol**
> *IEA*

Positive Impacts of Regulation:

- ✓ **Investor confidence and stability**: Establishing clear legal frameworks along with consistent and predictable policies safeguards investments and encourages long-term capital allocation. This stability assures investors that their investments will be protected from sudden regulatory changes, ultimately fostering a robust economic growth and innovation environment.
- ✓ **Environmental protection**: Implementing stringent regulations encourages companies to adopt industry-wide emissions reduction and sustainability benchmarks. This proactive approach leads to the development of cleaner technologies and practices, prompting industries to invest in renewable energy sources and improve operational efficiencies, thereby contributing to a healthier ecosystem (Figure 16.2).
- ✓ **Public trust and social license to operate**: Transparent and well-communicated policies are critical in earning public trust and acceptance of oil and gas operations. By actively engaging with communities and stakeholders and demonstrating a commitment to environmental stewardship and social responsibility, companies can bolster their reputation and secure the social license to operate, reducing the potential for conflicts and enhancing community relationships (Figure 16.3).

Challenges and Risks:

- **Regulatory uncertainty**: Frequent shifts in government policies, including abrupt changes to regulations and standards, create a climate of unpredictability that can deter potential investors. This lack of stability makes it challenging for businesses to devise long-term strategies, resulting in hesitancy to commit significant capital to projects that may face sudden policy alterations or shifts in regulatory expectations.
- **Bureaucratic delays**: Lengthy approval processes for essential permits and comprehensive environmental assessments often extend far beyond initial timelines. These delays can result from complex layers of bureaucratic red tape, leading to increased costs and project timelines. The protracted nature

Fiscal Regime Comparison Matrix

Country	Effective Tax Rate (%)	Royalty Structure	Profit Sharing	Policy Stability	Transparency
Norway	78%	Fixed	**Yes**	**High**	**High**
Nigeria	65%	Sliding Scale	**Yes**	Medium	Low
U.S.	45%	Lease-based	No	**High**	**High**
Qatar	35%	Fixed	**Yes**	**High**	Medium

FIGURE 16.2 Comparative attributes of fiscal regimes in oil-producing countries. Matrix comparing effective tax rates, royalty structures, profit-sharing mechanisms, and governance characteristics across Norway, Nigeria, the U.S., and Qatar. Supports fiscal scoring and strategic prioritization in upstream portfolio planning.

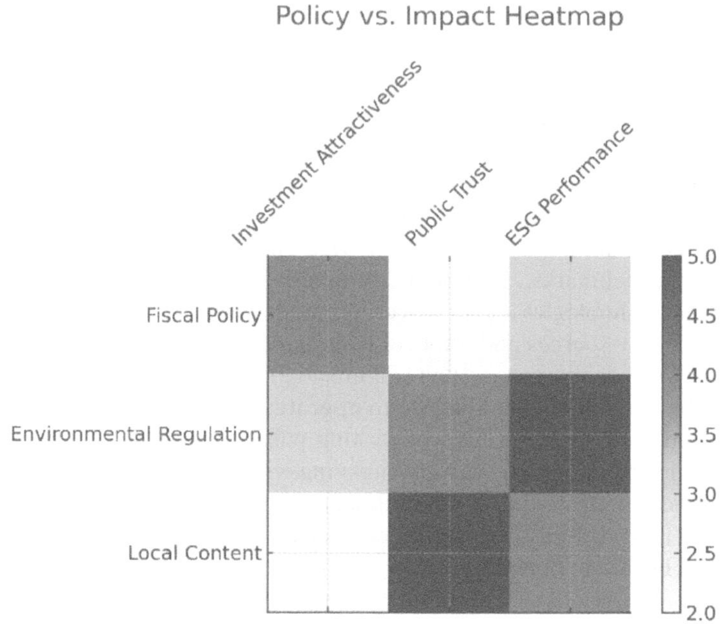

FIGURE 16.3 Policy vs. impact timeline and social license considerations. Assess the impact of key regulatory and social factors on upstream project execution. Transparent policy implementation and proactive stakeholder engagement enhance public trust and secure the social license to operate. Bureaucratic delays and regulatory uncertainty remain critical risks to timeline and capital efficiency.

of these procedures not only affects the immediate execution of projects but can also discourage stakeholders and reduce overall project viability.

- **Cost burdens**: The financial strain associated with compliance can be substantial, leading to high operational costs for businesses. These costs may include fees for obtaining necessary permits, expenses related to legal consultations, and investments in ensuring regulatory adherence. For many projects, particularly in sectors like renewable energy or construction, these burdens can render otherwise lucrative opportunities economically unviable, ultimately affecting their feasibility and potential contribution to economic growth.

16.4 REGULATORY TRENDS AND GLOBAL CASE STUDIES

1. **The European Union: A Push for Decarbonization**
The European Union is at the forefront of the transition toward low-carbon energy, spearheading initiatives to reduce greenhouse gas emissions across member states. To achieve this goal, the EU has implemented stringent regulations targeting emissions from various sectors, including transportation, industry, and agriculture. Additionally, the bloc has pledged to phase out fossil fuel investments, encouraging a shift toward renewable energy sources such as wind, solar, and hydroelectric power. This comprehensive approach not only aims to mitigate climate change but also seeks to enhance energy security and

promote sustainable economic growth throughout the region. Through policies such as the European Green Deal and the Fit for 55 package, the EU is committed to reducing carbon emissions by at least 55% by 2030 compared to 1990 levels, setting a clear roadmap for a greener and more sustainable future.

EU Taxonomy Freezes Oil Funding, Redirects Billions to Renewables

Financial Times, January 2023

- **Example:** The EU Taxonomy for Sustainable Finance restricts funding for upstream oil and gas projects, encouraging renewable energy investments instead.

2. **The United States: Balancing Growth and Environmental Protection**
The United States operates within a multifaceted regulatory framework that seeks to balance domestic energy independence with the imperative of environmental sustainability. This landscape encompasses a range of policies, including tax incentives for renewable energy development, stringent emissions standards for both transportation and industry, and comprehensive programs designed to protect natural resources. Additionally, the regulatory framework encompasses federal, state, and local regulations, reflecting the diverse energy needs and environmental priorities across different regions. As the nation grapples with the challenges of climate change, policymakers continuously navigate the interplay between fostering economic growth through energy production and ensuring the protection of ecosystems for future generations.

U.S. Halts Offshore Drilling After BP Disaster

New York Times, May 2010

- **Example:** The Inflation Reduction Act (IRA) of 2022 includes incentives for carbon capture but increases scrutiny on new drilling permits.

3. **The Middle East: National Oil Companies (NOCs) and State Control**
In many Middle Eastern countries, the oil industry is predominantly state-controlled, with National Oil Companies (NOCs) playing a central role in upstream activities such as exploration, drilling, and production. These NOCs, often backed by substantial government resources, have significant influence over global oil markets and production strategies. For instance, Saudi Aramco in Saudi Arabia and the Iraq National Oil Company are prime examples of how these entities not only drive domestic economic growth but also shape international energy policies. Their operations are closely intertwined with national interests, often leading to collaborations with foreign investors while maintaining strategic control over their rich hydrocarbon resources.
   - **Example:** Saudi Aramco operates under a highly regulated environment, aligning production with OPEC+ strategies to manage global oil supply.

4. **Africa: Regulatory Uncertainty and Investment Challenges**
Many African nations face significant challenges due to inconsistent energy policies, which hinder their ability to attract foreign investment. Despite possessing abundant hydrocarbon reserves, such as oil and natural gas, these countries often experience fluctuations in regulatory frameworks, a lack of

MCDA Scoring Matrix for Project Approval

Criteria	Weight (%)	Project A	Project B	Project C
Fiscal attractivness	30%	8.5	6.0	7.2
Regulatory Certainty	25%	9.0	5.5	6.8
EGS Alignment	20%	7.0	8.5	6.0
Community Support	15%	6.5	9.0	5.5
Infrastructure Fit	10%	8.0	7.0	6.5

FIGURE 16.4 Multi-criteria decision analysis (MCDA) scoring for upstream project selection. Weighted evaluation of three upstream projects across fiscal attractiveness, regulatory certainty, ESG alignment, community support, and infrastructure fit. Supports strategic prioritization by integrating financial and non-financial criteria into a unified decision framework.

infrastructure, and insufficient incentives for investors. This instability not only affects the confidence of potential investors but also limits the development of energy projects that could leverage these valuable resources for economic growth and energy security. As a result, nations rich in hydrocarbons find themselves struggling to capitalize on their natural wealth, which could otherwise transform their economies and improve living standards for their populations.

Nigeria's Oil Reform Hangs in the Balance

Reuters, November 2022

- **Example:** Nigeria's Petroleum Industry Act (PIA) aims to reform governance in the oil sector, but regulatory enforcement remains a challenge (Figure 16.4).

16.5 CASE STUDIES: THE ROLE OF REGULATION IN UPSTREAM DEVELOPMENT

16.5.1 Case Study 1: Norway's Balanced Approach to Oil Regulation

Background: Norway has one of the world's most stable regulatory environments, combining high taxation with investment-friendly policies.

Key Policies:

- 78% petroleum tax rate, but companies can deduct exploration costs.
- Strong environmental regulations and carbon pricing mechanisms.

Outcome:

- Norway remains a leading offshore producer while maintaining high ESG standards.

16.5.2 Case Study 2: U.S. Offshore Drilling Moratorium

Background: Following the Deepwater Horizon disaster, the U.S. temporarily banned offshore drilling in the Gulf of Mexico.

Key Policies:

- Stricter safety standards and well-control regulations were introduced.
- Increased financial responsibility for oil spill liabilities.

Outcome:

- While safety improved, the moratorium temporarily reduced investment in offshore oil projects.

16.5.3 Case Study 3: Mexico's Energy Reform and Market Liberalization

Background: In 2014, Mexico ended the state monopoly on oil and gas, allowing foreign investment for the first time in 75 years.

Key Policies:

- Compctitive bidding rounds for exploration and production.
- Fiscal incentives for deepwater and unconventional projects.

Outcome:

- Initial investment surged, but subsequent political changes led to regulatory uncertainty, slowing momentum.

16.5.4 Case Study 4: Canada's Carbon Pricing Strategy

Background: Canada introduced a national carbon pricing policy to reduce emissions while maintaining energy sector competitiveness (Figurc 16.5).

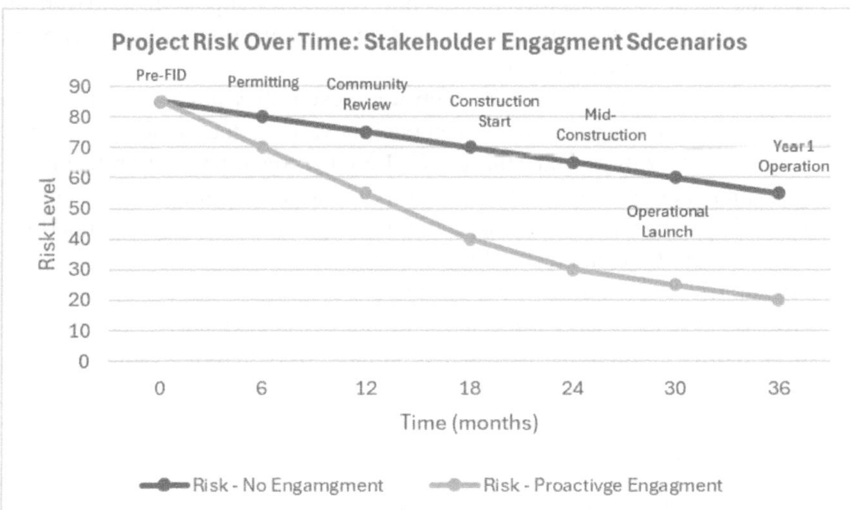

FIGURE 16.5 Project risk over time: Stakeholder engagement scenarios. Comparison of risk trajectories for upstream projects with and without proactive stakeholder engagement. Demonstrates how early transparency and community alignment reduce long-term execution risk and enhance social license to operate.

Key Policies:

- Carbon tax applies to upstream oil and gas production.
- Revenue is reinvested into clean energy projects.

Outcome:

- Mixed reactions from industry players, with concerns over competitiveness but recognition of sustainability benefits.

16.6 THE FUTURE OF ENERGY REGULATION IN UPSTREAM

The global regulatory landscape is undergoing significant transformations to effectively tackle the intertwined challenges of climate change, energy security, and rapid technological advancements. Key future trends are emerging in this arena

✓ **Carbon pricing and emissions trading:** There is an anticipated expansion of global carbon markets designed to incentivize the adoption of low-carbon technologies across various sectors. This includes not only traditional carbon trading systems in regions like the European Union but also the introduction of carbon pricing mechanisms in developing countries, which aim to create a unified approach to global emissions reduction.[3]

✓ **Stricter ESG compliance:** Companies in the oil and gas sector can expect to face increasingly stringent Environmental, Social, and Governance (ESG) requirements. Mandatory ESG reporting will become a standard practice, with regulatory bodies enforcing guidelines that necessitate the disclosure of sustainability targets and practices. This shift is driven by investor demand for transparency and accountability regarding corporate impacts on climate and society.

✓ **Regulatory support for new technologies:** Policymakers are likely to enact supportive regulations to promote innovative solutions such as Carbon Capture, Utilization, and Storage (CCUS), hydrogen production technologies, and alternative fuel sources. These initiatives will include government subsidies and grants for research and development, fostering a competitive environment for clean technology startups.

✓ **Nationalization and resource control:** In response to global market volatility and the increasing importance of energy independence, some nations may tighten regulations to enhance control over natural resources. This could manifest in nationalization or increased taxation on hydrocarbon production to bolster state revenues and prioritize domestic energy needs over export demands.

> The energy future will be won by those who adapt, not those who resist.
>
> *Christiana Figueres*

Example: The UK's "North Sea Transition Deal" exemplifies a balanced approach by seeking to harmonize ongoing oil production with the country's net-zero commitments. Under this deal, oil companies must invest substantial resources into decarbonization projects, emphasizing the transition to renewable energy sources while maintaining economic viability in the region. This strategic alignment aims to mitigate environmental impacts while securing jobs and energy supplies during the transition phase.

ESG MANDATES ARE COMING—READY OR NOT

Regulatory bodies from the EU, U.S., and Asia are rapidly requiring full ESG disclosures.

Companies without frameworks in place risk losing access to capital markets.

16.7 CONCLUSION: NAVIGATING ENERGY POLICIES FOR A SUSTAINABLE FUTURE

Energy policy and regulation will play a pivotal role in shaping the future landscape of upstream petroleum economics. As governments implement increasingly stringent regulations to address environmental concerns and promote sustainability, the petroleum industry faces significant challenges such as compliance costs and operational constraints. However, these regulations also catalyze innovation, compelling companies to invest in advanced technologies and methodologies that enhance efficiency in exploration and production. For instance, the adoption of renewable energy sources and improved extraction techniques can lead to reduced carbon footprints and increased economic viability. Ultimately, while the regulatory environment may pose short-term hurdles, it fosters a long-term shift toward more sustainable practices, ensuring that the industry can adapt and thrive in an evolving energy market. Oil and gas companies must proactively adapt to evolving policies by:

- **Engaging with regulators** to advocate for balanced policies.
- **Investing in compliance and ESG** to align with future regulations.
- **Diversifying portfolios** to hedge against policy shifts.

Companies that proactively embrace regulatory adaptation and implement sustainability initiatives are likely to maintain resilience and competitiveness in an increasingly complex and dynamic energy landscape. By staying ahead of evolving regulations, these organizations can mitigate risks, avoid potential penalties, and capitalize on new market opportunities. Furthermore, by investing in renewable energy sources, enhancing energy efficiency, and integrating sustainable practices into their operations, they not only contribute to environmental preservation but also appeal to a growing consumer base that prioritizes sustainability. This strategic approach positions them to thrive amid challenges such as fluctuating energy prices, technological advancements, and shifting consumer preferences.

16.7.1 Summary of Key Takeaways

✓ Energy policies impact investment decisions, environmental compliance, and industry sustainability.

✓ Upstream fiscal regimes vary globally, influencing profitability and risk.

- ✓ Regulatory trends include carbon pricing, ESG mandates, and market liberalization.
- ✓ Case studies highlight the benefits and challenges of different regulatory approaches.
- ✓ Future policies will drive decarbonization and technological advancements in upstream operations.

NOTES

1. The term “profit oil” refers to the portion of production remaining after cost recovery, which is then split between host governments and contractors under PSCs.
2. “Social license to operate” is a non-legal but powerful concept indicating community and stakeholder acceptance of a company’s activities.
3. The EU Emissions Trading System (EU ETS) is the world’s largest carbon market, covering over 10,000 installations across 30 countries.

17 AI, Automation, and Big Data in Upstream in Oil and Gas

AI won't replace petroleum engineers. But engineers using AI will replace those who don't.

Industry Trend Report, 2023

17.1 INTRODUCTION: THE DIGITAL TRANSFORMATION OF UPSTREAM OPERATIONS

The oil and gas industry is undergoing a significant digital transformation, with **artificial intelligence (AI), automation, and big data** playing a central role in optimizing exploration, production, and decision-making. These technologies are helping upstream companies improve efficiency, reduce operational risks, and enhance profitability in an increasingly complex and competitive landscape.

Historically, the upstream sector relied heavily on **manual data analysis, human-driven decision-making, and traditional exploration techniques**. As global energy markets become more volatile and regulatory pressures increase, companies invest in advanced digital solutions to remain competitive.

This chapter explores the key applications of AI, automation, and big data in upstream petroleum operations, highlighting how these technologies enhance efficiency, mitigate risks, and shape the industry's future (Figure 17.1).

17.2 THE ROLE OF AI IN UPSTREAM OIL AND GAS

AI is transforming upstream operations in the oil and gas industry by enabling advanced predictive analytics, which allows companies to anticipate equipment failures and optimize maintenance schedules. This reduces downtime and enhances operational efficiency. Additionally, AI facilitates autonomous decision-making processes, where algorithms can analyze vast amounts of data in real-time to determine optimal drilling locations and production strategies without human intervention. Furthermore, enhanced data interpretation through AI-driven tools provides deeper insights into geological and operational data, enabling better resource management, increased productivity, and reduced environmental impact. AI applications in the industry include:

1. **Predictive Maintenance and Equipment Monitoring**
   AI-powered predictive maintenance significantly reduces equipment failures by leveraging advanced analytics to interpret sensor data from machinery. By

DOI: 10.1201/9781003750345-20

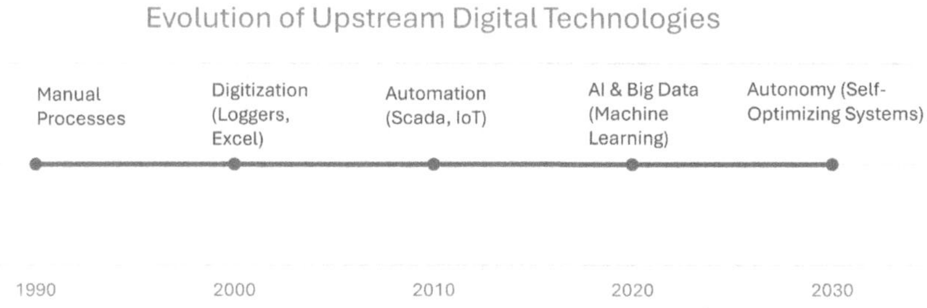

FIGURE 17.1 Evolution of upstream digital technologies (1990–2030). Timeline illustrating the progression from manual processes to autonomous systems in upstream oil and gas operations. Highlights key milestones: digitization (2000), automation (2010), integration of AI and big data (2020), and projected autonomy (2030). Frames the chapter's exploration of how digital technologies reshape exploration, production, and decision-making.

continuously monitoring parameters such as temperature, vibration, and pressure, these predictive maintenance systems can identify anomalies and predict potential breakdowns before they occur. This proactive approach minimizes unplanned downtime, enhancing operational efficiency and safety. Additionally, by anticipating maintenance needs, organizations can optimize repair schedules, which leads to reduced maintenance costs and extends the lifespan of equipment. Overall, integrating predictive maintenance creates a more reliable operational environment, allowing businesses to maintain productivity while reducing risks associated with equipment failures.

- **Example:** BP uses machine learning algorithms to monitor offshore platform equipment, identifying anomalies in pressure, temperature, and vibration data to prevent costly failures.

TOP AI USE CASES IN UPSTREAM

- Predictive maintenance
- Drilling optimization
- Reservoir modeling
- Seismic interpretation

2. **AI-Driven Exploration and Reservoir Characterization**
Machine learning models can analyze extensive datasets comprising seismic, geological, and petrophysical information, significantly enhancing the accuracy of reservoir characterization and exploration efforts. By leveraging advanced algorithms, AI can swiftly discern patterns and anomalies within these complex datasets, allowing for a more precise identification of hydrocarbon-rich formations. This efficiency surpasses traditional exploration methods, which often rely on manual interpretation and may overlook critical insights hidden within the data. Furthermore, machine learning techniques can continuously improve their predictive capabilities as they are exposed to new data, ultimately leading to more informed decision-making in resource extraction and management.
   - **Example:** ExxonMobil employs AI algorithms to analyze seismic images, helping geologists identify potential drilling sites with higher accuracy.
3. **Drilling Optimization and Automated Decision Support**
AI models leverage extensive datasets, including historical drilling data, real-time sensor inputs from the drilling rig, and various geological variables

such as rock formation properties and fluid characteristics. By integrating these data streams, the models can recommend optimal drilling parameters, such as weight on bit, rotation speed, and mud properties. This data-driven approach significantly reduces non-productive time (NPT) associated with equipment failures and operational delays, ultimately enhancing overall drilling efficiency and project cost-effectiveness. By continuously learning from ongoing operations, these AI systems adapt recommendations in real time, allowing for dynamic adjustments that further streamline the drilling process.

- **Example:** Shell's AI-driven **Drilling Advisory System** continuously adjusts drilling parameters to optimize performance and reduce operational risks.

4. **AI in Production Forecasting and Reservoir Management**
AI-driven reservoir models significantly enhance production forecasting by seamlessly integrating real-time well data, such as pressure, temperature, and fluid composition, with historical performance trends derived from past production data. By utilizing advanced machine learning algorithms, these models analyze intricate patterns and correlations within the data to predict future reservoir behavior more accurately. This approach enables dynamic reservoir management, allowing operators to make informed, timely adjustments to production strategies based on current conditions and predictive insights. As a result, operators can optimize output, improve resource allocation, and minimize operational costs while maximizing reservoir recovery efficiency (Figure 17.2).
   - **Example:** Saudi Aramco employs AI models to optimize water injection in mature fields, enhancing **enhanced oil recovery (EOR)** efforts.

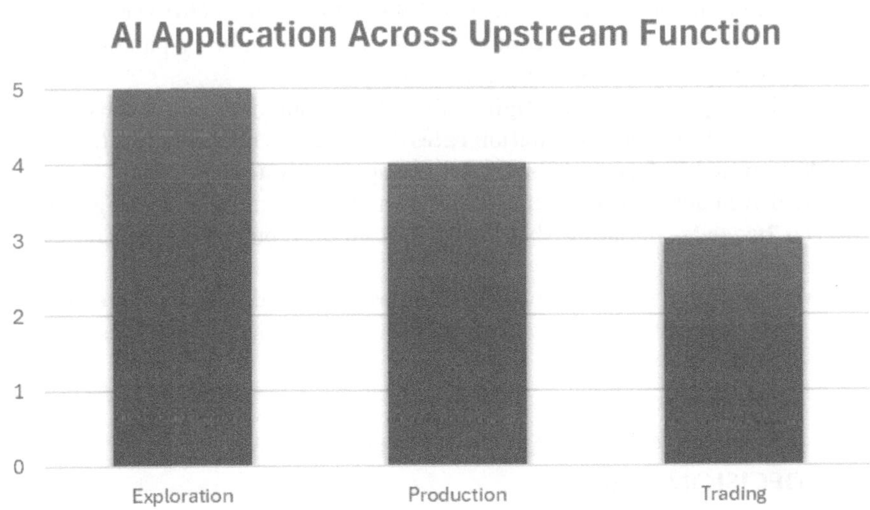

FIGURE 17.2 AI application across upstream functions. Comparative analysis of AI utilization in exploration, production, and trading. Exploration shows the highest adoption, driven by seismic interpretation and reservoir modeling. Production follows with strong use in predictive maintenance and drilling optimization, while trading reflects emerging applications in price forecasting and market analytics.

17.3 AUTOMATION IN UPSTREAM OPERATIONS

Automation plays a vital role in enhancing operational efficiency across various industries. By implementing automated systems, organizations can significantly minimize the need for human intervention, particularly in hazardous environments such as chemical plants or mining sites, where the risk to human life is elevated. This technology safeguards workers and streamlines processes, leading to faster production cycles and reduced downtime. Furthermore, automation contributes to greater process consistency by ensuring that tasks are performed with precision and repeatability, enhancing product quality and reliability. Overall, integrating automation into operational workflows is essential for driving productivity and safety while maintaining high performance standards. Key applications include:

1. **Autonomous Drilling and Remote Operations**
   Automated drilling systems reduce reliance on human drillers by using **real-time data analytics and robotic controls** to optimize drilling speed,[1] pressure, and trajectory. This increases efficiency and reduces the risk of human errors.
   - **Example:** Nabors Industries developed the **SmartRig** drilling system, which automates drilling operations and improves wellbore accuracy.
2. **Robotics and Drones in Offshore and Onshore Fields**
   Robots and drones are increasingly used for **equipment inspections, leak detection, and environmental monitoring** in remote or hazardous areas.
   - **Example:** TotalEnergies uses drones to inspect offshore platforms and detect methane leaks, reducing the need for human intervention and improving safety.
3. **Automated Production Facilities and Control Systems**
   Smart production systems integrate **Internet of Things (IoT) sensors** and AI-driven automation to control production rates, monitor pipeline integrity, and optimize flow rates in real-time.
   - **Example:** Chevron's **digital oilfields** use automation to **adjust valve operations and production rates** dynamically, maximizing efficiency.
4. **Automated Safety and Emergency Response Systems**
   AI-driven automation is transforming safety systems by **detecting potential hazards**, initiating shutdowns, and guiding emergency responses in case of accidents.
   - **Example:** BP has deployed **AI-powered gas detection systems** that automatically trigger alarms and shut down affected facilities in case of leaks.

17.4 BIG DATA AND ANALYTICS IN UPSTREAM DECISION-MAKING

Big data is revolutionizing the way upstream companies in the oil and gas sector process, store, and analyze vast amounts of information. By leveraging advanced analytics and machine learning algorithms, these companies can extract valuable insights from diverse data sources such as geological surveys, sensor data from drilling equipment, and historical production records. This shift toward data-driven decision-making enables more accurate risk assessments, optimizes resource allocation, enhances operational efficiency, and ultimately leads to improved profitability and sustainability in an increasingly competitive market. The key areas where big data is impacting the industry include:

1. **Data Integration and Real-Time Monitoring**
   Modern upstream operations in the oil and gas industry generate vast quantities of geological, geophysical, and production data. This includes detailed seismic surveys, well logs, and drilling reports that provide insights into subsurface formations and their potential for hydrocarbon extraction. Advanced big data platforms aggregate and process these diverse datasets, enabling real-time monitoring of drilling activities, reservoir performance, and equipment condition.[2] This integration facilitates informed decision-making, optimizing both production efficiency and resource management while minimizing operational risks and costs. By leveraging advanced analytics and machine learning, operators can identify trends, predict equipment failures, and enhance exploration efforts, thereby maximizing the economic potential of their assets (Figure 17.3).
   - **Example:** Equinor's **Integrated Operations Center** continuously processes real-time data from offshore platforms, allowing engineers to monitor and optimize production remotely.

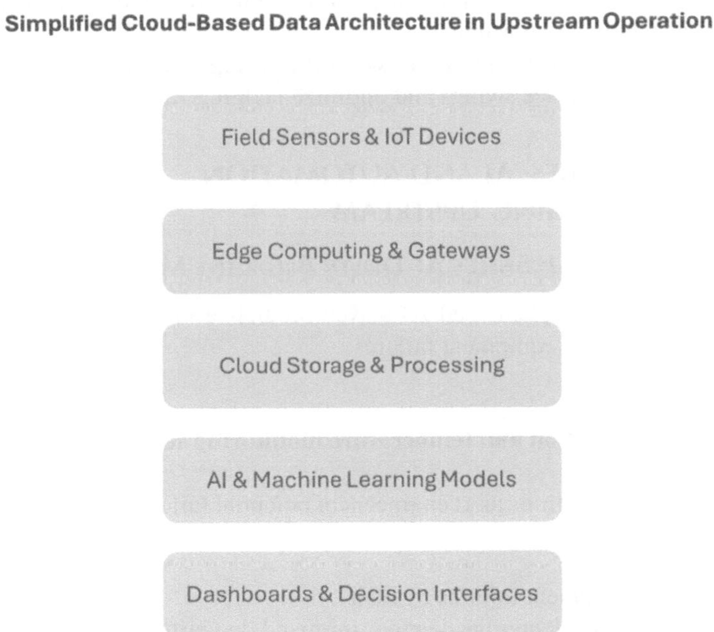

FIGURE 17.3 Simplified cloud-based data architecture in upstream operations. Illustrates the flow of operational data from field sensors and IoT devices through edge computing gateways, cloud processing, and AI models to decision dashboards. This architecture enables real-time monitoring, predictive analytics, and autonomous control across upstream assets.

2. **Advanced Analytics for Reservoir Management**
   Big data analytics empowers reservoir engineers to create sophisticated models that accurately simulate the intricate behaviors of oil and gas reservoirs. By analyzing vast amounts of data from geological, geophysical, and operational sources, these engineers can identify patterns and trends that were previously undetectable. This enhanced modeling capability allows for more effective production optimization strategies, such as improved drilling techniques and optimal resource extraction methods. Additionally, the insights gained from big data enable the development of tailored enhanced recovery strategies, ultimately leading to increased efficiency and yield in hydrocarbon production.
   - **Example:** Petrobras uses **machine learning-driven reservoir simulation models** to optimize offshore production in pre-salt formations.
3. **Market Analytics and Price Forecasting**
   Advanced AI-powered big data models meticulously analyze various facets of the global oil market, including supply and demand dynamics, geopolitical developments, and economic indicators. By leveraging vast datasets that include production levels, consumption trends, and international relations, these models can accurately predict price fluctuations.[3] This analytical capability not only aids investors in making informed decisions but also provides insights into potential market disruptions caused by political instability or shifts in energy policy. Consequently, stakeholders can better navigate the complexities of the oil market and position themselves strategically to capitalize on emerging opportunities.
   - **Example:** BP's trading division utilizes **big data analytics** to anticipate crude price swings and optimize hedging strategies.

17.5 CASE STUDIES: AI AND AUTOMATION TRANSFORMING UPSTREAM

17.5.1 Case Study 1: Shell's AI-Driven Predictive Maintenance

Background: Shell implemented AI-driven predictive maintenance across its offshore assets to prevent equipment failures.

Key Technologies:

- AI-based **vibration and temperature monitoring sensors** on compressors and turbines.
- Predictive algorithms alert engineers of potential failures before they occur.

Outcome:

- Reduced equipment failures by **30%**.
- **$1 billion in cost savings** due to minimized downtime and repairs.

> Shell Saves $1B Through Predictive AI
>
> *Financial Times, June 2022*

17.5.2 Case Study 2: Equinor's Automated Drilling Operations

Background: Equinor deployed **automated drilling control systems** on its offshore rigs to improve operational efficiency.

Key Technologies:

- AI-powered **drilling control systems** that adjust weight-on-bit and rotational speed in real time.
- Machine learning models analyzing **historical well data** to improve decision-making.

Outcome:

- Reduced **drilling time by 20%**.
- Increased wellbore stability and minimized drilling fluid losses.

> Equinor's Smart Rigs Cut Drilling Time by 20%
>
> *Upstream Online, April 2021*

17.5.3 Case Study 3: Chevron's Digital Oilfield and Big Data Integration

Background: Chevron launched its **Digital Oilfield Initiative** to improve production monitoring and operational efficiency.

Key Technologies:

- IoT-enabled **real-time sensors** on production wells.
- Big data analytics for **flow rate prediction and pipeline monitoring**.

Outcome:

- **15% increase in production efficiency**.
- Early detection of **pipeline leaks**, reducing environmental risks.

17.5.4 Case Study 4: BP's AI-Powered Seismic Data Interpretation

Background: BP implemented AI models to interpret **seismic imaging data**, improving exploration success rates.

Key Technologies:

- AI-based **deep learning models** trained on thousands of seismic datasets.
- Automated **fault and reservoir detection** algorithms.

Outcome:

- Reduced **seismic data processing time by 60%**.
- Improved **drilling accuracy**, leading to fewer dry wells.

17.6 THE FUTURE OF AI, AUTOMATION, AND BIG DATA IN UPSTREAM

The oil and gas industry is poised for significant transformation as it increasingly incorporates advancements in AI, automation, and big data analytics. This evolution is expected to manifest in several key areas:

- **More autonomous operations:** The industry is likely to see the deployment of fully automated drilling rigs and production facilities, minimizing human intervention. These autonomous systems can utilize real-time data to make instant decisions, enhancing operational safety and efficiency while significantly reducing the potential for human error.

> **AUTOMATION VS. AUTONOMY**
>
> Automation = Task-level replacement
>
> Autonomy = Full system-level decision-making

- **Advanced AI for reservoir simulation:** Innovations in AI will enable more precise subsurface imaging techniques, such as machine learning algorithms that analyze geological data to create accurate reservoir models.[4] This advancement will facilitate dynamic reservoir management, allowing for real-time adjustments in extraction strategies based on changing reservoir conditions, which can lead to increased recovery rates.
- **Expanded AI-driven trading and market forecasting:** Integrating advanced predictive analytics within trading platforms will provide market analysts with unprecedented insights into crude oil price trends. By analyzing historical data and monitoring real-time market signals, AI can enhance the accuracy of forecasts, helping companies make well-informed trading decisions and optimizing their financial outcomes.
- **Greater integration with renewable energy systems:** AI's role in optimizing hybrid energy models will become crucial as the industry moves toward sustainability. This includes managing offshore wind and gas projects, where AI can analyze performance metrics and environmental data to improve energy efficiency and reduce emissions. Such integration helps balance the fluctuations of renewable energy with traditional gas production, contributing to a more resilient energy grid.

Companies prioritizing investment in digital transformation initiatives will likely gain a substantial competitive edge. By leveraging AI and automation, they can achieve significant cost reductions, enhance operational efficiencies, and improve safety protocols in upstream operations, ultimately positioning themselves favorably in a rapidly evolving energy landscape.

17.7 CONCLUSION: THE DIGITAL FUTURE OF UPSTREAM

The upstream petroleum sector is undergoing a significant transformation driven by advancements in AI, automation, and big data analytics. Companies that adopt and integrate these digital technologies are poised to achieve greater operational efficiency, mitigate risks, and enhance profitability in an increasingly competitive and volatile energy market.

AI technologies, including machine learning algorithms, are being utilized to optimize exploration and production processes by analyzing geological data to identify potential drilling sites more accurately. Automation in drilling operations not only streamlines workflows but also enhances safety by reducing human exposure to hazardous environments.

Moreover, big data analytics empowers companies to harness vast amounts of information generated throughout the production cycle. By leveraging this data, operators can monitor equipment health in real-time, predict maintenance needs, and minimize downtime. The insights gained from data analysis are also crucial for making informed decisions regarding resource allocation and managing supply chain complexities.

As upstream operators continue to embrace these technological advancements, the petroleum industry is evolving toward a more data-driven, automated, and intelligent future. This shift not only enhances operational capabilities but also positions companies to adapt swiftly to changing regulatory landscapes and fluctuating market demands. Ultimately, those who invest in and prioritize these digital innovations will lead the charge in redefining efficiency and sustainability in energy production.

NOTES

1. Autonomous drilling systems integrate control loop automation with real-time formation feedback to dynamically adjust drilling parameters without human intervention.
2. Edge computing enables data processing near the data source, reducing latency and bandwidth requirements when transmitting data from remote oilfields to centralized systems.
3. Oil price models commonly incorporate time series forecasting methods such as ARIMA, LSTM neural networks, and Monte Carlo simulations using historical and macroeconomic variables.
4. Oil price models commonly incorporate time series forecasting methods such as ARIMA, LSTM neural networks, and Monte Carlo simulations using historical and macroeconomic variables.

Part IV

Market and Future Outlook

18 The Future of Upstream Economics

Navigating an Evolving Energy Landscape

The energy companies of tomorrow will not just drill deeper — they'll think cleaner.

Energy Strategy Journal, 2024

18.1 INTRODUCTION: A SECTOR IN TRANSITION

Technological advancements, such as automation, artificial intelligence, and advanced drilling techniques like hydraulic fracturing and horizontal drilling, have significantly transformed the upstream petroleum sector. Regulatory pressures are becoming increasingly stringent, with governments worldwide implementing policies to reduce carbon emissions and incentivize the adoption of cleaner energy sources. Simultaneously, shifting market dynamics driven by fluctuating oil prices, the influence of OPEC, and emerging competitors in renewable energy are reshaping investment strategies and operational priorities.

Moreover, growing concerns about climate change have led to a heightened social awareness regarding the environmental impacts of fossil fuels. This has prompted both public and private sectors to prioritize sustainability in their business models. While hydrocarbons remain a fundamental part of the global energy mix, accounting for a significant percentage of energy consumption, there is an undeniable shift toward renewable energy sources and technologies. This energy transition will not only influence the future of upstream economics, affecting capital investment and operational efficiencies, but will also redefine the roles and responsibilities of companies operating in the sector as they adapt to a rapidly changing energy landscape.

This chapter explores key **trends, challenges, and opportunities** defining the upstream sector over the coming decades. We will examine **how companies are adapting, what economic models are evolving, and how sustainability will integrate into upstream decision-making**.

18.2 KEY TRENDS SHAPING THE FUTURE OF UPSTREAM

Numerous significant trends are poised to shape economic decision-making, investment strategies, and operational frameworks in the upstream petroleum sector. These trends include the growing emphasis on sustainability and environmental responsibility, which is pushing companies to adopt cleaner technologies and reduce

DOI: 10.1201/9781003750345-22

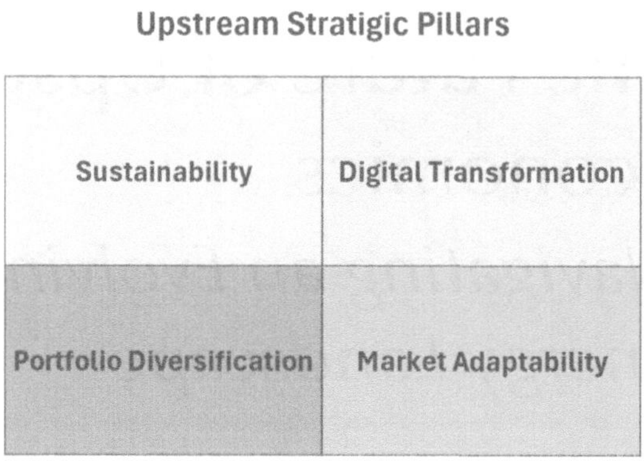

FIGURE 18.1 Strategic pillars for future-proofing upstream operations. Framework outlining four essential pillars for upstream resilience: sustainability, digital transformation, portfolio diversification, and market adaptability. These interconnected domains guide investment strategy, operational efficiency, and long-term competitiveness in a decarbonizing energy landscape.

their carbon footprints. Additionally, advancements in digital technology and data analytics are transforming exploration and production processes, enabling more efficient operations and enhanced resource management. Geopolitical factors, such as changes in regulatory environments, trade policies, and OPEC's production decisions, further complicate investment strategies by creating volatility in oil prices. Moreover, the increasing competition from renewable energy sources is prompting a reevaluation of long-term investments in fossil fuels. Finally, shifting consumer preferences and global demand patterns necessitate a more agile approach to operational frameworks, requiring companies to adapt quickly to market changes and innovate in their offerings. Collectively, these trends will significantly impact how stakeholders in the upstream petroleum industry navigate the complex landscape of modern energy production and investment (Figure 18.1).

1. **The Role of Oil and Gas in a Decarbonizing World**
   Despite the increasing investments in renewable energy sources, such as solar and wind, and advancements in hydrogen production and energy storage technologies, oil and gas are expected to remain integral components of the global energy mix for the foreseeable future. This persistence can be attributed to several factors, including the established infrastructure for fossil fuel extraction and distribution, the reliability of fossil fuels in meeting baseload energy demands, and the ongoing demand for petrochemical products in various industries. Furthermore, in regions where renewable technologies are still developing or energy transition policies are not yet fully implemented, oil and gas will likely continue to serve as crucial energy sources during the transition to a more sustainable energy landscape. However, their **share is expected to decline** as cleaner alternatives gain market traction.
   - **Short-term outlook (2025–2035)**: Continued demand for **natural gas as a transition fuel**, particularly in **Asia and Europe**.

- **Long-term outlook (2035–2050)**: Declining oil demand as **electric vehicles (EVs), renewable energy, and synthetic fuels gain market share**.
- **Implication for upstream:** Companies must adopt **cost-competitive strategies** and integrate **carbon management solutions** to remain viable.

2. **Technological Disruption and Digital Transformation**
Emerging technologies, such as artificial intelligence, machine learning, and advanced data analytics, are significantly transforming upstream operations in the oil and gas industry. These innovations enhance exploration by improving subsurface imaging and reservoir characterization, allowing for more accurate assessments of potential drilling sites. In production, technologies like automation, remote monitoring systems, and IoT devices streamline operations, increase efficiency, and reduce downtime. Additionally, sophisticated economic modeling tools leverage real-time data to provide more precise forecasts, optimize investment strategies, and assess the financial viability of projects, ultimately leading to more informed decision-making in the exploration and production phases.
    - **AI and machine learning**: Optimizing **reservoir management, drilling efficiency, and predictive maintenance**.
    - **Blockchain for supply chains**: Enhancing **transparency in oil trading, contract management, and regulatory compliance**.
    - **Automation and robotics**: Improving **drilling precision, reducing workforce risks, and optimizing safety protocols**.
3. **The Rise of ESG-Driven Investment and Carbon Markets**
    - **ESG investment shift:** Institutional investors are increasingly prioritizing companies with strong Environmental, Social, and Governance (ESG) performance. This shift is driven by a growing recognition that robust ESG practices can enhance long-term financial performance and mitigate risks. As a result, capital allocation is increasingly favoring firms that demonstrate responsible practices, sustainability initiatives, and transparency in their operations. Investment strategies now commonly incorporate ESG criteria, leading to a significant reallocation of funds toward companies that are committed to sustainable practices.
    - **Carbon markets[1] and pricing:** In response to climate change, governments worldwide are implementing carbon pricing mechanisms such as carbon taxes and cap-and-trade systems. These initiatives create financial incentives for businesses to reduce greenhouse gas emissions. The introduction of emissions penalties further raises compliance costs for upstream projects in sectors like oil and gas, compelling companies to adopt innovative technologies and practices that lower their carbon footprint. This regulatory landscape not only impacts operational costs but also influences investor confidence and market competitiveness.
    - **Net-zero commitments:** Major oil and gas companies, including BP, Shell, and TotalEnergies, are realigning their corporate strategies to meet ambitious net-zero emissions targets. These commitments

necessitate transformative changes in their investment portfolios, with a significant shift toward renewable energy sources and sustainable practices. As part of this transition, these companies are exploring new business models, including investments in electric vehicle infrastructure, carbon capture and storage technologies, and diversified energy solutions. The need for transparency in achieving these goals is paramount, with stakeholders seeking clarity on progress and actionable plans to ensure the credibility of net-zero pledges.

4. **Diversification and Portfolio Restructuring**
   Upstream companies are strategically diversifying their energy portfolios to navigate evolving market dynamics and enhance sustainability. Key initiatives include:
   - **Expansion into low-carbon technologies**: Companies are significantly increasing their investments in innovative low-carbon technologies. This includes the development of hydrogen production methods, such as electrolysis and steam methane reforming, alongside substantial funding in carbon capture, utilization, and storage (CCUS) technologies aimed at reducing greenhouse gas emissions. Additionally, there is a strong focus on integrating renewable energy sources, such as wind and solar power, to complement traditional oil and gas operations and align with global climate goals.
   - **Exit from high-risk assets**: There is a notable trend of divesting from marginal and high-cost oilfields, which often have unsustainable production economics and environmental concerns. Companies are shifting their focus toward short-cycle projects that promise faster returns on investment and are less exposed to volatility in oil prices. By choosing to divest, these companies aim to streamline operations and reduce exposure to geopolitical risks associated with high-risk regions[2].
   - **Strategic mergers and acquisitions**: To enhance operational efficiency and reduce costs, companies are increasingly engaging in strategic mergers and acquisitions. This consolidation allows firms to leverage synergies, optimize asset utilization, and achieve economies of scale. By acquiring complementary assets or merging with competitors, companies can strengthen their market position and improve resilience against market fluctuations.

WHAT MAKES AN UPSTREAM PROJECT FUTURE-PROOF?

- Low breakeven cost
- ESG compliance baked into economics
- Short-cycle recovery
- Technology-leveraged efficiency

These strategic moves are essential for upstream companies to remain competitive in a rapidly changing energy landscape while also committing to more sustainable practices.

18.3 CHALLENGES IN THE FUTURE OF UPSTREAM ECONOMICS

While new opportunities are emerging, the upstream industry faces **several fundamental challenges** that will define its **economic resilience and long-term sustainability**.

1. **Oil Price Volatility and Market Uncertainty**
   - The transition to **renewable energy and decarbonization policies** could lead to long-term **structural declines** in oil demand.
   - **Geopolitical risks** such as sanctions, trade restrictions, and supply chain disruptions can create price volatility.
   - **Economic implication**: Upstream projects require **strong hedging strategies and flexible investment models**.
2. **Regulatory and Policy Risks**
   - **Carbon taxation and emission limits:** Increasing regulation on **methane emissions, flaring restrictions, and carbon pricing**.
   - **Political instability in resource-rich nations:** Many oil-producing countries face governance issues that can **delay projects or change contract terms**.
   - **Economic implication**: Companies need **agile compliance strategies and political risk assessments**.
3. **Competition from Alternative Energy Sources**
   - The declining **levelized cost of energy (LCOE)** for renewables makes solar and wind **more competitive** than some oil projects.
   - The development of **advanced energy storage and synthetic fuels** may reduce dependence on hydrocarbons.
   - **Economic implication**: To remain relevant, upstream must focus on cost efficiency, technological adoption, and emissions reduction.
4. **Capital Discipline and Investment Constraints**
   - **Private equity and institutional investors** are shifting away from high-carbon assets, making it harder for upstream projects to secure financing.
   - Oil companies must **self-finance projects**, leading to stricter capital allocation and a focus on **high-return assets**.
   - **Economic implication**: Companies **prioritize short-cycle, low-breakeven projects** and reduce speculative exploration.

ESG METRICS TO WATCH IN UPSTREAM VALUATION

- Emission intensity (kg CO_2/boe)
- Capital % to low-carbon
- Renewable share of total output
- Transparency in Scope 1–3 reporting

18.4 FUTURE INVESTMENT AND ECONOMIC MODELS IN UPSTREAM

As upstream companies confront a myriad of challenges ranging from fluctuating commodity prices to increasing regulatory pressures, they are strategically adapting by adopting innovative investment strategies and economic models tailored for the evolving landscape.

1. **Flexible Investment Strategies**
   - **Short-cycle, high-return projects**: Companies are increasingly focusing on shale oil, liquefied natural gas (LNG), and deepwater drilling projects that promise rapid payback periods, often within a year or two. These projects are appealing due to their potential for quick revenue generation, allowing companies to respond swiftly to market changes.
   - **Long-term carbon-neutral investments**: Major oil companies are pivoting toward sustainable energy solutions, investing significantly in hydrogen technology, carbon capture, utilization and storage (CCUS[3]), and biofuels. These investments align with global decarbonization goals and are anticipated to enhance long-term operational resilience.
   - **Hybrid financing models**: Environmental, Social, and Governance (ESG) criteria into financing is becoming commonplace. Companies are turning to ESG-linked loans that offer favorable terms as a reward for sustainable practices, green bonds dedicated to funding eco-friendly initiatives, and innovative carbon credit financing methods that incentivize carbon reduction.
2. **Integration of Carbon Pricing and Sustainability Metrics**
   - Firms are increasingly recognizing the importance of incorporating carbon intensity metrics within their financial models. This change not only facilitates compliance with regulatory requirements but also appeals to investors who prioritize sustainability.
   - Economic models are evolving to reflect the costs associated with ESG compliance, including investments in carbon offset programs. This shift acknowledges the growing demand for transparent sustainability practices from stakeholders.
   - Projects that demonstrate carbon neutrality are anticipated to secure price premiums in regulated markets, signaling a market trend that favors environmentally responsible operations (Figure 18.2).
3. **Data-Driven Decision-Making and AI-Based Forecasting**
   - The adoption of AI-driven economic modeling is revolutionizing how companies approach production planning, cost forecasting, and risk assessment. These advanced tools analyze vast datasets to optimize decision-making processes and improve operational efficiency.
   - Scenario analysis tools are now essential for dynamic adjustments in response to volatile oil price swings, policy reforms, and stringent ESG mandates. This proactive approach enables companies to remain agile and well-prepared for various market scenarios, ensuring sustainability and profitability in a challenging environment[4].

These enhanced investment strategies and economic models reflect the upstream sector's commitment to navigating a complex future while embracing sustainability and technological advancements.

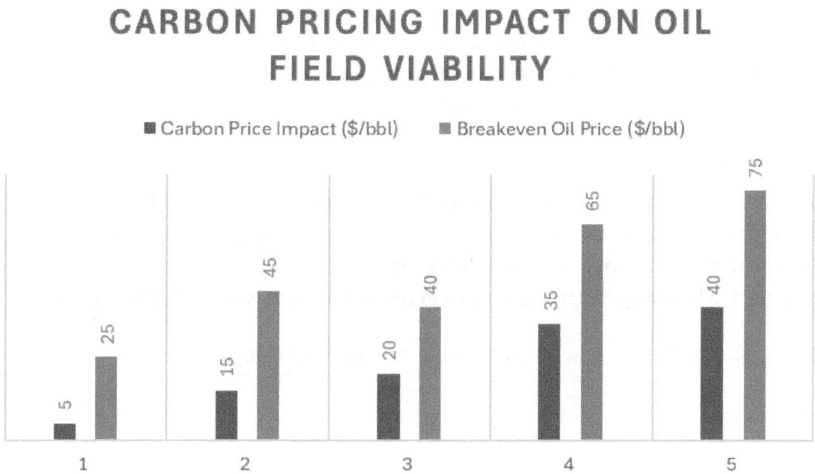

FIGURE 18.2 Carbon pricing impact on oil field viability. Comparison of breakeven oil prices and carbon price burdens across five upstream assets. Highlights how rising carbon costs disproportionately affect high-emission fields, increasing economic pressure and accelerating portfolio restructuring toward low-carbon, short-cycle projects.

18.5 CASE STUDIES: COMPANIES NAVIGATING THE FUTURE

The energy transition is reshaping investment strategies in the upstream sector. Major oil and gas companies are adopting diverse approaches to balance profitability with sustainability while positioning themselves for long-term success in a low carbon future. The following case studies illustrate how industry leaders integrate new energy technologies, restructure their portfolios, and leverage **sustainability initiatives**.

18.5.1 Case Study 1: BP's Transformation into an Integrated Energy Company

Background and Strategy:

British Petroleum (BP) has transformed strategically, shifting from a traditional oil company to a diversified **integrated energy provider**. Recognizing the changing market dynamics and the pressure to align with **global decarbonization goals**, BP announced a plan to:

- **Cut upstream oil and gas production by 40% by 2030** while expanding its renewables and low-carbon energy portfolio.
- Invest **$5 billion annually in low-carbon energy**, including **wind, solar, and bioenergy** projects.
- Develop **carbon capture and storage (CCS) facilities** and **hydrogen production hubs**.

Major Investments:

- **Offshore wind:** BP acquired a **50% stake in Equinor's U.S. offshore wind assets**, strengthening its position in renewable energy.

- **Solar and battery storage:** Partnering with **Lightsource BP**, the company is expanding **solar projects** across multiple markets.
- **Hydrogen and CCS:** BP invests in **blue and green hydrogen**, targeting **major European industrial hubs**.

Economic Impact:

BP's transition **reflects a fundamental shift in capital allocation**, with **30% of CAPEX** directed toward renewables and clean technologies. The company expects **low-carbon investments to drive long-term growth**, compensating for declining oil revenues while **capitalizing on government incentives for green energy**.

> BP Slashes Oil Production by 40% – Bets Big on Renewables
>
> *Financial Times, 2023*

18.5.2 Case Study 2: ExxonMobil's Investment in Carbon Capture and Storage (CCUS)

Background and Strategy:

ExxonMobil has historically been a leader in **traditional oil and gas production**. Still facing increasing pressure from **investors, regulators, and climate advocacy groups**, the company is **expanding its commitment to carbon management**.

- ExxonMobil aims to **become the world's leading CCUS company**, leveraging its existing infrastructure and expertise in **carbon sequestration**.
- The company committed **over \$15 billion to CCUS projects** between 2022 and 2027.
- ExxonMobil focuses on **reducing emissions from its existing operations** rather than exiting the oil and gas business.

Major Investments:

- **Houston carbon hub:** A multi-billion-dollar initiative designed to capture and store **100 million metric tons of CO_2 annually** from industrial sources.
- **LaBarge CCUS facility:** The project in Wyoming captures **7 million metric tons of CO_2 annually**, making it one of the most significant carbon sequestration projects globally.
- **Strategic partnerships:** ExxonMobil has formed alliances with **heavy industry players**, including **steel, cement, and petrochemical manufacturers**, to offer carbon storage solutions.

Economic Impact:

- **Carbon credits and monetization:** ExxonMobil is positioning CCUS as a **profitable business model**, creating **long-term revenue streams** from **carbon trading markets**.
- **Regulatory compliance and incentives:** The company is taking advantage of **U.S. tax credits (45Q)**, which **financially reward** large-scale CO_2 sequestration.

- **Competitive edge:** By investing in CCUS, ExxonMobil ensures its **long-term role in a decarbonizing economy**, maintaining the viability of **hard-to-abate sectors like refining and petrochemicals**.

Exxon's $15B Carbon Gamble: A Future in CO_2 Storage?

Bloomberg, 2022

18.5.3 Case Study 3: TotalEnergies' Expansion into Hydrogen and Biofuels

Background and Strategy:
TotalEnergies, formerly **Total**, is accelerating its shift toward **renewable energy and low-carbon fuels** while **maintaining a balanced hydrocarbon portfolio**. The company has committed to:

- **Reducing oil product sales by 30% by 2030.**
- Investing in **green hydrogen, biofuels, and sustainable mobility solutions**.
- Expanding its **renewable energy portfolio to 100 GW by 2030**.

Major Investments:

- **Green hydrogen expansion:** In partnership with Air Liquide, TotalEnergies is constructing Europe's largest green hydrogen plant in France.
- **Biofuels leadership:** The company is **converting its La Mède refinery** into a **biofuel production hub**, focusing on **sustainable aviation fuel (SAF)**.
- **Solar and wind:** Major investments include **solar projects in India** and **offshore wind farms in the North Sea**.

Economic Impact:

- **Diversification and risk management:** By expanding into **hydrogen and biofuels**, TotalEnergies is reducing its dependence on **volatile oil markets**.
- **Government support and regulations:** TotalEnergies benefits from **EU carbon pricing mechanisms and subsidies** promoting clean energy.
- **Competitive positioning:** The company's investments in SAF and hydrogen fuel position it as a **leader in alternative energy markets**.

TotalEnergies Converts Refineries to Biofuel Hubs

Reuters, 2023

18.5.4 Case Study 4: Saudi Aramco's Strategy to Expand Oil While Embracing Energy Transition

Background and Strategy:
Saudi Aramco, the world's **largest oil producer**, has taken a **dual approach**:

- Expanding its **oil and gas production capacity to safeguard market share**.
- Investing in **carbon-neutral solutions and emerging energy markets**, including **CCUS, hydrogen, and renewables**.

While many Western oil companies **scale down their fossil fuel operations**, Aramco remains **bullish on long-term oil demand**, particularly in **Asia and emerging markets**.

Major Investments:

- **Oil capacity expansion:** Aramco is investing **\$150 billion** to increase its **maximum sustainable oil production capacity to 13 million barrels per day (bpd)**.
- **Hydrogen leadership:** The company is developing **one of the world's largest blue hydrogen plants** to **export hydrogen to Asia and Europe**.
- **Carbon capture and enhanced oil recovery (EOR):** Aramco integrates **CCUS into its reservoirs**, enabling **carbon-neutral oil production**.
- **Renewable energy development:** The company has launched **solar and wind projects** across Saudi Arabia, supporting the nation's **Vision 2030** sustainability goals.

Economic Impact:

- **Strategic market control:** Aramco aims to capture market share in a shrinking supply environment by expanding oil capacity while competitors reduce output.
- **Long-term profitability:** Aramco's **low-cost production model (\$4 per barrel) ensures economic viability**, even in price downturns.
- **Diversified energy future:** Investments in **hydrogen, CCUS, and renewables** balance its traditional oil business with **emerging energy opportunities** (Figure 18.3).

18.5.5 Key Takeaways from Case Studies

See Figure 18.3.

18.6 CONCLUSION: THE FUTURE OF UPSTREAM STRATEGY

The global upstream industry is undergoing a significant transformation, driven by the dual pressures of profitability and sustainability amid shifting market dynamics. Companies operating in this space must navigate these complexities while responding to consumer demand for cleaner energy solutions. Each corporation highlighted

Case Study Matrix

Company	Strategic Focus	Key Tech/Assets	Impact
BP	Integrated energy pivot	Wind, Solar, Bioenergy	Long-term renewables growth
ExxonMobil	Carbon capture leader	CCUS, LaBarge, 45Q	Revenue from CO_2 markets
TotalEnergies	Hydrogen & Biofuel transition	SAF, Green H_2	Risk diversification
Aramco	Expand oil + lead in H_2 & CCUS	\$150B capex, Blue H_2	Low-cost dominance + energy export

FIGURE 18.3 Strategic responses to the energy transition: Case study matrix. Comparison of four leading upstream companies—BP, ExxonMobil, TotalEnergies, and Aramco—based on strategic focus, key technologies, and economic impact. Highlights diverse approaches to balancing profitability, sustainability, and long-term competitiveness in a decarbonizing energy landscape.

in this report has adopted a distinct strategy to ensure resilience and adaptability in the face of industry evolution. This includes proactive measures such as implementing advanced carbon capture technologies, expanding their portfolios to include renewable energy sources, and optimizing the efficiency and performance of existing fossil fuel assets.

Adaptability in the upstream sector isn't just an asset — it's survival.

Final Thoughts:

- **Integration of sustainability**: Companies that incorporate sustainability metrics into their financial planning, such as lifecycle analyses and carbon footprint assessments, will enhance their appeal to investors who prioritize environmental, social, and governance (ESG) criteria.
- **Impact of carbon pricing and ESG-linked capital markets:** The establishment of robust carbon pricing mechanisms, combined with ESG-linked financial instruments, is poised to significantly influence investment patterns over the next decade, encouraging upstream operators to adopt greener practices and technologies.
- **Role of technology in competitiveness:** Continued advancements in technology that lead to cost reductions in renewable energy production and hydrogen fuel generation will be pivotal in determining the long-term competitiveness of oil and gas companies. The adaptation of these technologies will likely dictate the strategic direction of the industry going forward.

The following chapter will delve deeper into the collaborative efforts between governments, investors, and industry leaders as they work together to shape a future for the upstream sector that is not only sustainable but also economically viable for all stakeholders involved.

NOTES

1. Cap-and-trade programs like the EU ETS assign emission caps and allow market-based trading of allowances, incentivizing lower emissions.
2. High-cost deepwater and Arctic fields often face breakeven prices exceeding $70/bbl, making them vulnerable to transition risks
3. U.S. 45Q tax credits offer up to $85/ton for captured CO_2 permanently stored, making large-scale CCUS projects economically viable.
4. Modern economic models increasingly rely on Monte Carlo simulations and real options valuation to reflect volatility and uncertainty.

19 Late-Life Asset Economics

19.1 INTRODUCTION: THE SHIFT TOWARD MATURE FIELDS

The oil and gas industry has always been driven by the search for new frontiers—wildcat wells, untapped basins, and high-impact exploration. But in today's capital-constrained and carbon-conscious world, a quiet revolution is underway: **late-life assets (LLAs)** are taking center stage.

These are the mature oil and gas fields nearing the end of their production lives, often dismissed in the past as liabilities or decommissioning burdens. Yet increasingly, they are being re-evaluated through a more pragmatic lens—not as declining assets, but as **cash-flow generators**, **tax shields**, and in some cases, **low-risk investment opportunities**.

The growing importance of LLAs is not accidental. It reflects a broader shift in industry economics and investor sentiment. Exploration budgets have shrunk. Mega-projects take too long. Energy transition pressures are mounting. In this environment, mature fields—with known reservoirs, existing infrastructure, and minimal permitting hurdles—offer something rare: **predictable returns** in a volatile world.

Across the globe, governments and operators are scrambling to redefine how to extract value from aging fields. In the UK North Sea, fiscal reforms aim to extend field life and delay abandonment. In Southeast Asia, companies like Petronas and MedcoEnergi are acquiring IOCs' brownfields and squeezing new value from old wells. In Canada, abandonment liabilities have become a key criterion in asset divestment. Everywhere, **late-life economics** are being rethought.

This chapter explores that shift—how economic frameworks, fiscal policies, and technology are converging to **turn aging assets into strategic levers**. We'll examine the unique financial profile of LLAs, analyze rejuvenation strategies, and evaluate the environmental, social, and governance (ESG) and abandonment challenges they carry. Case studies and recent news will illustrate how national oil companies (NOCs), private equity-backed firms, and small independents are leveraging LLAs for growth—while others walk away.

> We used to chase barrels; now we chase margin and cash. That's why late-life assets matter.
>
> *Anonymous CFO, Mid-Cap Operator*

DOI: 10.1201/9781003750345-23

Far from being the industry's afterthought, LLAs may just become its **future-proof core**—if managed with clarity, precision, and flexibility.

19.2 ECONOMIC CHARACTERISTICS OF LATE-LIFE ASSETS

LLAs have a distinct economic profile that sets them apart from both greenfield and peak-producing brownfield projects. These fields sit in a **narrow economic corridor**—where small changes in production, cost, or fiscal terms can tip the scale between viability and abandonment.

19.2.1 Declining Production, Stable Costs

The defining feature of LLAs is their **natural decline in production**. Reservoir pressure is depleted, water cut is rising, and artificial lift becomes essential. Yet, unlike greenfield projects, **fixed operating costs remain relatively high** due to aging infrastructure, safety requirements, and staff retention.

- **Operating costs per barrel often rise** as production declines.
- Decommissioning obligations loom, but OPEX must be spent until then.
- Marginal fields may generate **positive cash flow but negative NPV**, especially under full-cycle cost assumptions.[1]

19.2.2 The Cash Cow Paradox

Despite weak long-term economics, many LLAs **generate strong short-term cash flow**. Infrastructure is already in place. Drilling is done. CAPEX is minimal. If OPEX is managed tightly and fiscal terms are favorable, LLAs can function as **low-risk cash cows** (Figure 19.1).

Metric	Early-Life Assets	Late-Life Assets
Capex Intensity	High	Low
Cash Flow Volatility	High (esp. price-driven)	Moderate
Decline Rate	Shallow to moderate	Steep
Tax Shield Potential	Low	High
ESG Risk (perception)	Low	High (flaring, leaks)
Decommissioning Liability	Low	High

FIGURE 19.1 Comparative economic and ESG profile: Early-life vs. late-life assets. Side-by-side comparison of upstream asset characteristics across six dimensions: CAPEX intensity, cash flow volatility, decline rate, tax shield potential, ESG risk perception, and decommissioning liability. Highlights how LLAs offer lower capital exposure and higher tax efficiency, but carry elevated ESG and abandonment risks.

19.2.3 Fiscal Levers and Tax Shields

Governments often provide **fiscal incentives** to extend the life of marginal fields. These may include:

- **Accelerated depreciation or uplift allowances**
- **Reduced abandonment tax exposure**
- **Cost recovery enhancements or ring-fencing flexibility**
- **Decommissioning tax rebates** (e.g., UK's PRT reliefs)

For some operators, LLAs offer **tax shields** against profits from other fields, especially in ring-fenced corporate tax structures. This makes LLAs attractive as **portfolio-balancing tools—despite their limited standalone appeal**.

19.2.4 Reinvestment vs. Harvest Mode

Economic decisions in LLAs often revolve around whether to:

- **Reinvest** (e.g., infill drilling, workovers, tie-backs)
- **Harvest** (reduce OPEX, extract remaining value, prepare for decommissioning)

This trade-off is deeply context-specific:

- In **low oil price** environments: harvesting dominates.
- In **high oil price** windows: selective reinvestment becomes viable—especially if costs are already sunk.

Figure 19.2 offers a simplified framework for determining late-life asset strategy based on oil price scenarios and remaining reserves.

19.3 FISCAL INCENTIVES AND DECOMMISSIONING TREATMENT

The economics of LLAs are inseparable from the tax regimes and fiscal treatments that shape their final years. As reserves dwindle and operating costs rise, governments

Oil Price	Remaining Reserves	Strategy
High	Moderate–High	Infill/Extend
High	Low	Harvest & Exit
Low	High	Wait / Delay
Low	Low	Decommission

FIGURE 19.2 Strategic decision matrix for late-life asset management. Framework for determining optimal strategy based on oil price environment and remaining reserves. High-price scenarios favor reinvestment or harvesting, while low-price conditions prompt delay or decommissioning. Supports dynamic portfolio planning and capital discipline in mature fields.

are increasingly challenged to keep mature fields productive without compromising long-term decommissioning responsibilities. Strategic fiscal incentives—particularly those related to abandonment cost recovery—have become essential tools for both extending field life and enabling smooth operator transitions.

19.3.1 Abandonment Cost Recovery Mechanisms

Decommissioning is often the most significant single expenditure in a late-life asset's lifecycle, and how these costs are treated financially can make or break an operator's decision to divest, farm out, or retain operations. Countries with mature basins have responded with varying approaches:

- **United Kingdom**
  The UK provides **tax relief on decommissioning** through a system of tax deductions and **Decommissioning Relief Deeds (DRDs).**[2] Under the DRDs, the government ensures that companies can claim tax relief for 40–75% of qualified decommissioning costs—even if their tax liability is insufficient. This certainty reduces financial risk and encourages asset transfers.
- **Malaysia**
  PETRONAS allows cost recovery for decommissioning in PSC Cost Oil components, subject to budget approval and contract terms. In recent years, regulators have explored incentivizing reuse and partial decommissioning to reduce abandonment liabilities.
- **Canada**
  In Alberta and British Columbia, decommissioning costs can be deducted under **capital cost allowance (CCA)** rules, and federal programs such as the **Site Rehabilitation Program (SRP)** have provided billions in grants and credits to accelerate cleanup while supporting economic activity.

> Clarity on decommissioning tax relief is essential to unlocking transactions in mature basins. Without it, the market for late-life assets freezes.
>
> *UK Oil & Gas Authority (OGA), 2021*

19.3.2 Incentives to Defer Decommissioning

Beyond cost recovery, some governments deploy **incentives to defer decommissioning**, hoping to extract additional value while delaying the creation of massive abandonment liabilities.

- **UK: The MER UK Strategy**
  The **Maximizing Economic Recovery (MER) UK** policy has promoted late-life asset stewardship by advancing **tax stability**, **license flexibility**, and **infrastructure sharing**. Operators who take over LLAs are encouraged to continue production through incremental investments, with tax allowances tailored to brownfield upgrades.

- **Norway**
  Norway allows accelerated depreciation and refund mechanisms to support enhanced oil recovery (EOR) projects in mature fields—helping defer abandonment and increasing recovery rates.
- **Indonesia**
  The Indonesian government has offered **tax holidays** and extended **PSC tenures** for operators willing to invest in mature fields, especially when tied to EOR pilots or modular re-development.

19.3.3 R&D Incentives for Brownfield Innovation

Brownfield technologies—such as compact subsea systems, mobile plug-and-abandon units, or AI-driven predictive maintenance—can radically improve late-life economics. To encourage such innovation:

- **Australia** and **Canada** offer **R&D tax credits** for the development of technologies applicable to marginal and mature fields.
- The UK provides a **Patent Box regime**, reducing corporate tax on profits derived from patented technologies, which is often relevant in late-life field re-engineering.
- The **Netherlands** has integrated **brownfield incentives** into its SDE++ subsidy scheme, indirectly supporting emissions-reduction technologies used in legacy assets.

19.3.4 Case Study: UK North Sea and the Late-Life Fiscal Reset

In 2015, the UK launched a multi-pronged fiscal initiative targeting mature assets:

- Introduction of the **Investment Allowance** to replace previous brownfield allowances, offering streamlined incentives for all new capital.
- Establishment of the **OGA's Late Life Asset Stewardship team**, which worked with industry to identify commercial barriers.
- Formalization of **Decommissioning Relief Deeds**, giving M&A confidence for late-life asset buyers.

The result was a renewed wave of late-life asset transfers from supermajors to independent operators, including Shell's sale of its Brent field stakes and BP's divestment of its Andrew Area to Premier Oil.

> By 2020, over 40 North Sea fields had changed hands under new late-life terms—with several extending production by 5–10 years.
>
> *OGA, 2021 North Sea Stewardship Report*[3]

19.4 BROWNFIELD REJUVENATION STRATEGIES

While LLAs are often perceived as declining liabilities, targeted rejuvenation techniques can unlock previously stranded barrels, delay abandonment, and restore economic viability. The key lies in deploying **cost-effective interventions** that enhance recovery without incurring the capital burden of new field developments.

19.4.1 Conventional Rejuvenation: Workovers, Sidetracks, Infill Drilling

These "first-response" options remain foundational to brownfield economics:

- **Workovers** involve repairing or replacing downhole equipment to restore or improve flow rates—often at a fraction of the cost of new drilling. Common interventions include sand control, tubing replacement, or artificial lift upgrades.
- **Sidetracks** allow re-entry into the reservoir via a new wellbore path, bypassing damaged or depleted zones. They are particularly valuable in reservoirs with compartmentalization or bypassed pay.
- **Infill drilling** targets undrained zones between existing wells. While requiring new surface or subsea infrastructure, infill campaigns benefit from existing data and infrastructure and shorter payback periods.

> In the mature Gulf of Mexico shelf, infill wells and sidetracks have yielded IRRs >30% in select fields
>
> *Wood Mackenzie, 2022*

19.4.2 AI and Digital Twins: Enhancing Operational Precision

In recent years, digital technologies have transformed brownfield optimization:

- **Digital twins** replicate the physical asset virtually, allowing real-time simulation of flow, pressure, equipment integrity, and production outcomes.
- **AI-driven surveillance** tools use historical and live data to predict equipment failures, optimize lift performance, and identify hidden production opportunities.
- **Predictive maintenance** powered by machine learning helps reduce unplanned downtime and extend equipment lifespan, particularly for legacy offshore facilities with aging systems.

> A 2021 SPE study found that AI deployment in mature North Sea assets reduced downtime by 15% and improved OPEX efficiency by 12%.[4]

19.4.3 Cost-Effective EOR in Mature Fields

Enhanced Oil Recovery (EOR) methods—long considered expensive—are being re-evaluated for mature assets through modular, scalable deployment:

- **Polymer flooding** has proven successful in sandstone fields where mobility control is critical. Indonesia's Tertiary Recovery initiatives include dozens of such pilots.

Rejuvenation Strategy	Typical CAPEX (USD/well)	OPEX Impact	Payback Period	Best Use Case
Workovers	$250k–$1M	↓ Minor OPEX savings	<1–2 years	Damaged completions, artificial lift failures
Sidetracks	$2M–$6M	↑ Modest OPEX	2–4 years	Bypassed pay, compartmentalized reservoirs
Infill Drilling	$5M–$15M	↑↑ New facilities	3–5 years	High-POR zones between producers
Digital Twin & AI Monitoring	$1M–$3M (platform-wide)	↓↓ Significant OPEX	<2 years (platform)	Legacy offshore platforms
EOR (Polymer, LSW, CO_2)	$5M–$50M (field-wide)	↑↑↑ Depends on method	5–10 years	Selective mature fields with strong reservoir data

FIGURE 19.3 Comparative economics of brownfield rejuvenation strategies. Evaluation of five rejuvenation options—workovers, sidetracks, infill drilling, digital twin/AI monitoring, and enhanced oil recovery (EOR)—across CAPEX, OPEX impact, payback period, and strategic applicability. Supports tactical decision-making in mature fields by aligning intervention type with reservoir conditions and economic constraints.

Note: **Cost and payback figures vary significantly by basin, regulatory regime, and reservoir complexity. Values represent indicative industry ranges.**

- **Low-salinity waterflooding** (LSW) and **surfactant injection** offer relatively low-cost recovery boosts, especially in carbonate-rich or oil-wet reservoirs.
- **Gas injection (CO_2, N_2)** is often tied to decarbonization efforts, enabling dual benefits: increased oil recovery and carbon sequestration. Several UAE and Norwegian projects combine CO_2-EOR with emissions offsetting (Figure 19.3).

19.5 SUMMARY

Brownfield rejuvenation strategies represent a vital middle ground between passive decline and high-cost re-development. When tailored appropriately, they can extend field life, enhance recovery, and improve cash flow—all while delaying abandonment costs. The integration of **engineering**, **data science**, and **economics** is increasingly the hallmark of success in this phase of the asset lifecycle.

19.6 M&A AND TRANSFER OF LATE-LIFE ASSETS

As upstream portfolios mature, LLAs often become economic deadweight for International Oil Companies (IOCs), even when technically recoverable reserves remain. However, these same assets can represent value-creation opportunities for smaller operators with leaner structures, flexible risk profiles, and specialized brownfield expertise.

19.6.1 Divestment by IOCs: Strategic and Financial Drivers

Global IOCs have been actively reshaping portfolios to align with shareholder expectations, energy transition goals, and capital efficiency:

- **Capital discipline**: IOCs seek higher returns by focusing on large-scale, low-cost, and short-cycle developments (e.g., deepwater, LNG, shale).
- **Carbon intensity and ESG pressures**: Mature assets tend to have higher emissions and lower margins—making them unattractive under ESG scrutiny.
- **Decommissioning liabilities**: IOCs face massive end-of-life costs. Selling assets before these burdens mature reduces long-term obligations.

> Between 2017 and 2023, IOCs divested over $140B in mature upstream assets globally
>
> *Wood Mackenzie, 2024*[5]

19.6.2 Acquisition by Smaller Players: A Strategic Upside

Smaller, often regional or privately backed E&Ps see value where IOCs see decline:

- **Lower overhead models**: Independents operate with lower fixed costs and faster decision-making cycles.
- **Asset familiarity**: Acquirers may already be operating in similar basins or have local insight that IOCs lack.
- **Extended recovery potential**: Through tailored investments—sidetracks, workovers, new lift systems—smaller firms can unlock additional barrels economically.
- **Tax incentives**: Some jurisdictions allow buyers to offset decommissioning costs or benefit from fiscal packages for brownfield reinvestment.

> For independents, even a 10–15% recovery uplift on a mature asset can represent transformational cash flow.

19.6.3 Value Creation Post-Acquisition

Success hinges on what happens **after** the deal:

- **Lean operational turnarounds**: Buyers often introduce cost-efficient processes, renegotiate contracts, and optimize manpower.
- **Reservoir re-evaluation**: Updating models with modern software and integrating new surveillance data helps redefine development plans.
- **Digitalization and automation**: Applying AI-driven monitoring or automated well testing can improve production while reducing OPEX.
- **Modular Capex**: Focus on capital-light projects—such as mobile production units, leased facilities, or modular compression—to minimize exposure.

19.6.4 Case Study: Petronas Carigali's Acquisition from Shell (Sarawak, Malaysia)

In 2021, **Petronas Carigali** acquired Shell's entire upstream portfolio in **Sarawak's shallow waters**, covering eight producing fields and associated facilities.

- **Shell's motivation**: Strategic refocus on deepwater and energy transition-aligned assets. The mature Sarawak fields had declining production and high emissions intensity.
- **Petronas' strategy**:
  - Already operating in the area with existing infrastructure.
  - Plans to **extend field life** using low-cost re-development, gas monetization, and AI-driven production optimization.
  - Local operator advantage enabled faster regulatory engagement and resource allocation.

Result: Petronas reported a **15% improvement in asset uptime** and a 12% reduction in unit OPEX within 18 months.[6]

19.7 SUMMARY

Late-life asset transactions illustrate a key trend in upstream realignment: **value is not in the barrels alone, but in who operates them and how**. Where IOCs see ESG exposure and capital inefficiency, lean independents and NOCs may see opportunity. Proper post-acquisition integration, technical upgrades, and cost discipline are essential to turning aging assets into cash-flow generators.

19.8 RISK EVALUATION AND REAL OPTIONS IN LATE-LIFE ASSETS

LLAs present a unique risk-return profile. Declining reserves, rising costs, and increasing regulatory scrutiny raise the stakes for investment decisions. Traditional net present value (NPV) metrics often fall short of capturing the embedded flexibility that asset holders retain—such as deferring abandonment or restarting production later. This is where **real options** analysis becomes a critical tool.

19.8.1 Abandonment Cost Uncertainty

Decommissioning costs are notoriously variable and **subject to regulatory, environmental, and technical uncertainty**. This poses financial and strategic challenges:

- **Range of costs**: Plug and abandonment (P&A) for subsea wells can range from $2M to $20M per well, depending on depth, wellbore complexity, and local standards.
- **Escalation risk**: Delay may result in stricter environmental rules or rising service costs, increasing the future burden.
- **Accounting and liability**: Companies are required to estimate abandonment liabilities years in advance, often leading to balance sheet distortions and capital constraints.

> In the UK North Sea alone, decommissioning liabilities exceed £45 billion
>
> *OGUK, 2023*

19.8.2 The Real Option to Delay or Suspend

Unlike greenfield projects, LLAs often carry a built-in **option to pause or defer**—especially in the absence of abandonment deadlines. This flexibility has real economic value.

Options Embedded in LLAs:

- **Option to suspend**: Temporary shutdowns during low-price periods can preserve cash and reduce losses.
- **Option to re-enter**: Reactivating idle wells when prices improve, or new tech reduces OPEX.
- **Option to defer decommissioning**: Waiting for government incentives, partnerships, or cost-reducing technologies.

Example: In 2020, several mature Gulf of Mexico platforms suspended production rather than decommission, anticipating a price recovery and regulatory relief.[7]

19.8.3 Strategic Decommissioning Postponement

Governments increasingly recognize the **national value of deferring abandonment** when technically and economically justified.

- **Malaysia** and **Australia** offer fiscal incentives to delay abandonment if viable production or secondary recovery is still possible.
- The **UK's MER UK strategy** explicitly promotes maximizing economic recovery before field decommissioning, urging operators to consider asset transfer before shutdown.

The cost of early abandonment is not only financial—it eliminates future revenue potential, energy security contribution, and employment (Figure 19.4).

This **decision tree** can be expanded to include:

- Probability of oil price recovery.
- Expected cost inflation.
- Policy incentives.
- Break-even shifts via OPEX reduction or tech adoption.

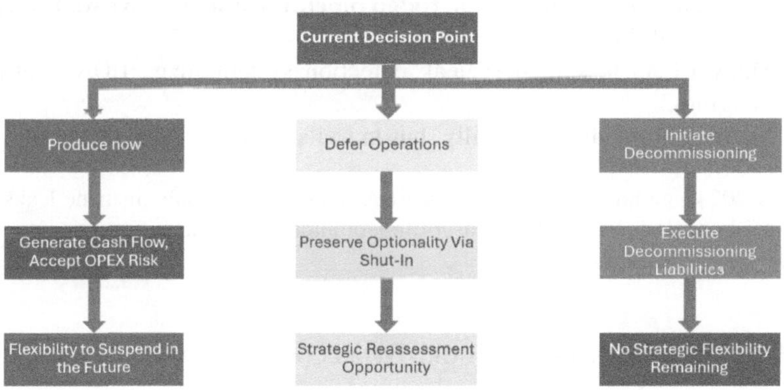

Decision Point	Short-Term Action	Strategic Implication
Continue Production	Maintain operations and cash flow	Flexibility to suspend if conditions worsen
Defer Operations	Temporarily shut-in asset	Preserve asset value for future reassessment
Initiate Decommissioning	Begin plug & abandonment activities	Terminal outcome — no residual flexibility

FIGURE 19.4 Strategic options for a mature offshore asset nearing economic limit. Decision tree outlining three pathways for a late-life offshore asset under current oil price conditions: continue production to generate cash flow, defer operations to preserve optionality, or initiate decommissioning. Highlights the embedded flexibility in LLAs and the economic value of delaying irreversible decisions.

19.9 SUMMARY

Late-life asset economics are not just about current NPV—they're about **embedded strategic flexibility**. Decommissioning can be deferred, paused, or optimized. Real options analysis helps decision-makers recognize when "doing nothing now" is a value-maximizing strategy. Ignoring these flexibilities leads to premature abandonment, missed upside, and stranded value.

19.10 ESG, METHANE, AND AGING INFRASTRUCTURE

As ESG frameworks gain momentum globally, late-life oil and gas assets face heightened scrutiny. Aging infrastructure not only poses technical and financial challenges but also emerging **ESG liabilities** that operators can no longer ignore. Methane leakage, decommissioning delays, and inspection non-compliance are fast becoming reputational and regulatory flashpoints.

19.10.1 Risk of Methane Leaks from Late-Life Assets

Methane (CH_4) is over **80 times more potent than CO_2** over 20 years, making it a major contributor to near-term climate risk. Older offshore and onshore facilities are disproportionately responsible for fugitive emissions.

- **Poor seals, aging flanges, corroded pipelines, and inactive wells** increase leak probability.
- Many LLAs lack modern leak detection systems (e.g., LDAR, infrared imaging).
- Methane from idle or partially shut-in wells often goes undetected for years.

> A 2024 satellite study over Southeast Asia detected chronic methane leaks from offshore platforms nearing end-of-life—prompting regulatory concern
>
> *IEA, 2024*[8]

19.10.2 ESG Liabilities vs. Short-Term Economic Gains

Operators holding LLAs must now **balance declining production economics with ESG risk exposure.**

- **Short-term view**: Squeeze remaining value through extended production.
- **ESG view**: Aging fields are ticking time bombs for emissions, spills, and headline risk.

Investor pressure is mounting:

- ESG-focused funds increasingly **screen out** companies with deferred decommissioning liabilities.
- **Credit ratings** may be affected by environmental non-compliance risk.

- **Pension funds** and **sovereign wealth funds** demand demonstrable ESG strategies, not just cash flow.

> Every delay in decommissioning becomes an ESG footnote in annual reports.
>
> *WoodMac Energy Transitions Group, 2025*

19.10.3 Regulatory and Reputation Risk from Decommissioning Delays

Delays in shutting down aging infrastructure carry more than cost implications—they now attract **policy crackdowns** and public backlash.

Recent Developments (2024–2025)

- **Indonesia**: Delayed decommissioning of multiple offshore fields led to visible gas sheens and international headlines. Activists now demand international oversight of abandonment plans.
- **Australia's NOPSEMA**[9]: Announced stricter inspection regimes and compulsory ESG reporting for aging platforms following multiple methane leak incidents in the Timor Sea.

> In 2025, NOPSEMA shut down a platform after non-compliance with structural and emissions inspections—marking the first enforcement of its 2024 ESG compliance rules.

19.11 SUMMARY

The ESG lens magnifies every weakness of aging petroleum infrastructure. What was once a cost decision is now an issue of **reputation, regulatory compliance, and future access to capital**. Methane emissions, in particular, represent both an environmental liability and a measurement challenge—requiring proactive investment in detection, repair, and responsible asset retirement.

Late-life asset strategy is no longer a technical discussion alone—it is now a **boardroom and investor concern**, shaping capital access, license renewals, and societal license to operate.

19.12 CONCLUSION: STRATEGIC ROLE IN PORTFOLIO MANAGEMENT

LLAs have long carried a reputation of decline, risk, and eventual abandonment. But that's only part of the picture. For strategic operators and investors, LLAs offer **cash flow, tax efficiency, infrastructure leverage, and even upside optionality**—if managed correctly.

The role of these aging assets has shifted from being **passive liabilities** to becoming active instruments in **portfolio rebalancing, strategic exits, and ESG-aligned rejuvenation**.

19.12.1 When LLAs Are Cash Cows

Mature fields with sunk capital and depreciated infrastructure often deliver **strong operating margins**, even as output declines.

- Low remaining CAPEX and relatively predictable OPEX
- High uptime and operational familiarity
- Opportunity to defer abandonment through low-risk infill drilling or workovers

These assets become **capital-light cash contributors**, helping fund growth projects or absorb price shocks elsewhere in the portfolio.

> Not every old well is a burden. Some are the economic anchors of entire portfolios.
>
> *Portfolio Strategist, 2024*

19.12.2 When They Become ESG Traps

However, the longer LLAs stay online, the more **ESG and decommissioning risks compound.**

- Methane emissions, structural integrity failures, or regulatory non-compliance can erase years of profit in a single incident.
- Negative ESG ratings affect borrowing costs, JV attractiveness, and M&A deal flow.
- Delays in abandonment accrue future liabilities, often underestimated on the books.

Strategic inflection points arise when marginal returns fall below the reputational cost of maintaining the asset.

19.12.3 LLAs as Tools in Exit and Transition Strategy

Rather than seeing LLAs as the end of the road, advanced portfolio managers treat them as **strategic levers**:

- **Divestment** to ESG-resilient niche operators (with lower hurdle rates)
- **Joint abandonment** cost-sharing among legacy JV partners
- **Carbon offset pilots or CCS repurposing** for aging reservoirs
- **Option to delay or mothball** production pending market signals

This positions LLAs not as obstacles, but as **strategic chess pieces** in long-term capital allocation and energy transition planning.

> Late-life assets are not liabilities — they're optionality in disguise.
>
> *Upstream Portfolio Advisor, 2025*

NOTES

1. In some PSC-based regimes, such as Indonesia and Malaysia, the operator may still achieve cost recovery on LLAs even when the economic metrics (IRR, NPV) are weak under traditional DCF models.
2. OGA (2021), *UKCS Stewardship Survey.*
   HMRC guidance on **Decommissioning Relief Deeds.**
3. OGA, 2021 North Sea Stewardship Report.
4. SPE-204567-MS (2021) or general SPE literature on AI in mature fields.
5. Wood Mackenzie's Global M&A Tracker (or similar summary from 2024/2025).
6. Industry news articles (e.g., *Offshore Technology*, *Upstream Online*) confirming Shell's exit and Petronas's reported OPEX reduction.
7. BOEM (Bureau of Ocean Energy Management) 2020–2021 field data reports.
8. Environmental Defense Fund (EDF) satellite leak studies.
9. NOPSEMA ESG Compliance Policy Update (2024–2025).

20 The Economics of Offshore vs. Onshore Development

In oil and gas, the choice between offshore and onshore isn't about right or wrong — it's about risk versus readiness.

20.1 INTRODUCTION: OFFSHORE VS. ONSHORE—UNDERSTANDING THE TRADE-OFFS

The development of oil and gas reserves, whether offshore or onshore, is influenced by a complex interplay of economic, technical, and regulatory factors. Offshore projects often benefit from access to larger and potentially more productive reserves, which can lead to higher long-term returns. However, they come with significant challenges, including higher initial capital expenditures, advanced technological requirements for drilling and production, and logistical complexities due to remote locations.

Conversely, onshore projects typically have lower upfront costs and often face less stringent logistical challenges, making them more accessible for exploration and production. Yet, they are not without their drawbacks—onshore operations may encounter greater regulatory scrutiny, environmental concerns from local communities, and variations in land access that can affect project timelines and costs.

Both offshore and onshore developments also contend with fluctuating market prices for crude oil and natural gas, which influence the overall economic viability and investment decisions. Moreover, technological advancements—such as horizontal drilling and hydraulic fracturing—have revolutionized both sectors, enhancing recovery rates and lowering operational costs. These advancements contribute to evolving risk profiles, as the industry must continually adapt to changing market conditions, regulatory landscapes, and environmental considerations. Ultimately, the choice between offshore and onshore development hinges on a thorough assessment of these distinctive advantages and challenges, which together determine the cost structures, risk profiles, and long-term profitability of each project.

This chapter examines the fundamental economic differences between offshore and onshore oil and gas development, highlighting the factors influencing investment decisions, cost considerations, technological advancements, and risk mitigation strategies. Through real-world case studies, we analyze how companies optimize their portfolios by balancing offshore and onshore assets.

 DOI: 10.1201/9781003750345-24

20.2 COST STRUCTURE ANALYSIS: OFFSHORE VS. ONSHORE DEVELOPMENT

Cost structure is a fundamental element in determining the economics of oil and gas projects, as it influences both feasibility and profitability. When comparing offshore and onshore developments, they exhibit markedly different characteristics in terms of capital expenditure (CAPEX), operating expenditure (OPEX), and breakeven price[1] considerations.

Offshore projects typically involve higher CAPEX due to the complexities of deep-water drilling, the need for specialized equipment, and the construction of platforms and infrastructure that can withstand harsh marine environments. This can lead to initial investments reaching billions of dollars. In contrast, onshore developments generally require lower CAPEX, benefiting from more accessible drilling sites, established infrastructure, and less regulatory complexity.

Operating expenditure also differs significantly between the two. Offshore operations tend to incur higher OPEX due to increased maintenance costs, logistical challenges associated with transporting personnel and materials, and the need for ongoing technological support to manage operations in isolated and demanding conditions. Onshore projects, while they still incur OPEX, often enjoy lower costs from easier access and more straightforward operational management.

Additionally, the breakeven price, which is the oil or gas price at which a project becomes economically viable, varies between offshore and onshore developments. Offshore projects often require a higher breakeven price due to their elevated CAPEX and OPEX, whereas onshore projects, especially those in established and prolific basins, can sustain lower breakeven prices, giving them a competitive advantage in a fluctuating market.

Understanding these nuances in cost structure is crucial for investors, operators, and policymakers as they navigate the complexities of oil and gas exploration and production decisions (Figure 20.1).

Capital Expenditures (CAPEX[2]):

- **Offshore:** There are high upfront costs due to the need for drilling platforms, subsea infrastructure, and complex transportation networks. Deepwater developments require floating production storage and offloading (FPSO) units or long-distance pipelines to transport hydrocarbons.[3]

Offshore vs. Onshore Cost Breakdown Table (CAPEX, OPEX, Breakeven)

Cost Type	Offshore (Deepwater)	Onshore (Shale/Others)
CAPEX	Very High (>$5B)	Low to Moderate ($20M–$300M)
OPEX	High	Moderate
Breakeven Price	$50–$70/bbl	$30–$50/bbl

FIGURE 20.1 Offshore vs. Onshore Cost Breakdown: CAPEX, OPEX, and breakeven price. Side-by-side comparison of offshore (deepwater) and onshore (shale/others) developments across capital expenditure, operating costs, and breakeven price thresholds. Highlights the economic contrast between high-cost, long-cycle offshore projects and lower-cost, flexible onshore operations.

- **Onshore:** Lower initial costs as drilling and production infrastructure are more straightforward to deploy. Equipment, rigs, and transportation networks are more accessible, lowering development costs.

Operating Expenditures (OPEX):

- **Offshore:** Higher OPEX due to complex logistics, expensive maintenance, and challenging working conditions. Regular inspections, repairs, and well interventions require specialized offshore vessels and highly trained personnel.
- **Onshore:** Lower OPEX because maintenance and transportation are more straightforward and less costly. Production facilities are more accessible, reducing labor and logistical expenses.

Breakeven Price Considerations:

- **Offshore:** Deepwater projects typically require oil prices above **$50–$70 per barrel** to remain profitable due to higher CAPEX and OPEX.
- **Onshore:** Shale and conventional onshore oil fields have lower breakeven prices, ranging between **$30 and $50 per barrel**, making them more resilient to price fluctuations.

20.3 RISK AND UNCERTAINTY IN OFFSHORE AND ONSHORE PROJECTS

Offshore Risks:

1. **Technical and engineering challenges:** The harsh marine environments where offshore projects are situated necessitate developing complex and highly specialized engineering solutions. This includes the need for advanced materials and durability against corrosion, which significantly escalates initial capital expenditure and ongoing operational risks. Continuous technological innovations, such as dynamic positioning systems, are crucial but often come with steep learning curves and associated costs.
2. **Geopolitical risks**: Many offshore oil and gas fields are situated in areas with overlapping territorial claims or disputed maritime boundaries. This geopolitical volatility leads to regulatory uncertainties and potential legal disputes, which can halt operations or alter project timelines. For instance, sanctions imposed due to geopolitical tensions can affect companies' ability to operate in certain regions, complicating supply chains and access to necessary resources.
3. **Weather-related risks**: Offshore rigs and platforms are exposed to the full force of oceanic storms, hurricanes, and unpredictable sea conditions. These weather-related risks not only threaten the structural integrity of the facilities but can also lead to costly shutdowns that impact production schedules. Advanced weather forecasting and real-time monitoring systems are vital, but can add to project costs.
4. **Regulatory and environmental risks**: The catastrophic consequences of events such as the Deepwater Horizon oil spill have prompted stringent regulatory changes across the industry. This has resulted in heightened compliance costs, rigid operational protocols, and extended timelines for project

approvals due to environmental assessments and stakeholder consultations. Companies must prioritize ecological impact mitigation and invest in robust disaster response strategies to navigate these complexities.

Onshore Risks:

1. **Geological risks**: The production profiles of onshore oil and gas fields, especially in shale formations, require ongoing and often extensive drilling programs to sustain output levels. Geological uncertainties can lead to unexpected dry holes or resource depletion, increasing operational costs and requiring adjustments to extraction strategies and technologies.
2. **Land Use and community relations**: Onshore drilling projects frequently encounter opposition from landowners, indigenous communities, and environmental activists. These conflicts can result in protracted negotiations and public relations challenges, delaying project timelines and inflating costs. Building strong relationships with local stakeholders is essential for smooth project execution and can mitigate potential roadblocks.
3. **Water usage and environmental impact**: The process of hydraulic fracturing (fracking) necessitates substantial volumes of water, which raises concerns over groundwater contamination and potential seismic activity. Regulatory scrutiny over water management practices and environmental assessments is intensifying, with many regions imposing limits on water usage and requiring comprehensive studies to evaluate the broader ecological impact.
4. **Regulatory and taxation issues**: The regulatory landscape for onshore projects is continuously evolving, with governments frequently revising taxation policies, royalty rates, and environmental regulations. These changes can significantly impact projects' economic viability, necessitating agile financial planning and risk assessment strategies to adapt to new fiscal environments. Companies must invest in compliance monitoring and an adaptive approach to navigate this dynamic regulatory framework effectively (Figure 20.2).

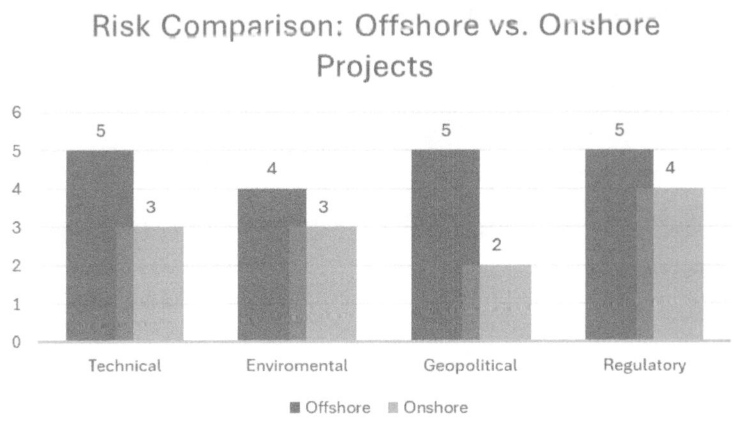

FIGURE 20.2 Risk comparison across offshore and onshore projects. Bar chart comparing risk levels in four categories—Technical, environmental, geopolitical, and regulatory—For offshore and onshore developments. Offshore projects consistently show higher risk exposure, reinforcing the need for advanced mitigation strategies and robust portfolio balancing.

20.4 TECHNOLOGICAL ADVANCEMENTS IN DRIVING COST REDUCTIONS

Technology has played a crucial role in significantly narrowing the cost gap between offshore and onshore developments in the energy sector.

Advances in Offshore Technology:

- **Subsea processing and tiebacks**: This innovative approach processes hydrocarbons directly on the seabed, substantially reducing infrastructure costs and eliminating the need for complex surface processing facilities. Operators can reduce overall development costs and improve efficiency by employing subsea tiebacks to transport hydrocarbons to existing platforms or directly to shore.
- **Floating production storage and offloading (FPSO) units**: FPSOs have revolutionized offshore operations by eliminating the necessity for expensive fixed platforms. These floating units can process and store hydrocarbons, significantly reducing capital expenditures (CAPEX) while offering enhanced flexibility to relocate according to market conditions and field development needs.
- **Digital twin and AI monitoring**: Using digital twin technology combined with AI-driven monitoring solutions enhances the predictive maintenance capabilities of offshore platforms. By creating a virtual replica of physical assets, operators can harness real-time data analytics to identify potential failures before they occur, minimizing downtime and optimizing operational efficiency (Figure 20.3).

TOP EMERGING TECHNOLOGIES IN UPSTREAM

- Digital twin monitoring
- Automated drilling platforms
- Multistage fracturing
- Subsea tiebacks

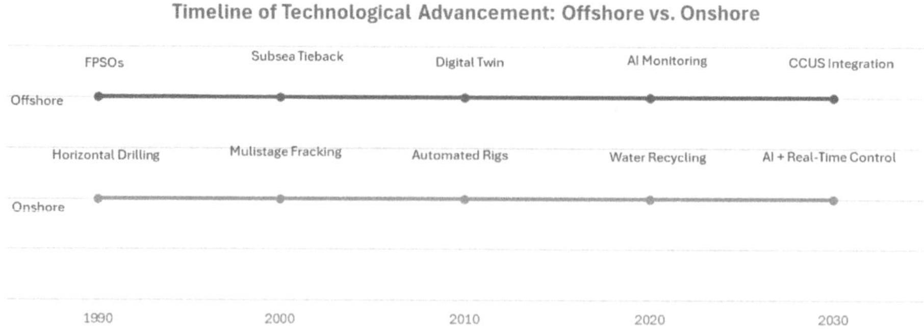

FIGURE 20.3 Economics of brownfield rejuvenation options. Comparison of five strategies by CAPEX, OPEX impact, payback period, and Best use case. Highlights cost-effective paths to extend field life and delay abandonment.

Advances in Onshore Technology:

- **Horizontal drilling and multistage fracturing**: These advanced drilling techniques have transformed the shale oil and gas sector by allowing access to previously inaccessible hydrocarbons. Horizontal drilling enhances recovery rates by maximizing exposure to oil and gas reservoirs. At the same time, multistage fracturing optimizes the creation of fractures within the rock, leading to improved flow rates and economics in production.
- **Automated drilling rigs**: The introduction of automated drilling rigs has significantly reduced drilling times and operational costs. These rigs leverage automation and advanced robotics to increase drilling precision and efficiency, reduce labor costs, and improve safety for personnel involved in the drilling process.
- **Enhanced oil recovery (EOR)**: EOR techniques, such as CO_2 injection and steam flooding, extend the lifespan of mature onshore fields by improving recovery rates. By injecting CO_2 or steam into reservoirs, operators can unlock additional hydrocarbons that would otherwise remain trapped, thus maximizing the total extraction from existing fields and ensuring a more sustainable resource management approach.

These technological advancements illustrate how innovation drives cost reductions and improves efficiency across offshore and onshore oil and gas developments.

20.5 INVESTMENT STRATEGIES: BALANCING OFFSHORE AND ONSHORE ASSETS

Oil and gas companies adopt a strategic approach to balancing their offshore and onshore investments, aiming to manage risks while effectively maximizing returns on their capital.

Advantages of Offshore Investments:

- **Large reserves with high long-term production potential:** Offshore fields often contain substantial reserves supporting prolonged production cycles, ensuring sustained revenue streams. For instance, deepwater projects in the Gulf of Mexico are known for their significant yield capacity.
- **Longer project lifespans:** Offshore investments typically have extended lifespans, benefiting companies from long-term cash flow stability. This is particularly true for fields with established infrastructure and technology that can extend their operational life.
- **Technological advancements**: Innovations such as subsea processing and enhanced oil recovery techniques have dramatically reduced operational costs and increased recovery rates. Companies like Shell and BP continually invest in R&D to push these technological boundaries further.

WHEN DOES OFFSHORE MAKE ECONOMIC SENSE

- Long reserve life
- Stable geopolitics
- Supportive carbon pricing (CCUS integration)
- Technological edge

Advantages of Onshore Investments:

- **Lower capital expenditure (breakeven) and quicker development timelines**: Onshore projects generally require less initial investment and can be brought online much more rapidly than their offshore counterparts. For example, the North American shale boom has shown how quickly companies can drill and complete wells, enabling rapid responses to market fluctuations.
- **More effortless scalability and adaptability to market conditions**: Onshore operations allow for more flexible scaling. Companies can adjust the drilling and production pace in response to real-time market demands without the extensive logistical challenges offshore projects face.
- **Lower geopolitical and regulatory risks**: Onshore investments typically involve fewer regulatory hurdles and are less susceptible to geopolitical tensions, making them more stable in volatile regions. Countries with stable regulatory frameworks, like the United States, provide safer environments for these investments (Figure 20.4).

Example:
ExxonMobil and Chevron have demonstrated a strategic balance in their investment portfolios. Both companies have aggressively expanded their onshore shale assets in the Permian Basin, renowned for its prolific output and relatively lower production costs. Concurrently, they continue to invest in offshore projects in the deepwater Gulf of Mexico and offshore Guyana, capitalizing on the high potential returns of offshore reserves while ensuring their portfolios remain diversified and resilient against market fluctuations.

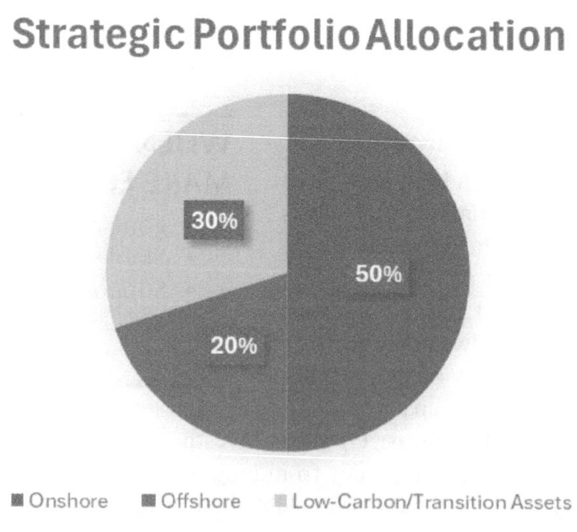

FIGURE 20.4 Strategic portfolio allocation across asset types. Illustrates a representative upstream portfolio Split: 50% low-carbon/transition assets, 30% offshore, and 20% onshore. Highlights the growing emphasis on sustainability while retaining legacy production and infrastructure leverage.

20.6 CASE STUDIES: OFFSHORE VS. ONSHORE DEVELOPMENT

20.6.1 Case Study 1: Deepwater Guyana—ExxonMobil's Offshore Success

- ExxonMobil's offshore discoveries in Guyana's **Stabroek Block** have positioned the country as a major oil producer.
- Development of FPSOs has enabled rapid commercialization of offshore reserves.
- Despite high initial CAPEX, production costs have decreased over time, making Guyana a key part of ExxonMobil's growth strategy.
- **Key takeaway:** Offshore megaprojects, when executed efficiently, can deliver substantial long-term value.

Guyana Becomes a Deepwater Giant as ExxonMobil Hits Fifth Discovery

Bloomberg, 2022

20.6.2 Case Study 2: The Permian Basin—Onshore Shale Boom

- The **Permian Basin** in the U.S. has become the world's most prolific shale oil-producing region.
- Companies like **Pioneer Natural Resources, ExxonMobil, and Chevron** have invested heavily in horizontal drilling and fracking to maximize output.
- Low CAPEX and flexible drilling programs allow rapid adaptation to oil price changes.
- **Key takeaway:** Onshore shale plays offer quick returns and scalability, making them attractive in volatile markets.

Permian Rigs Back Online as Prices Stabilize – Shale Rebounds Again

Reuters, 2023

20.6.3 Case Study 3: The North Sea—Balancing Offshore Maturity and Efficiency

- The **North Sea** remains a key offshore oil region despite aging fields.
- Companies like **BP and Equinor** have focused on subsea tiebacks and enhanced recovery techniques to extend field life.
- High OPEX and decommissioning liabilities remain challenges.
- **Key takeaway:** Mature offshore basins require cost-efficient solutions to maintain production viability.

UK North Sea Fields Face $25B Decommissioning Bill

Financial Times, 2022

Case Study Matrix Table

Project/Region	Type	Key Strengths	Risks/Limitations
Deepwater Guyana	Offshore	Large reserves, high long-term ROI	High CAPEX, offshore exposure
Permian Basin	Onshore	Fast returns, scalable	Water issues, price sensitivity
North Sea	Offshore	Mature infra, tiebacks	Decommissioning, aging fields
Ghawar Field	Onshore	Ultra-low cost, high reliability	Long-term reservoir decline risk

FIGURE 20.5 Offshore vs. onshore case study matrix. Comparison of four key oil and gas regions—Guyana, Permian Basin, North Sea, and Ghawar—across asset type, strengths, and limitations. Highlights how reserve size, cost structure, and maturity shape investment strategy and risk exposure.

20.6.4 Case Study 4: Saudi Arabia's Ghawar Field—The World's Largest Onshore Oilfield

- Saudi Aramco's **Ghawar field** has been producing for over 70 years and remains a cornerstone of the global oil supply.
- Advanced water injection and reservoir management have maintained high production levels.
- **Key takeaway:** Large onshore conventional fields remain highly cost-effective and reliable sources of oil production (Figure 20.5).

20.7 CONCLUSION: THE FUTURE OF OFFSHORE AND ONSHORE DEVELOPMENT

Various factors, including prevailing market conditions, technological advancements, and the evolving regulatory landscape, influence the economic viability of offshore versus onshore oil and gas development.

- **Offshore projects**: These developments are often viewed as attractive opportunities due to their access to large-scale hydrocarbon reserves and the potential for long-term production stability. However, they demand significant capital expenditures (CAPEX), often exceeding billions of dollars. Moreover, offshore projects face challenges such as complex logistical operations, longer lead times, and heightened strategic risk management due to their isolation and exposure to harsh environmental conditions. Practical risk assessment and mitigation strategies are essential for navigating these challenges and ensuring project feasibility.
- **Onshore developments**: In contrast, onshore projects generally entail lower upfront costs and shorter project cycles, allowing companies to react quickly to fluctuations in market demand. This flexibility enables operators to easily shut down or ramp up production, thus enhancing resilience during periods of volatility. Onshore developments can be executed with various technologies that are often more readily implemented and adapted, resulting in a more agile resource extraction and management approach.
- **Technological innovations**: The continual evolution of technology remains a pivotal factor for offshore and onshore operations. Innovations such as

enhanced drilling techniques, improved reservoir modeling, and automated systems are reducing operational costs and improving overall efficiency. Integrating big data analytics and artificial intelligence is also crucial in optimizing production and minimizing downtime across all types of projects.

- **Sustainability and carbon management**: Increasingly, sustainability concerns and the imperative for effective carbon management are shaping investment decisions in the oil and gas sector. Companies are under growing pressure to incorporate emissions reduction strategies, implement practices such as carbon capture[4] and storage (CCS), and transition to cleaner energy sources. The demand for sustainable practices will likely affect how investments are allocated between offshore and onshore projects, with stakeholders expecting transparent reporting and accountability on environmental impacts.

In summary, by employing a balanced portfolio strategy encompassing both offshore and onshore investments, oil and gas companies can optimize their profitability and effectively manage risks associated with fluctuating market dynamics and regulatory changes in an ever-evolving energy landscape.

Technology turns what used to be impossible into today's baseline.

Key Takeaways:

✓ Offshore and onshore developments present distinct cost structures, operational risks, and investment dynamics that require careful consideration.

✓ Offshore projects necessitate substantial CAPEX but can yield long-term, stable production if managed effectively.

✓ Onshore projects are generally more cost-effective and offer greater operational flexibility, making them remarkably resilient in unstable market conditions.

✓ Ongoing technological advancements are essential for reducing project costs and enhancing the feasibility of both offshore and onshore operations.

✓ Implementing a balanced portfolio strategy allows companies to manage risks proactively while maximizing profitability in an increasingly complex energy market.

NOTES

1. Onshore shale projects, particularly in the Permian Basin, can maintain breakeven below $40/bbl due to drilling efficiency and infrastructure proximity.
2. Deepwater offshore developments often require investments of $5–15 billion before first oil, especially in ultra-deep water environments.
3. FPSOs reduce infrastructure costs by eliminating fixed platforms, especially valuable in remote or harsh offshore basins.
4. Carbon capture integration into offshore tiebacks can reduce net emissions and help meet ESG-linked funding criteria.

21 The Future of Upstream Petroleum Economics

21.1 INTRODUCTION: THE EVOLVING ENERGY LANDSCAPE

> Tomorrow's upstream winners will be those who solve today's uncertainty with data, discipline, and direction.

The upstream petroleum sector stands at a critical juncture. The convergence of the **energy transition, rapid technological advancements, evolving geopolitical dynamics, and a growing emphasis on environmental, social, and governance (ESG) factors** is reshaping the industry profoundly. As the global energy landscape shifts, upstream operators must adapt to new market realities, regulatory pressures, and consumer expectations.

Key Challenges and Opportunities:

- **Balancing hydrocarbon demand with carbon reduction:** While hydrocarbons remain essential for global energy security, reducing carbon emissions is becoming a key priority. Governments and industry players must navigate policies aimed at decarbonization while ensuring energy availability.
- **Leveraging technological innovations:** Advancements in automation, artificial intelligence, digital twins, and enhanced oil recovery (EOR) techniques provide opportunities to optimize production, reduce costs, and minimize environmental impact.
- **Managing geopolitical risks and supply chain disruptions:** The volatility of energy markets, international conflicts, and trade policies creates uncertainty. Companies must develop resilient supply chain strategies and assess geopolitical risks in investment decisions.
- **Integrating ESG factors into investment and operations:** Investors, regulators, and consumers increasingly focus on sustainability and corporate responsibility. Upstream operators must align their strategies with ESG principles to maintain access to capital and build long-term resilience.

As the industry evolves, navigating these challenges will require **adaptability, innovation, and strategic foresight**. The following sections will explore the **future of upstream economics**, examining emerging trends, disruptive forces, and potential pathways for the sector.

 DOI: 10.1201/9781003750345-25

21.2 KEY DRIVERS OF CHANGE

Several key forces are shaping the future of the upstream petroleum sector. These drivers will determine how companies adapt to evolving market conditions, regulatory pressures, and technological advancements.

1. **Energy Transition**
   The global shift toward **lower-carbon energy sources** is accelerating, creating challenges and opportunities for the oil and gas industry. The pace and direction of this transition will significantly impact future demand for hydrocarbons. Different scenarios project **varying oil and gas consumption levels**, depending on the **speed of renewable energy adoption, policy choices, and advancements in alternative energy technologies.**
   - **Example:** The **International Energy Agency (IEA[1])** predicts that global oil demand could peak by 2025 and decline steadily under its Sustainable Development Scenario. However, **natural gas may remain a key transition fuel** due to its lower carbon footprint than coal and oil.

   > IEA Says Oil Demand May Peak by 2025 Under Net Zero Push
   >
   > *Financial Times, 2023*

2. **Technological Disruption**
   Continued exploration, drilling, production, and digitalization innovation are transforming upstream operations. These advancements are **driving down costs, improving efficiency, and unlocking access to previously uneconomic resources.**
   - **Example: Autonomous drilling rigs, AI-driven predictive maintenance, and digital twins** enhance operational efficiency and reduce downtime. AI-powered **reservoir modeling** and **machine learning algorithms** are optimizing exploration and production strategies.
3. **Geopolitical Uncertainty**
   International relations, **resource nationalism, trade policies, and potential conflicts** create uncertainty and volatility in global energy markets. These geopolitical risks can **impact investment decisions, disrupt supply chains, and restrict access to key resources.**
   - **Example: Sanctions on oil-producing countries** like **Iran and Venezuela** have disrupted global oil supply chains, leading to price volatility and prompting shifts in energy trade routes. Similarly, conflicts in energy-rich regions like the Middle East have historically influenced global oil prices.
4. **ESG Factors**
   Growing societal and investor pressure for **sustainability, environmental protection, and social responsibility** pushes upstream companies to **integrate ESG principles into their operations and investment strategies.** Companies that fail to address **climate risks, emissions reductions, and**

corporate governance issues may struggle to access capital and maintain stakeholder trust.

- **Example: Major oil and gas companies** are setting ambitious **net-zero carbon** targets, **investing in renewable energy projects**, and enhancing **ESG transparency and reporting** to meet investor and regulatory expectations.

Implications of Key Drivers:

- Upstream companies must **adopt more sustainable practices** and **invest in cleaner technologies** to remain competitive.
- **Technological innovation** will continue to **improve efficiency, reduce costs, and lower environmental impact.**
- Companies must develop **resilient supply chains** to **mitigate geopolitical risks** and **ensure stable production.**
- **Integrating ESG factors** into business decisions will **enhance resilience** and **attract long-term investment.**

21.3 SCENARIOS FOR THE FUTURE OF UPSTREAM

Global energy demand, policy decisions, technological advancements, and climate action will shape the future of the upstream petroleum sector. Different scenarios outline potential pathways for the industry, ranging from continued reliance on hydrocarbons to a rapid shift toward renewable energy sources.

21.3.1 High-Demand Scenario

In this scenario, **global energy demand continues to grow,** primarily driven by population growth, economic expansion, and industrialization, particularly in **developing countries.** Oil and gas remain **dominant energy sources** despite the rising share of renewables.

Key Characteristics:

- Strong demand for **transportation fuels, petrochemicals, and electricity generation from natural gas.**
- **Developing economies in Asia and Africa** experience rapid industrialization, driving energy consumption.
- Governments **prioritize economic growth over aggressive decarbonization policies.**
- **Exploration and production** continue at high levels to meet rising energy demand.

Example:

- Countries like **India, Indonesia, and Nigeria** expand infrastructure, transportation, and manufacturing sectors, **increasing their reliance on oil and gas.**
- **Liquefied natural gas (LNG) demand surges** as emerging markets transition from **biomass and coal to gas-fired power generation.**

Impact on Upstream:

- Companies continue to **invest in conventional and unconventional oil and gas projects** to meet demand.
- **Operational efficiency, cost competitiveness, and technological advancements** in extraction become **key to maintaining market share**.
- Oil and gas prices remain **strong, attracting continued investment in exploration and production.**

21.3.2 Transition Scenario

This scenario envisions a **gradual shift** to **lower-carbon energy sources,** with oil and gas demand **declining over time** but still playing a **substantial role,** particularly in sectors **with limited alternatives.** Natural gas is a **transition fuel,** displacing coal in **power generation** and supporting **industrial applications.**

Key Characteristics:

- Governments implement **carbon pricing mechanisms, emissions regulations, and incentives for cleaner energy.**
- Oil and gas consumption **gradually declines** but **remains essential** for aviation, shipping, and petrochemical industries.
- **Natural gas demand grows** as it replaces coal in **electricity generation** and **industrial applications.**
- Investments in **carbon capture and storage (CCS) and hydrogen technologies** increase.[2]

Example:

- The **EU and China phase out coal-fired power plants**, replacing them with a mix of **natural gas, nuclear, and renewables**.
- **Oil demand for passenger vehicles declines** as electric vehicles (EVs) become mainstream, but **petrochemical demand remains strong**.

Impact on Upstream:

- Companies **balance traditional hydrocarbons** with **investments in natural gas, hydrogen, and low-carbon technologies.**
- **Diversification and portfolio optimization** become **critical** for adapting to shifting energy demands.
- **Regulatory compliance and ESG performance** influence investment decisions and financing availability.

21.3.3 Low-Carbon Scenario

This scenario assumes a **rapid global transition** to **renewable energy,** driven by **strong policy support, carbon pricing, and technological breakthroughs.** Oil and gas demand **declines sharply,** and their role in the energy mix **diminishes significantly.**

Comparing the Three Scenarios

Factor	High-Demand Scenario	Transition Scenario	Low-Carbon Scenario
Oil Demand	Continues to grow	Declines gradually	Declines rapidly
Natural Gas Demand	High	Strong as a transition fuel	Declines with the rise of hydrogen
Renewable Energy Growth	Slow but expanding	Rapid expansion	Dominant energy source
Upstream Investment	Strong	Diversified between hydrocarbons and low-carbon tech	Shift toward renewables and clean energy
Role of ESG & Policy	Limited influence	Increasing influence	Decisive government action & decarbonization policies

FIGURE 21.1 Comparing future scenarios for upstream petroleum. Side-by-side comparison of three energy transition pathways—high demand, transition, and low carbon—across oil and gas demand, renewables growth, upstream investment, and ESG/policy influence. Highlights how upstream strategy must adapt to divergent futures.

Key Characteristics:

- **Aggressive climate policies** and international agreements drive **decarbonization efforts.**
- **Significant investments** in **renewable energy, battery storage, and electrification.**
- **Fossil fuel consumption declines** in **transportation, power generation, and heating.**
- Carbon pricing, emissions regulations, and subsidies for clean energy **accelerate the shift away from hydrocarbons.**

Example:

- **Electric vehicle penetration exceeds 70% globally,** reducing oil demand for transportation.
- **Hydrogen and biofuels replace** fossil fuels in **aviation, shipping, and heavy industry.**
- **Major economies impose strict emissions caps,** forcing oil and gas producers to decarbonize or exit certain markets.

Impact on Upstream:

- **Significant oil and gas demand reduction** leads to **asset write-downs and stranded resources.**
- Companies must **pivot rapidly** to **renewable energy, hydrogen, and carbon capture technologies.**
- **Mergers, acquisitions, and divestments accelerate** as firms **shift away from traditional hydrocarbons** (Figure 21.1).

WHAT KEEPS UPSTREAM ECONOMICALLY VIABLE IN THE 2030S?

- Sub-$40 breakeven
- ESG-linked finance access
- Carbon-neutral pathways
- Tech-enhanced recovery

21.3.4 Strategic Considerations for Upstream Companies

Regardless of which scenario unfolds, **upstream companies must prepare for multiple possibilities** and position themselves for long-term success:

- **Enhance operational efficiency:** Lower costs and improve recovery rates to remain competitive.

- **Diversify investments:** Balance hydrocarbons with low-carbon technologies, renewables, and hydrogen.
- **Embrace digitalization:** Leverage AI, automation, and data analytics to optimize operations.
- **Strengthen ESG performance:** Align with regulatory requirements and investor expectations.
- **Monitor policy and market trends:** Adapt to evolving energy demand and decarbonization policies.

These scenarios **highlight the uncertainty facing the upstream sector.** While hydrocarbons **will remain part of the global energy mix** for decades, companies must **adapt their strategies to thrive in a changing landscape.**

21.4 IMPLICATIONS FOR UPSTREAM PETROLEUM ECONOMICS

The evolving energy landscape presents **challenges and opportunities** for the upstream petroleum sector. Companies must navigate cost pressures, energy transition policies, geopolitical risks, and shifting investor expectations. This section explores the key **economic implications** of these trends and how upstream companies can adapt to remain competitive.

21.4.1 Cost Competitiveness

To thrive in a changing market, **upstream companies must prioritize cost efficiency** and optimize operations. As oil price volatility increases and competition from renewables grows, cost reduction will be a **critical determinant of profitability.**

Key Strategies for Cost Competitiveness:

- **Automation and AI:** Using **AI-driven analytics, machine learning, and automation** to optimize production and reduce downtime.
- **Advanced drilling and recovery techniques:** Enhancing **well productivity and recovery rates** through better drilling and completions.
- **Supply chain optimization:** Improving **logistics, procurement strategies, and supplier negotiations** to lower operating expenses.
- **Lean operations:** Streamlining business processes and **minimizing waste** for long-term cost efficiency.

Example:

- **BP and ExxonMobil** have invested heavily in **AI-powered predictive maintenance** to improve efficiency and reduce unplanned downtime, lowering production costs.
- **Chevron** has optimized offshore operations using **autonomous drilling rigs** to enhance well productivity.

Economic Impact:

- **Lower breakeven prices**, making projects viable even in low-price environments.
- **Higher profit margins** and greater resilience during price downturns.

21.4.2 Portfolio Diversification

As the energy transition accelerates, upstream companies must **balance their portfolios** between conventional hydrocarbons, unconventional resources, and cleaner energy alternatives.

Key Diversification Strategies:

- **Investing in natural gas:** Gas is seen as a **transition fuel** that helps reduce emissions while maintaining energy security.
- **Expanding unconventional resources:** Shale, tight oil, and deepwater exploration will remain key growth areas.
- **Venturing into renewables:** Some upstream companies are **expanding into solar, wind, and biofuels** to future-proof their businesses.
- **Carbon capture and hydrogen projects:** Investing in **carbon capture, utilization, storage (CCUS),** and **hydrogen production** to align with decarbonization goals.

Example:

- **TotalEnergies** and **Shell** have diversified into offshore wind and hydrogen projects, reducing dependence on crude oil.
- **Equinor** has shifted from a pure oil producer to an **integrated energy company**, developing large-scale carbon capture and hydrogen facilities.

Economic Impact:

- **Reduces reliance on oil price fluctuations** by having multiple revenue streams.
- **Positions companies for long-term growth** in emerging energy markets.

21.4.3 Risk Management in a Volatile Market

Companies must adapt their risk management frameworks to increased uncertainty in energy policies, geopolitical events, and technological shifts.

Key Risk Management Strategies:

- **Scenario planning and stress testing:** Evaluating various future energy transition pathways and their financial impact.
- **Financial hedging:** Using **commodity derivatives and futures contracts** to mitigate price fluctuations.
- **Geopolitical risk assessment:** Diversifying investments across stable and high-potential markets to balance exposure.
- **Flexible capital allocation:** Ensuring **capital expenditures (CAPEX) can be adjusted quickly** based on market conditions.

Example:

- **Saudi Aramco** adopts **long-term hedging strategies** to stabilize revenue amid oil price volatility.
- **BP** uses dynamic scenario planning to adjust its investment strategies **based on energy transition policies.**

Economic Impact:

- **Protects revenue streams** from oil price swings and market uncertainties.
- **Enables companies to remain agile** and pivot strategies as needed.

21.4.4 Sustainability and ESG Integration

ESG factors increasingly shape investment decisions and regulatory policies. **Companies that fail to address ESG concerns may face capital constraints, regulatory penalties, and reputational damage.**

Key ESG Strategies:

- **Greenhouse gas (GHG) emission reductions:** Investing in **low-carbon technologies and carbon offset programs.**
- **Water and waste management:** Enhancing **environmental stewardship** in drilling and production.
- **Community engagement and social responsibility:** Building strong relationships with **host communities** to maintain a positive social license to operate.
- **Transparent ESG reporting:** Meeting **investor and regulatory expectations** for sustainability disclosures.

Example:

- **ExxonMobil** is investing **$15 billion in carbon capture and storage (CCS)** projects.
- **TotalEnergies** has committed to achieving **carbon neutrality by 2050** and is actively shifting toward renewable energy.

Economic Impact:

- **Attracts ESG-conscious investors** and enhances access to sustainable financing.
- **Reduces regulatory risks** and strengthens corporate reputation (Figure 21.2).

21.4.5 Final Thoughts

Upstream industries must evolve their **financial strategies** to succeed in the rapidly changing energy environment. **Companies that prioritize cost efficiency, portfolio**

Summary: The New Economics of Upstream

Key Area	Strategic Focus	Economic Benefits
Cost Competitiveness	AI, automation, digitalization	Lower breakeven prices, higher margins
Portfolio Diversification	Balancing hydrocarbons with renewables & gas	Reduced price exposure, new revenue streams
Risk Management	Scenario planning, hedging, flexible CAPEX	Financial stability, adaptability in uncertain markets
ESG & Sustainability	Carbon reduction, social responsibility	Access to capital, long-term regulatory compliance

FIGURE 21.2 The new economics of upstream: Strategic levers and benefits. Summary of four key focus areas—cost competitiveness, portfolio diversification, risk management, and ESG integration—alongside their strategic priorities and economic impacts. Highlights how upstream companies can remain viable amid transition and volatility.

diversification, proactive risk management, and ESG integration will be better positioned for long-term success.

21.5 THE ROLE OF INNOVATION AND TECHNOLOGY

Technological advancements are revolutionizing **upstream petroleum operations, making them more efficient, cost-effective, and sustainable.** Integrating **automation, AI, digitalization, and enhanced recovery techniques** reshapes **how oil and gas companies explore, develop, and produce hydrocarbons.**

21.5.1 Improving Efficiency and Reducing Costs

The upstream sector is under continuous pressure to **optimize costs and enhance operational efficiency.** Advances in **automation, robotics, and AI-driven analytics** are transforming operations by reducing downtime, improving drilling precision, and streamlining workflows.

> Equinor and TotalEnergies Bet Big on Hydrogen Future
>
> *Bloomberg, 2024*

Key Technological Innovations:

- **Autonomous drilling rigs:** Self-operating rigs minimize human intervention, reduce costs, and improve safety.
- **Predictive maintenance:** AI-driven **real-time equipment monitoring** prevents unexpected failures and costly repairs.
- **Robotic inspections:** Robots equipped with sensors can inspect offshore platforms and pipelines, improving safety and efficiency.

INNOVATION TIMELINE IN UPSTREAM (2000–2040)

Key tech waves:

- 2000–10: Horizontal drilling, fracking
- 2010–20: AI, digital twins
- 2020–30: CCS, autonomous rigs, blockchain
- 2030–40: AI economic modeling, hydrogen optimization

Example:

- **BP and Shell** use **real-time drilling automation and AI-driven predictive analytics** to improve drilling efficiency and prevent downtime.
- **Equinor** deployed an **autonomous subsea drone** for offshore inspections, reducing operational risks.

Economic Impact:

- **Lower operational costs** through automation and predictive analytics.
- **Higher production efficiency** with real-time performance optimization.

21.5.2 Enhanced Oil Recovery (EOR) and Unconventional Resources

Upstream companies invest in advanced recovery technologies to maximize production from mature fields and unlock previously uneconomic resources.

Key Technologies in EOR and Unconventional Resources:

- **CO_2 injection:** Injecting carbon dioxide into reservoirs increases oil recovery rates.
- **Microbial EOR:** Using bacteria to break down hydrocarbons and improve flow rates.
- **Hydraulic fracturing (fracking):** Advancements in fracturing fluids and proppant technologies enhance production in **shale and tight oil formations.**

Example:

- **Chevron and ExxonMobil** use **supercritical CO_2 injection** to boost recovery in depleted reservoirs.
- **Pioneer natural resources** has improved **fracking techniques** to enhance Permian Basin production.

Economic Impact:

- **Extends the life of mature fields** and maximizes oil recovery.
- **Unlocks additional reserves** in unconventional formations.

21.5.3 Minimizing Environmental Footprint

Sustainability and carbon management are increasingly important for **regulatory compliance and investor confidence.** Clean technologies are being deployed to **reduce emissions, manage waste, and improve water use efficiency.**

Key Technologies in Environmental Management:

- **Carbon capture, utilization, and storage (CCUS):** Captures CO_2 emissions from production sites and stores them underground.
- **Water recycling systems:** Advanced treatment processes **reuse water** in drilling and hydraulic fracturing.
- **Methane leak detection:** Using **satellites, sensors, and infrared cameras** to detect and mitigate methane emissions.

Example:

- **Saudi Aramco** invests in **CCUS to capture and store 9 million metric tons of CO_2 annually.**
- **ExxonMobil** uses **advanced leak detection systems** to minimize methane emissions from operations.

Economic Impact:

- **Regulatory compliance** with strict emissions standards.
- **Lower operational costs** by reusing water and reducing carbon liabilities.

21.5.4 Digitalization and Data Analytics

Big data, AI, and **cloud computing are transforming upstream decision-making. Digital twins, blockchain, and machine learning algorithms** enhance reservoir modeling, predictive analytics, and production optimization.

Innovation and Technology in Action

Technology	Application	Economic Benefits
Automation & AI	Autonomous drilling, predictive maintenance	Reduced downtime, lower costs
Enhanced Oil Recovery (EOR)	CO2 injection, microbial EOR	Increased hydrocarbon recovery
Environmental Tech	CCUS, methane leak detection	Reduced emissions, regulatory compliance
Digitalization	Digital twins, blockchain, AI analytics	Optimized operations, better ESG tracking

FIGURE 21.3 Innovation and technology in upstream: Applications and economic impact. Summary of key technologies—automation, EOR, environmental tech, and digitalization—alongside their operational applications and economic benefits. Highlights how innovation drives cost reduction, recovery efficiency, and ESG performance.

Key Digital Technologies in Upstream Operations:

- **Digital twins:** Virtual replicas of physical assets provide **real-time monitoring and predictive analytics.**
- **Machine learning algorithms:** AI optimizes **reservoir modeling, production forecasting, and maintenance schedules.**[3]
- **Blockchain for ESG compliance:** Blockchain enhances **data integrity in emissions tracking and reporting.**

Example:

- **TotalEnergies** has deployed **digital twins** across offshore assets to **optimize maintenance schedules and improve efficiency.**
- **Chevron and IBM** use **blockchain technology** to **ensure data transparency in emissions tracking.**

Economic Impact:

- **Reduces downtime** and improves decision-making.
- **Enhances ESG reporting transparency**, attracting sustainable investments (Figure 21.3).

21.5.5 Final Thoughts

Technology is reshaping the future of upstream petroleum economics. Companies that embrace automation, AI, digitalization, and environmental sustainability will achieve **cost efficiency, resilience, and long-term profitability.**

21.6 THE HUMAN ELEMENT

The **future of upstream petroleum economics** depends not solely on technology and market dynamics—**people also shape it.** A highly skilled and adaptable workforce is crucial for navigating industry changes, adopting new technologies, and addressing

THREE TYPES OF UPSTREAM INNOVATORS

- The ReSharper: TotalEnergies, Equinor
- The Tech-Optimizer: BP, Shell
- The Carbon-Giant: ExxonMobil, Aramco

sustainability challenges. Companies that invest in **training, adaptability, diversity, and safety** will position themselves for long-term success.

21.6.1 Skill Development

As automation, digitalization, and environmental regulations reshape the industry, upstream professionals must develop **new competencies** in **data analytics, AI, automation, and sustainability.** Companies must invest in **training programs** to ensure employees have the skills for future operations.

Key Areas for Skill Development:

- **AI and machine learning:** Understanding data-driven decision-making for reservoir modeling, predictive maintenance, and automation.
- **Sustainability and carbon management:** Training employees in **emission reduction strategies, carbon capture, and environmental impact assessments.**
- **Advanced robotics and automation:** Operators must learn to manage **robotic drilling rigs, autonomous inspections, and real-time remote monitoring systems.**

Example:

- **BP and Shell** partner with universities to **offer AI and digitalization training** for their engineers and geologists.
- **Equinor** launched an internal **machine learning and sustainability certification** to equip employees with digital and environmental skills.

Economic Impact:

- **Enhances efficiency** and productivity through data-driven decision-making.
- **Reduces operational risks** with AI-powered predictive maintenance.

21.6.2 Workforce Adaptability

The **energy transition and technological advancements** require a culture of **continuous learning.** Employees must **adapt to industry shifts** and **upskill** to stay relevant.

Strategies for Workforce Adaptability:

- **Lifelong learning programs:** Companies should offer **on-the-job training, reskilling initiatives, and access to certification programs.**
- **Cross-disciplinary collaboration:** Engineers, data scientists, and environmental experts should collaborate to **integrate sustainability with upstream operations.**
- **Flexible career paths:** Employees should be encouraged to explore **roles in digitalization, renewable energy, and energy efficiency.**

Example:

- **TotalEnergies** allows employees to transition from traditional oil and gas roles into **renewable energy projects and data science teams.**
- **Saudi Aramco** provides an **online learning platform** for workers to access AI, automation, and sustainability courses.

Economic Impact:

- **Increases job security** by equipping workers with future-proof skills.
- **Boosts innovation** through cross-disciplinary knowledge exchange.

21.6.3 Diversity and Inclusion

A **diverse workforce** fosters innovation and enhances decision-making by incorporating **varied perspectives** on technological development, sustainability, and geopolitical risks.

Key Diversity Initiatives:

- **Inclusive hiring practices:** Companies should recruit from **underrepresented groups, including women, minorities, and local talent.**
- **Leadership development programs:** Offering mentorship and career development opportunities to ensure **diverse representation in leadership roles.**
- **Cultural competency training:** Training employees to **work effectively in global teams** and respect local cultures in project regions.

Example:

- **Chevron** has launched an **initiative to increase female representation in engineering and leadership roles.**
- **ExxonMobil** has diversity-focused recruitment to **enhance gender and cultural representation** in upstream operations.

Economic Impact:

- **Improves decision-making** by incorporating diverse perspectives.
- **Strengthens local partnerships** by hiring local talent and respecting cultural nuances.

21.6.4 Health and Safety in Upstream Operations

Upstream oil and gas operations pose significant safety risks, especially in offshore drilling, deepwater projects, and high-pressure environments. Protecting workers **remains a top priority.**

Key Health and Safety Innovations:

- **Wearable safety technology:** Smart helmets, biometric sensors, and real-time monitoring devices **track worker health and detect fatigue or hazardous exposure.**

Examples of Workforce Development Initiatives

Initiative	Purpose	Example
Digital Training Programs	Teach AI, automation, and data analytics skills	BP partners with MIT to train engineers in AI for drilling operations
Sustainability Education	Equip employees with carbon management and ESG expertise	TotalEnergies includes sustainability modules in employee training
Diversity and Inclusion Programs	Create a more innovative and inclusive workforce	Chevron increases female hiring in engineering roles
Wearable Safety Tech	Monitor worker health and prevent accidents	Shell deploys biometric wearables on offshore platforms

FIGURE 21.4 Workforce development initiatives in upstream petroleum. Examples of training, inclusion, and safety programs that support digital transformation, ESG integration, and operational resilience. Highlights how companies equip their workforce for a changing energy landscape.

- **AI-powered risk assessment:** AI-driven safety analytics **identify high-risk scenarios and recommend preventive measures.**
- **Virtual reality (VR) training:** VR simulations **train workers in hazardous environments, enhancing preparedness for real-life situations.**

Example:

- **Equinor and BP** use **AI-driven predictive safety systems** to **reduce offshore accidents.**
- **Shell** has implemented **VR training for emergency response simulations,** helping workers practice high-risk procedures in a controlled environment.

Economic Impact:

- **Reduces workplace injuries and downtime.**
- **Enhances safety compliance and lowers insurance costs** (Figure 21.4).

FUTURE-FOCUSED UPSTREAM STRATEGY FRAMEWORK

Four stacked pillars:

- Innovation
- ESG integration
- Portfolio diversification
- Resilience and risk management

21.6.5 Final Thoughts

The **future of upstream petroleum economics** depends on technology, market dynamics, and **people.** Investing in **education, adaptability, diversity, and safety** will be essential for companies to thrive in an evolving industry.

21.7 CONCLUSION: NAVIGATING THE UNCERTAIN FUTURE

The future of **upstream petroleum economics** is uncertain, but **change brings opportunity.** The **energy transition, technological advancements, geopolitical shifts, and ESG factors** will determine how companies succeed in this evolving landscape.

Companies that **embrace innovation, prioritize sustainability, diversify their portfolios, and develop robust risk management strategies** will be best positioned to **remain competitive and resilient** in the decades ahead.

21.7.1 Strategies for Success

1. **Embrace Innovation**
   Continuously invest in **technological advancements, digital transformation, and automation** to improve efficiency, reduce costs, and minimize environmental impact.

Example:

- An oil company invests in **AI-driven predictive maintenance**, reducing downtime and optimizing production.
- A company pioneers **automated drilling** and **carbon capture** to stay ahead of the competition.

Impact:

- Enhances efficiency and cost-effectiveness.
- Unlocks new resource potential through **advanced EOR** and **automation.**

2. **Integrate Sustainability (ESG Focus)**
   Incorporate **ESG factors** into all aspects of operations. Setting sustainability goals, engaging stakeholders, and adopting transparent reporting will be **crucial for long-term success.**

Example:

- A major energy company commits to **net-zero emissions by 2050**, investing in **renewables, CCS, and community programs**.

Impact:

- Attracts investors prioritizing **sustainable energy.**
- Reduces regulatory risks and enhances corporate reputation.

3. **Diversify Portfolios**
   A balanced portfolio—**combining conventional and unconventional oil and gas with renewable energy and clean technology investments**—will provide **resilience and long-term growth.**

Example:

- A company invests in **offshore wind farms, solar projects, and hydrogen** alongside traditional oil and gas assets.

Impact:

- Creates **multiple revenue streams** across energy sectors.
- Reduces reliance on a single commodity.

4. **Manage Risks**
   With increased uncertainty, companies must develop **robust risk management frameworks** using **scenario planning, stress testing, and geopolitical analysis** to prepare for market shifts.

Example:

- A company plans scenarios for different energy transition pathways and adjusts investment strategies accordingly.
- An oil major diversifies **supply chains** to avoid geopolitical bottlenecks.

Impact:

- Minimizes exposure to **sudden price shocks** and regulatory changes.
- Ensures business continuity during **market disruptions.**

21.7.2 Opportunities for Growth

1. **Renewable Energy Investments**
The global push for **clean energy** presents opportunities for upstream companies to **expand into renewables, including offshore wind, solar, and hydrogen.**

Example:

- A major oil company partners with a renewable energy firm to **develop a large-scale green hydrogen plant.**

Impact:

- Opens new revenue streams in the **clean energy sector.**
- Positions the company as an **energy leader beyond fossil fuels.**

2. **Carbon Capture and Storage (CCS)**
CCS technologies are **critical for emissions reduction**, and upstream com panies are uniquely positioned to **develop large-scale CCS projects.**

Example:

- A company captures CO_2 from **natural gas processing facilities** and stores it in **depleted reservoirs.**

Impact:

- Aligns with **net-zero targets and regulatory incentives**.
- Creates **new revenue opportunities** in carbon markets.

3. **Digital Transformation**
Digitalization will **revolutionize** exploration, production, and asset management. **AI, data analytics, and blockchain** will drive efficiencies and improve decision-making.

Example:

- A company implements a **digital twin** for a major offshore field, optimizing real-time reservoir performance.

Impact:

- Reduces operational costs and enhances productivity.
- Improves safety and risk management through **real-time data analysis.**

21.7.3 Final Thought: The Future Belongs to Adaptable, Innovative, and Strategic Companies

The **upstream sector is evolving** rapidly. **Companies that adapt, innovate, and embrace sustainability** will emerge as **leaders in the next phase of the energy industry.**

By balancing hydrocarbon production with clean energy investments, leveraging technology, and managing risks effectively, upstream companies will continue to play a crucial role in meeting global energy needs while transitioning toward a more sustainable future.

The key to success? Stay flexible, invest in innovation, and build resilience.

21.8 CONCLUSION AND FUTURE OUTLOOK

21.8.1 Final Thoughts on Upstream Petroleum Economics

The upstream petroleum industry is a dynamic and high-stakes sector that requires a careful balance of technical expertise, economic analysis, and strategic decision-making. This book has explored key aspects of petroleum economics, from exploration and production cost structures to risk assessment, financial modeling, and investment strategies. Through detailed case studies, economic evaluations, and real-world decision-making frameworks, we have demonstrated how companies navigate the complexities of upstream operations to maximize profitability while mitigating risks.

Several core insights have emerged from our analysis:

- **Project viability hinges on cost efficiency and economic resilience**: Projects with lower breakeven prices and optimized CAPEX/OPEX structures remain competitive even in volatile markets.
- **Data-driven decision-making enhances investment returns**: Integrating financial modeling, Monte Carlo simulations, and sensitivity analysis strengthens investment strategies and risk mitigation.
- **Technology and digitalization are reshaping oil economics**: Digital advancements are revolutionizing how oil companies manage assets from AI-driven reservoir modeling to real-time production optimization.
- **ESG considerations are redefining success**: Regulatory frameworks, carbon emissions policies, and sustainability goals are becoming central to project feasibility and capital allocation.

21.8.2 The Future of Upstream Petroleum Economics

The coming decades will witness profound shifts in the upstream oil and gas industry, driven by technological advancements, evolving market dynamics, and global energy transition policies. Some key trends shaping the future include:

1. **The Growing Role of Digitalization and AI in Oilfield Economics**
   The future of upstream decision-making will be **increasingly data-driven**, leveraging:
   - **Machine learning and predictive analytics**: AI-driven subsurface imaging, well placement optimization, and reservoir management will reduce exploration risks and improve recovery factors.

- **Digital twins and real-time monitoring**: Companies will use virtual simulations of oilfields to optimize production strategies and enhance efficiency.
- **Automated economic forecasting models**: AI-powered pricing algorithms will provide better market predictions and financial risk assessments.

2. **Capital Discipline and Cost Optimization in a Volatile Market**
   Oil price volatility is expected to persist, forcing companies to:
   - **Adopt lean project execution strategies**: Modular infrastructure, phased development, and CAPEX-light models will dominate.
   - **Enhance supply chain efficiencies**: Companies will prioritize local sourcing, streamlined logistics, and just-in-time inventory management.
   - **Diversify asset portfolios**: Supermajors and NOCs will balance high-risk deepwater assets with low-cost onshore developments to hedge against economic downturns.
3. **The Integration of Carbon Management and ESG-Driven Investments**
   Climate policies and investor expectations will reshape upstream investment decisions:
   - **Carbon capture and storage (CCS)**: **Large-scale deployment of CCS technology will become necessary** for projects to remain economically and environmentally viable.
   - **ESG-linked project financing**: Access to capital will increasingly depend on sustainability metrics, with banks and institutional investors favoring low-carbon projects.
   - **Regulatory pressures and carbon pricing**: Governments will enforce stricter emissions targets, making carbon pricing mechanisms a key factor in economic evaluations.
4. **Geopolitical Shifts and Market Realignments**
   The global energy landscape is evolving, and geopolitical dynamics will influence the upstream sector:
   - **OPEC+ and global supply control**: OPEC+ policies will continue shaping oil price stability, with periodic production cuts or increases affecting project economics.
   - **The rise of non-traditional oil players**: Emerging producers such as Guyana and Namibia are attracting major investment and altering traditional supply chains.
   - **Energy security and strategic reserves**: Nations will prioritize energy independence, leading to increased investments in domestic exploration and production.
5. **The Role of Upstream in the Energy Transition**
   While oil and gas will remain critical to global energy supply for decades, the industry must adapt to shifting demand patterns:
   - **Hydrogen and low-carbon alternatives**: Companies invest in blue hydrogen (derived from natural gas with CCS) and green hydrogen as part of their diversification strategy.
   - **Renewable integration in oil operations**: Offshore wind, solar power for EOR, and hybrid energy solutions will improve the sustainability of upstream projects.

- **Declining long-term demand for crude oil**: Peak oil demand forecasts suggest that by the 2030s, demand may plateau, requiring companies to optimize assets and extend profitability through efficiency gains.

21.8.3 Strategic Imperatives for Industry Leaders

For oil and gas companies to remain competitive in the evolving energy landscape, they must:

- ✓ **Strengthen resilience against market volatility** by Focusing on low-breakeven assets, flexible cost structures, and price hedging strategies.
- ✓ **Leverage technology for competitive advantage**: Invest in AI-driven analytics, automation, and digital oilfield innovations.
- ✓ **Integrate ESG and carbon management into core strategy**: Develop low-carbon solutions, adopt CCS, and align projects with sustainability goals.
- ✓ **Diversify portfolios for long-term growth**: Explore opportunities in LNG, hydrogen, and renewable energy while optimizing conventional assets.
- ✓ **Monitor geopolitical trends and adapt strategies accordingly**: Understand global energy realignments and navigate regulatory shifts to secure investments and market positions.

21.9 FINAL WORDS

The upstream oil and gas industry faces challenges and opportunities at a crossroads. Economic volatility, technological advancements, and regulatory pressures will continue shaping investment decisions. However, companies that embrace **innovation, financial discipline, sustainability, and strategic foresight** will survive and thrive in the future energy landscape.

This book has provided the tools and insights to navigate these complexities, ensuring that decision-makers can make informed, data-driven, and strategic choices in the ever-evolving world of petroleum exploration and production.

The energy transition is not a threat — it's a test of imagination and execution.

NOTES

1. The IEA Sustainable Development Scenario assumes a 50% decline in oil use by 2040 and aggressive adoption of low-carbon technologies.
2. Blue hydrogen is produced from natural gas with carbon capture; green hydrogen is made from water electrolysis using renewable electricity.
3. Digital oilfield economics increasingly use machine learning algorithms to forecast CAPEX, OPEX, and breakeven sensitivity under multiple scenarios.

Bibliography

Ahmed, T., & McKinney, P. D. (2011). *Advanced reservoir engineering.* Gulf Professional Publishing.

Alruqimi, M., & Di Persio, L. (2024a). Enhancing multistep Brent oil price forecasting with a multi-aspect metaheuristic optimization approach and ensemble deep learning models. *arXiv preprint* arXiv:2407.12062. Retrieved from https://arxiv.org/abs/2407.12062

Alruqimi, M., & Di Persio, L. (2024b). Enhancing multi-step Brent oil price forecasting with ensemble multi-scenario Bi-GRU networks. *arXiv preprint* arXiv:2407.11267. Retrieved from https://arxiv.org/abs/2407.11267

Berrisch, J., Pappert, S., Ziel, F., & Arsova, A. (2022). Modeling volatility and dependence of European carbon and energy prices. *arXiv preprint* arXiv:2208.14311. Retrieved from https://arxiv.org/abs/2208.14311

BP. (2020). *Thunder Horse Field.* Retrieved from https://www.bp.com/en/global/corporate/what-we-do/bp-worldwide/gulf-of-mexico/thunder-horse.html

Chevron. (2019). *Gorgon LNG Project.* Retrieved from https://www.chevron.com/projects/gorgon

Chevron. (2021). *Gorgon Project.* Retrieved from https://australia.chevron.com/our-businesses/gorgon-project

Economides, M. J., Hill, A. D., & Ehlig-Economides, C. (2012). *Petroleum production systems* (2nd ed.). Pearson.

Equinor. (2020). *Johan Sverdrup Field.* Retrieved from https://www.equinor.com/en/what-we-do/johan-sverdrup.html

Howarth, R. (2024). Is LNG worse for the climate than coal? *Financial Times.* Retrieved from https://www.ft.com/content/ac59729d-48f8-4b10-be3d-a41f7d13295b

International Energy Agency. (2020). *World Energy Outlook 2020.* Retrieved from https://www.iea.org/reports/world-energy-outlook-2020

Mian, M. A. (2011). *Project economics and decision analysis* (Vols. 1 & 2). PennWell Corporation.

Occupational Safety and Health Administration. (2021). *Oil and Gas Extraction - Safety and Health.* Retrieved from https://www.osha.gov/oil-and-gas-extraction

Pang, T., Tan, K., & Fan, C. (2023). Carbon price forecasting with quantile regression and feature selection. *arXiv preprint* arXiv:2305.03224. Retrieved from https://arxiv.org/abs/2305.03224

Rystad, J. (2024). *A top energy strategist is optimistic about climate change.* Associated Press. Retrieved from https://apnews.com/article/7927a9ac8172b0f278d0db35d5f19f0c

Shell. (2020). *Prelude FLNG.* Retrieved from https://www.shell.com/about-us/major-projects/prelude-flng.html

Smith, J., & Rajan, R. (2015). Advanced drilling techniques for unconventional oil and gas reservoirs. *Journal of Petroleum Technology, 67*(4), 56–67. https://doi.org/10.2118/178992-MS

Thompson, P., & Green, D. (2018). Impact of enhanced oil recovery on reservoir performance. *Energy Journal, 39*(7), 34–49. https://doi.org/10.1016/j.eneco.2015.01.022

United States Environmental Protection Agency. (2021). *Clean Air Act Overview.* Retrieved from https://www.epa.gov/clean-air-act-overview

Zhao, X., & Zhang, X. (2019). Carbon price forecasting models based on big data analytics. *Journal of Industrial and Business Economics, 46*(3), 453–472. https://doi.org/10.1080/17583004.2019.1568138

Glossary of Upstream Petroleum Economics

Abandonment: Permanently shutting down a well, plugging it, and restoring the surrounding environment after production ceases.

Appraisal Well: A well drilled after a discovery to determine a hydrocarbon reservoir's size, extent, and productivity.

Artificial Lift: Any method used to increase production when natural reservoir pressure is insufficient. Examples include gas lift, electric submersible pumps (ESPs), and sucker rod pumps.

Asset Retirement Obligation (ARO): Oil and gas companies have a legal and financial responsibility to decommission and restore drilling sites after production ceases properly.

Associated Gas: Natural gas found in crude oil reservoirs, either dissolved in oil or as a separate gas cap.

Barrel of Oil Equivalent (BOE): A standardized unit of energy measurement representing the energy content of one barrel of crude oil (~5.8 million BTUs of natural gas).

Blowout: An uncontrolled release of crude oil or natural gas from a well due to a pressure imbalance.

Blowout Preventer (BOP): A critical safety system that prevents the uncontrolled release of oil or gas during drilling.

Break-Even Price: The oil price at which an upstream project becomes economically viable, covering all operational and capital costs.

Brent Crude: A global oil price benchmark derived from a blend of North Sea crude oils.

Brownfield: An existing, mature oil or gas field that is already producing.

Capital Expenditures (CAPEX): Funds a company uses to acquire, maintain, or improve fixed assets such as drilling rigs, production facilities, and infrastructure.

Carbon Capture and Storage (CCS): A method of reducing CO_2 emissions by capturing them at industrial facilities and storing them underground.

Carbon Intensity: The amount of CO_2 emitted per unit of energy produced, used as a metric for environmental impact.

Cash Flow at Risk (CFaR): A financial risk management metric used to estimate the potential adverse impact of market volatility on future cash flows.

Casing: Steel pipes inserted into a wellbore to prevent the walls from collapsing and to isolate fluids between different rock layers.

CO_2 Injection: A method of enhanced oil recovery (EOR) involving carbon dioxide injection to boost output and store CO_2.

Completion: Making a drilled well ready for production by installing necessary equipment.

Concession Agreement: An upstream contract where an oil company gains full rights to explore and develop a hydrocarbon block in exchange for taxes and royalties paid to the host government.

Cost Oil: The portion of production allocated to an oil company under a Production Sharing Agreement (PSA) to recover exploration and development costs.

Decline Curve Analysis: is a method for predicting the future production rate of a well based on historical data.

Decline Rate: The rate at which a reservoir's production naturally decreases over time due to depletion.

Decommissioning Relief Deed (DRD): A UK fiscal mechanism allowing tax relief on decommissioning costs, even after asset transfers.

Depletion Allowance: A tax deduction that accounts for the declining production of an oil or gas field over time.

Development Well: A well drilled in a proven field to extract hydrocarbons instead of an exploratory well.

Directional Drilling: A technique that allows drilling wells at multiple angles, rather than just vertically, to reach specific targets.

Downstream: The oil and gas value chain's refining, distribution, and marketing segment.

Dry Hole: A well that is found to be non-commercial, meaning it does not contain recoverable hydrocarbons.

Economic Limit: The point at which an oil or gas well's operating expenses exceed the revenue generated, making continued production uneconomical.

Enhanced Oil Recovery (EOR): Techniques like CO_2 injection, steam flooding, and chemical flooding that increase oil recovery from reservoirs.

Environmental Impact Assessment (EIA): A study conducted before project development to evaluate the potential environmental effects of exploration and production activities.

EOR (Enhanced Oil Recovery): Techniques such as gas, water, or polymer injection are used to extract additional oil from mature fields.

E&P (Exploration & Production): The upstream sector of the oil and gas industry, focusing on finding and extracting hydrocarbons.

Exploration and Production (E&P): The upstream sector activities include discovering and extracting hydrocarbons.

Exploration License: A permit issued by a government allowing a company to explore for hydrocarbons in a defined area.

Farm-Out Agreement: One oil company transfers part of its working interest in a project to another company in exchange for funding exploration or development activities.

Field Development Plan (FDP): A detailed strategy for developing an oil or gas field, including drilling schedules, infrastructure, and production targets.

Fiscal Regime: The royalties, taxes, and profit-sharing arrangements governing oil and gas contracts.

Fixed vs. Floating Contracts: Fixed contracts have stable pricing terms, while floating contracts allow pricing to fluctuate based on market conditions.

Flare Gas: Excess natural gas that is burned off rather than captured, often due to a lack of infrastructure.

Gas Injection: A form of enhanced oil recovery where gas (such as CO_2 or natural gas) is injected into a reservoir to improve oil recovery.

Gas-to-Liquids (GTL): A technology that converts natural gas into liquid fuels like diesel and jet fuel.

Geological Risk: The uncertainty associated with hydrocarbons' presence, quantity, and recoverability in a reservoir.

Geophysical Survey: Using seismic, gravity, and magnetic methods to locate potential hydrocarbon deposits.

Gross Production: The total amount of hydrocarbons produced from a well before accounting for ownership splits or royalties.

Hedging: A financial strategy used by oil and gas companies to manage price risk by using futures, options, and swaps.

Horizontal Drilling: is a technique in which the wellbore is turned horizontally within the reservoir to maximize contact with the hydrocarbon-bearing formation.

Hydraulic Fracturing (Fracking): A healthy stimulation technique that injects high-pressure fluid into rock formations to increase oil and gas flow.

Indicated vs. Inferred Resources: Indicated resources have been geologically tested and estimated with high confidence, while inferred resources are estimated with less certainty.

Infill Drilling: Drilling additional wells in a field to enhance recovery and maximize production.

Integrated Oil Company: A company that operates across multiple sectors of the oil and gas industry, including upstream, midstream, and downstream.

Internal Rate of Return (IRR): The discount rate at which an investment's Net Present Value (NPV) equals zero, measuring its profitability.

Jack-up Rig: A mobile offshore drilling unit with legs that can be extended to the seabed for stability during drilling.

Joint Operating Agreement (JOA): A contract between multiple companies participating in a joint venture to define operational and financial responsibilities.

Lease Agreement: A contract granting an oil company the right to explore and produce hydrocarbons from a specific area.

Lifting Cost: The cost of extracting and transporting a barrel of oil or a cubic foot of gas to the surface.

Liquefied Natural Gas (LNG): Natural gas cooled to -162°C for easier storage and transport.

Local Content Requirements: Regulations mandating that some goods, services, or labor in an oil and gas project be sourced locally.

Marginal Field: A hydrocarbon field with low production potential that may become economically viable with improved technology or higher oil prices.

MER UK Strategy: The UK's "Maximizing Economic Recovery" framework for extending the life of North Sea assets.

Midstream: The oil and gas sector deals with hydrocarbon transportation, storage, and processing.

Monetization Strategy: A company's plan to convert its hydrocarbon reserves into financial returns, which may involve direct sales, joint ventures, or partnerships.

Netback Price: The price received for oil or gas after subtracting transportation and refining costs.

Net Present Value (NPV): The discounted sum of future cash flows from an investment, used to assess its financial viability.

Non-Associated Gas: Natural gas that exists independently in a reservoir and is not dissolved in crude oil.

Offshore Drilling: Extracting oil and gas from beneath the ocean floor.

Oil Reserves (Proven, Probable, Possible):

- **Proven (1P):** Reserves with a 90% probability of commercial extraction.
- **Probable (2P):** Reserves with a 50% extraction probability.
- **Possible (3P):** Reserves with a 10% extraction probability.

Oil Sands: A mixture of sand, water, clay, and bitumen that requires specialized extraction techniques.

Operator: The company responsible for managing and executing oil and gas exploration and production activities in a designated area.

OPEX (Operating Expenditures): The ongoing costs of running an oil and gas operation, including maintenance, wages, and utilities.

P&A (Plug and Abandonment): The final step in a well's life—sealing the well and restoring the site.

Pay Zone: The reservoir part where hydrocarbons can be extracted economically.

Petroleum Fiscal Terms: The financial arrangements, including royalties, taxes, and cost-recovery mechanisms, define how revenue is shared between governments and companies.

Primary Recovery: The first stage of oil extraction, relying on natural reservoir pressure to drive hydrocarbons to the surface.

Production Plateau: A period during which an oil field maintains peak production levels before starting to decline.

Production Sharing Agreement (PSA): A contract where the government allows an oil company to explore and produce hydrocarbons in exchange for a share of the production.

Proved Reserves (1P): Estimated quantities of hydrocarbons that are recoverable with reasonable certainty under current conditions.

PSC (Production Sharing Contract): A contract model where the operator recovers costs and shares profits with the host country.

Qualified Interest: A company's percentage ownership in a lease or contract area.

Reserves-to-Production (R/P) Ratio: The number of years a company or country can continue producing hydrocarbons at current rates before reserves are depleted.

Reservoir Simulation: Using computer models to predict reservoir behavior under different production scenarios.

Risk Service Contract (RSC): An oil company is paid a fixed fee for services rather than sharing production revenue.

Royalty: A percentage of production paid to the government or landowner as compensation for resource extraction.

Seismic Survey: A technique to map underground rock formations using controlled sound waves.

Sidetrack: A secondary wellbore drilled from an existing well to reach untapped reserves.

Spot Market vs. Long-Term Contracts: The **spot market** involves immediate purchase and delivery of crude oil, while **long-term contracts** lock in prices and volumes over extended periods.

Stranded Assets: Oil and gas assets may become uneconomical due to regulatory, market, or technological changes.

Thermal EOR: An enhanced oil recovery method that uses heat to reduce oil viscosity and improve flow.

Tight Oil: Oil trapped in dense, low-permeability rock formations that require advanced extraction techniques.

Turnkey Contract: A contract where a contractor agrees to deliver a fully operational project without further input from the client.

Unconventional Resources: Hydrocarbons found in formations requiring advanced extraction techniques, such as shale oil and tar sands.

Upstream: The industry involves finding and producing crude oil and natural gas.

Vapor Recovery Unit (VRU): Equipment used to capture and recover vaporized hydrocarbons to reduce emissions.

Well Abandonment: The process of safely sealing a well at the end of its productive life to prevent environmental damage.

Well Completion: Preparing a well for production, including installing casing, perforating, and stimulating the formation.

Workover: Maintenance or intervention performed on an existing well to restore or improve production.

Zero Routine Flaring: An initiative to reduce emissions by eliminating routine gas flaring in upstream operations.

Index

O

P

R

S

T